D1116346

Technical Illustration

Techniques and Applications

John A. Dennison
Instructor, Industrial Technology
North Kansas City High School
North Kansas City, Missouri
Former Professor and Chair
Department of Power and Transportation
Central Missouri State University
Warrensburg, Missouri

Charles D. Johnson
Professor, Communication Technology
Department of Industrial Technology
University of Northern Iowa
Cedar Falls, Iowa

Publisher
The Goodheart-Willcox Company, Inc.
Tinley Park, Illinois

Library of Congress Catalog Card Number 2001033409
International Standard Book Number 1-56637-871-0
1 2 3 4 5 6 7 8 9 10 – 03 – 06 05 04 03

The Goodheart-Willcox Company, Inc. Brand Disclaimer: Brand names, company names, and illustrations
for products and services included in this text are provided for educational purposes only, and do not represent or imply endorsement or recommendation by the author or the publisher.

The Goodheart-Willcox Company, Inc. Safety Notice: The reader is expressly advised to carefully read,
understand, and apply all safety precautions and warnings described in this book or that might also be indicated in undertaking the activities and exercises described herein to minimize risk of personal injury or
injury to others. Common sense and good judgment should also be exercised and applied to help avoid all
potential hazards. The reader should always refer to the appropriate manufacturer's technical information,
directions, and recommendations; then proceed with care to follow specific equipment operating instructions. The reader should understand these notices and cautions are not exhaustive.

The publisher makes no warranty or representation whatsoever, either expressed or implied, including but
not limited to equipment, procedures, and applications described or referred to herein, their quality, performance, merchantability, or fitness for a particular purpose. The publisher assumes no responsibility for
any changes, errors, or omissions in this book. The publisher specifically disclaims any liability whatsoever,
including any direct, indirect, incidental, consequential, special, or exemplary damages resulting, in whole
or in part, from the reader's use or reliance upon the information, instructions, procedures, warnings, cautions, applications, or other matter contained in this book. The publisher assumes no responsibility for the
activities of the reader.

Library of Congress Cataloging-in-Publication Data

Dennison, John A.
  Technical illustration : techniques and applications / by John A.
Dennison, Charles D. Johnson.
    p. cm.
  ISBN 1-56637-871-0
  1. Technical illustration—Technique. I. Johnson, Charles D.
(Charles David), 1949- II. Title.
T11.8 .D46 2003
604.2—dc21   2001033409

Introduction

Technical Illustration—Techniques and Applications is a textbook designed to introduce you to the field of technical illustration. You will learn about the drawing tools used by technical illustrators and the basic techniques for using these tools. Both manual drawing and computer-based illustration methods are addressed in this text. In addition to these hands-on skills, this text also discusses illustration composition and design.

Technical Illustration—Techniques and Applications includes thorough instruction for developing the three predominant types of pictorial drawings—axonometric, oblique, and perspective. The text also provides coverage of computer-aided drafting (CAD) methods with instruction on 2D-based drafting in addition to 3D-based surface modeling and solid modeling.

After learning basic technical illustration drawing techniques, you will advance to methods that cause illustrations to "spring to life," such as shading, rendering, coloring, and airbrushing.

Examples of technical illustrations are shown throughout the text to reinforce the content. You will learn techniques in a wide variety of illustration applications—from pictorial assembly drawings to three-dimensional models, renderings, and advertisement graphics. *Technical Illustration—Techniques and Applications* also includes chapters devoted to technical manuals and publication design and production.

Features of the Text

This text is divided into four sections. The chapters in each section are intended to build on the skills taught in previous chapters. The first section, *Introduction to Technical Illustration*, provides an overview of the field and the common tools and equipment used by technical illustrators. Basic sketching and drawing techniques are also covered. In the second section, *Pictorial Generation*, the axonometric, oblique, and perspective pictorial drawing methods are discussed. This section also introduces CAD-based drawing, including surface modeling and solid modeling. Section Three, *Illustration Enhancement*, covers manual and computer-based shading and rendering and airbrushing methods. The final section, *Illustration Applications and Production*, covers design and production methods for technical manuals and similar publications.

Each chapter of the text begins with a *Learning Objectives* section and an *Introduction*. A foundation for learning the concepts in the chapter is developed in these sections. Important concepts are emphasized in the introduction. The chapter objectives listed provide you with an idea of what you will be learning as you read the chapter. The objectives also outline what you should be able to accomplish upon completing the chapter and alert you to key terms.

The content at the end of each chapter provides a number of ways to evaluate your progress. Each chapter contains a section of review questions. In addition, some chapters provide activities that are designed to be completed in class or as outside assignments. For the chapters that teach specific drawing skills, drawing problems are provided. You are instructed to complete the drawings by hand or by using computer-based methods, depending on the chapter. The review questions, activities, and drawing problems will help you test your skills and knowledge as you progress through the text.

Different fonts for type are used throughout the text to identify key terms and other special features. Important terms in each chapter appear in ***bold italic*** type. Many of these terms are referenced with definitions in the *Glossary*. Common commands and command options used in computer-aided drafting systems are capitalized and designated with **bold helvetica** type. This font helps you identify the typical name for a command.

Where necessary in the text, cautions appear to alert you to safety information in relation to tools, equipment, or procedures. These are printed in red.

Using the Text

There are three basic principles that need to be applied in using this text to learn technical illustration techniques. You will have to read, practice, and experiment. All three of these principles are very important to mastering the skills and techniques involved in technical illustration. Elimination or relaxation of any one of them will significantly reduce your ability to develop expertise in illustration.

Reading the Text

Although many figures in this text can provide considerable instruction on their own, *Technical Illustration—Techniques and Applications* will require that you read the written material, integrate the concepts with your prior drawing or drafting skills, analyze the figures that relate to the text you are reading, and then incorporate that instruction in practicing your illustration skills. As you are introduced to new skills and concepts, you will find it helpful to apply them on your own.

Learning through Practice

Reading the text alone will not allow you to apply an understanding of the techniques used in technical illustration. Practice is necessary to reinforce the instruction provided. You will find that some technical illustration skills require much more practice to learn than others. For example, learning how to apply shading techniques may appear relatively easy while reading and looking at the examples in the text. If you are typical of most learners, when you begin practicing those techniques, you will not be satisfied with your projects in the beginning. As you continue to practice and learn from prior mistakes, your abilities will improve significantly.

Practice is vital for refining a new skill so that it becomes a habit or an automatic response that does not require time-consuming conscious thought. Your practice of a technique should always involve self-evaluation during and after the project. By identifying what you did wrong as well as what you did correctly, you can begin to avoid making the same mistakes over and over. Practice provides the best opportunity to correct errors in technique made early in the learning process. Also, you will find that practice is necessary to fully understand how each illustration technique can be applied to a variety of different situations.

Learning through Experimentation

Simple completion of assigned projects can teach you the fundamental illustration skills. Repetition will tend to make you more proficient as you gain more and more practice. After learning a technical illustration skill, you will need to experiment with it. Unless every situation you encounter is very similar to the projects you learned on, you may have difficulty applying your skills to different design situations. Major growth in illustration skill typically takes place when students have developed enough basic skills to begin experimenting with how they may apply these techniques to other applications, or when they start blending various techniques in new situations. Experimentation also involves more practice. As you experiment with different applications and techniques, you will be gaining more practice at the same time.

About the Authors

John A. Dennison has an extensive background in drafting and design technology for industry and education. His career in industry includes drafting for aircraft and spacecraft projects for McDonnell-Douglas Corp. and technical illustration work for the manufacturing engineering departments of a number of manufacturing firms. Many examples of his work appear in this text. Dr. Dennison's career in education includes 10 years in Missouri community colleges as a vocational drafting technology instructor, Coordinator of Applied Sciences, and Dean of Vocational, Technical and Adult Education. Dr. Dennison taught and administered technology programs at Central Missouri State University for 24 years prior to retirement and returned to teaching at the high school level to reside closer to his family. He received undergraduate and master's degrees from Central Missouri State University and a doctorate from the University of Missouri-Columbia.

Charles D. Johnson's professional background includes over 20 years of teaching communication courses at the junior high, high school, and college levels. Dr. Johnson is active in Technology Education at both state and national levels. He has served on a variety of state and national committees and has authored articles for national publications, in addition to serving on the Editorial Review Board for *The Technology Teacher*. He has also coauthored a guide on integrating special needs students into technology classes. He received his undergraduate degree from Florida State University, a master's degree from Western Carolina University, and a doctorate from North Carolina State University.

Acknowledgments

The authors wish to thank the following individuals and companies for contributing graphic images and other materials to this textbook.

A.B. Dick
Adobe Systems, Inc.
Air Nouveau
AM International, Inc.
ARKLA Products Company
Autodesk, Inc.
Badger Air-Brush Co.
Cerwin-Vega, Inc.
Cimatron Ltd.
CNC Software, Inc.
Computer Design, Inc.
John Deere & Co.
Domino Amjet, Inc.
Eastman Kodak Co.
ECRM Imaging Systems
Engle Homes
Epson America, Inc.
Fender Musical Instruments
    Corp.
Ford
Fuji Photo Film USA, Inc.
Graphtec
Heidelberg Eastern, Inc.
Hewlett-Packard Co.
Honda
Alan T Horwell (Alan's Image
    Factory; www.aifweb.com)
IBM Corp.
I-DEAS Artisan Series
Intertec Publishing Corp.
JUN-AIR USA, Inc.
Kenwood USA Corporation

David Kimble
Jack Klasey
Koh-I-Noor, Inc.
Macromedia, Inc.
The Martin Guitar Company
Jo-Anne Mason
    (www.joannemason.com)
MechSoft.com
Mitsubishi Electronics
National Computer Graphics
    Association
NexGen Ergonomics, Inc.
Olson Manufacturing and
    Distribution, Inc.
Paasche Airbrush Co.
Pioneer Graphics
Polaroid Corp.
PTC
Roland DGA
Ryobi Outdoor Products, Inc.
Staedtler, Inc.
SURFCAM
Trek Bicycle Corp.
U.S. Department of Commerce,
    Patent and Trademark Office
The Utley Company, Inc.
Vemco Corp.
WAM!NET, Inc.
Jason Weiesnbach
W.E.T. Studios
X-Rite, Inc.
Xyvision, Inc.

Special thanks to Jack Klasey for his contribution of photographs and technical assistance.

About the Cover

The jukebox on the cover is a manual illustration by David Kimble. He is a highly talented illustrator who specializes in "see-throughs," which show the exterior of an object as well as interior details. His specialty is cutaway illustrations of high-performance cars, but he often creates other types of illustrations as well. His work includes a cutaway of the Starship Enterprise that sold over 1 million copies in poster form.

Rock-Ola Manufacturing Corporation, a company that builds high-quality reproduction jukeboxes, contracted with Mr. Kimble to create the cover illustration. This illustration, and similar "see-throughs" that he creates, are done in a series of precise steps. First, he photographs the stages of assembly and studies the lighting of the object. He then traces a large photograph of the item using drafting film and pencil, and artistically renders all the interior detail. As an aid, he develops a grid using the perspective of the photograph. Next, an inking is completed using drafting film and a fineline technical pen. A film positive is made from the inking, which makes the inked lines even thinner so that they will not be visible on the final illustration. He completes the illustration by airbrushing the film on both sides, a technique he developed to help achieve transparent effects.

When Mr. Kimble works on a project, he works seven days a week, for 14 to 16 hours per day. The jukebox required about seven weeks for completion.

Special Recognition

Special thanks to PTC and Ducati North America for contributing the image of the motorcycle used in this text. It illustrates the sequence of four phases of a drawing project: A basic sketch, CAD wireframe representation, rendered three-dimensional model, and completed model with special effects applied.

The image represents the product development process—from concept to delivery. PTC's Product Lifecycle Management solutions empower manufacturers to bring superior products to market faster by optimizing the process. Windchill PDMLink from PTC enables manufacturers to control and leverage product data throughout each stage. (Motorcycle image courtesy of Ducati North America)

Trademarks

AutoCAD is a registered trademark of Autodesk, Inc. Photoshop is a registered trademark of Adobe Systems, Inc. Freehand is a registered trademark of Macromedia, Inc. Other trademarks are registered by their respective owners.

Contacting Goodheart-Willcox

Your feedback is important. If you have any suggestions or comments about the text, please direct your findings to:

Managing Editor—Technology
Goodheart-Willcox Publisher
18604 West Creek Dr.
Tinley Park, IL 60477

Brief Contents

Contents

Section Two
Pictorial Generation

Section Four
Illustration Applications and Production

Introduction to Technical Illustration

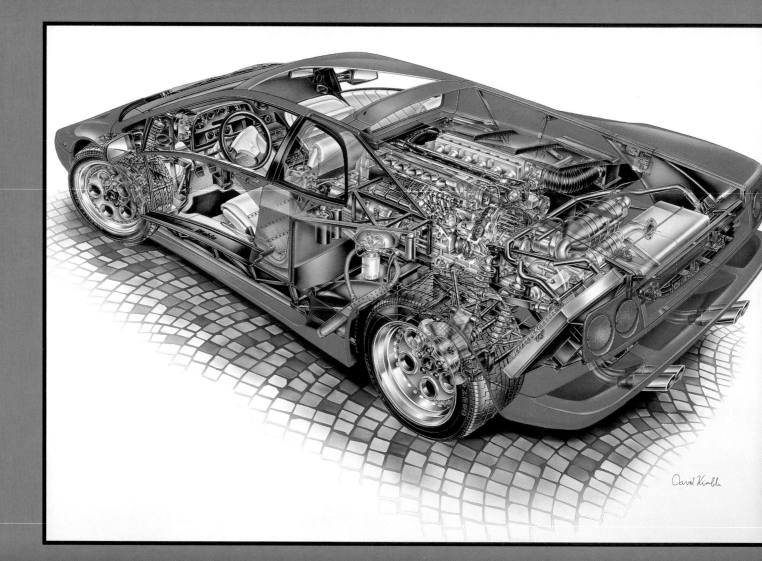

Chapter 1

Technical Illustration Applications and Careers

Learning Objectives

At the conclusion of this chapter, you will be able to:

☐ Define technical illustration.

☐ Identify the principal applications of technical illustration.

☐ List the principal careers and job titles associated with technical illustration.

☐ Identify the skills and personal characteristics important to becoming a competent technical illustrator.

Introduction

The purpose of this textbook is to introduce you to the many tools, skills, and graphic techniques necessary to produce technical illustrations. This chapter provides an introduction to the field of technical illustration.

Technical illustration is a specialized process used to develop graphic materials for a specific audience. Nearly all technical illustrations include some type of pictorial view of a product or process. The technical illustration communicates a desired message to an identified group of people.

Technical illustration has a wide range of applications. When you look around, you can see that we are surrounded by technical illustration on a daily basis. Materials produced by technical illustrators are visible in many forms and have a wide variety of uses.

Many job classifications in the field of technical illustration require the same core skills. To become a successful technical illustrator, you must possess specific abilities. Developing drafting skills, basic artistic skills, computer-aided drafting (CAD) ability, and positive personal characteristics will lead to a bright future in the technical illustration field.

Technical Illustration Overview

A general awareness of the meaning and purpose of technical illustration should be acquired to fully understand its applications. Many images and illustrations encountered in daily life are a result of technical illustration. Technical illustrations describe, explain, or clarify some aspect of a product or process. Therefore, they must be technically accurate and visually effective.

The Meaning and Purposes of Technical Illustration

Analyzing the words "technical" and "illustration" may help you gain a better understanding of the practice. The term "technical" originated with the Greek word *technikos,* which means "to have an art or be skillful in something." Today, we often use the term *technical* to indicate having skills or abilities in a specialized field. The term is also used as a reference to something that relates to technology and the applied sciences. "Illustration" is the process of showing or illustrating. To "illustrate" is to develop a visual or graphic representation that clarifies or explains something. *Technical illustration* can be defined as the process of applying graphic skills to produce visual materials that explain or clarify some aspect of a product or process.

Why do we need technical illustration? Among the most important reasons is the need to communicate a specific message to an extremely diverse population. Technical illustrations are produced for audiences with a wide range of ages, educational backgrounds, and social and economic classes.

For example, users of gas barbecue grills need to know how to connect a pressure regulator for a gas barbecue grill to a gas pressure tank. Although this is a relatively simple procedure, it is important to make sure that the users can perform it in a safe manner. The people buying this product may include automotive technicians, accountants, retired history teachers, and engineers. Some of these people may never have heard of a gas pressure regulator, while others may already know how to connect one.

This audience diversity requires a communication method that does not rely totally on a written description. Including a technical illustration showing the key concepts will help make sure that anyone can perform the task in a safe and correct manner.

Figure 1-1 shows an example of how this communication task may be presented in an owner's manual. Many individuals may be able to perform the task using only the written instructions. However, a person who cannot read English can see from the drawing that the POL fitting on the regulator is inserted into the opening on the propane tank valve. The graphic also shows that the fitting must turn counterclockwise.

Do not assume that an illustration alone will communicate what the audience needs to know. The illustration alone in **Figure 1-1** communicates how the regulator attaches to the tank, but it does not tell the entire process. The user is referred to a "Leak Testing" section. This must be included as a written instruction. The reference also appears in all uppercase, bold letters to capture visual attention.

Another purpose of technical illustration is to create an emotional or mental response from the audience. For example, illustrations of products in a catalog need to look appealing to the potential customer. To create visual appeal, technical illustrators often use techniques that are meant to appear attractive and impressive. Attractive illustrations are typically used to inform potential consumers of new or innovative applications, and generally aim at developing a feeling of desire or interest. Advertising and marketing departments devote great effort to develop illustrations that leave a favorable mental impression upon viewers.

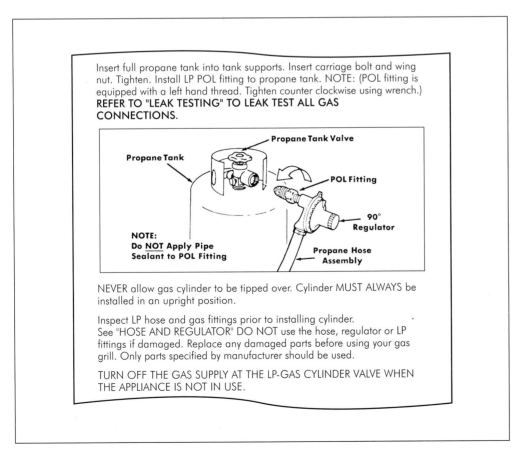

Figure 1-1. Both a written description and an illustration are used to show how to connect a propane tank to a propane hose assembly. (ARKLA Products Company)

Figure 1-2. This advertisement illustration is designed to generate an emotional response. (Image of car courtesy of Ford)

The example in **Figure 1-2** shows how technical illustrations are used in advertising to generate viewer desire. Using an illustration in these applications can be much more effective than using a photograph.

Although professional photographers can produce excellent photographs of products, a technical illustrator often needs to modify the image. The image may need to be altered to make the product appear attractive or easier to use, or it may be necessary to remove distracting background and surface feature details.

An illustrator will often develop visual images of an object that does not even exist. A drawing of a proposed architectural structure may be developed to sell a visual concept and gain acceptance from investors or zoning officials. The illustration may also be used to evaluate and compare designs.

Applications of Technical Illustration

Examples of the many applications of technical illustration are all around each of us at home, at school, and on the job. Nearly all technical illustrations are three-dimensional pictorial drawings. These drawings are easily understood by people who do not have the technical training necessary to interpret multiview orthographic drawings.

Without the pictorial drawings that accompany the written description, many home assembly projects would never be completed. You have heard the saying "a picture is worth a thousand words." This is quite often true. Technical illustrations can show how items go together by conveying useful details. They may provide full-size drawings of the various fasteners to help the user find the correct one, and written instructions to go with the visual images.

There are many examples of technical illustration found in the home. The owner's manual for a DVD player or video cassette recorder includes illustrations showing connection and operation details. A vacuum cleaner may include instruction sheets showing how to replace the belts. Photograph-based illustrations in catalogs, magazines, and sales brochures are also examples of technical illustrations.

Look under the hood of most cars and you will find a printed sticker showing how the hoses and drive belts are routed. Look in the trunk and you will probably find a decal showing where to locate and how to operate the jack. The vehicle owner's manual also contains technical illustrations showing vehicle operating controls and electrical schematics of the fuse box.

There are so many examples of technical illustration in our daily lives that we seldom recognize them as products of the illustration process. However, these illustrations were carefully designed and created by technical illustrators.

Examples of school-oriented technical illustration include instructional charts in classrooms, instructional slides, overhead projector transparencies, and textbook illustrations. **Figure 1-3** shows a typical transparency combining a variety of technical illustration techniques. Notice that the pictorial images provide a three-dimensional appearance. Also, the views are used to communicate the procedures shown. The graphic images and text both explain the desired concepts to the audience.

Figure 1-4 shows an example of an illustration you might find in any drafting textbook. Notice that it does not attempt to achieve the realism of an actual metric scale. Realism might distract the audience from focusing on the important features of metric divisions. Instead of absolute realism, this illustration simply uses the basic rectangular shape of one surface of a ruler.

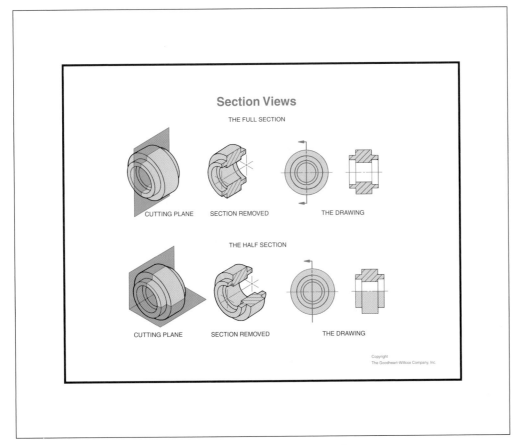

Figure 1-3. A typical transparency image for instructional purposes.

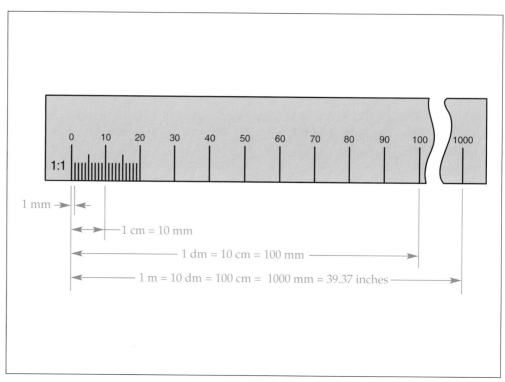

Figure 1-4. This textbook illustration shows how to read a metric ruler. The focus of the illustration is the numbered divisions. To avoid distraction, the scale is drawn very simply, rather than realistically.

Figure 1-5 is another example of something that a technical illustrator might produce. The illustration does not have a pictorial image. However, the illustrator needed to use a variety of design concepts to make the chart effectively communicate its message and be pleasing to the eye. A single font is used to help develop a sense of unity in the column-style layout. The size and boldness of the text is changed to achieve variety. Unnecessary narrative has been eliminated. Only the basic information is included.

Common Metric Prefixes		
Prefix	**Multiplication factor**	**Symbol**
mega	$1\ 000\ 000 = 10^6$	M
kilo	$1\ 000 = 10^3$	k
hecto	$100 = 10^2$	h
deka	$10 = 10$	da
deci	$0.1 = 10^{-1}$	d
centi	$0.01 = 10^{-2}$	c
milli	$0.001 = 10^{-3}$	m
micro	$0.0001 = 10^{-6}$	μ

Figure 1-5. This chart is an example of a technical illustration. Although there is no pictorial image, the appearance and usefulness of the table is determined by its design.

Figure 1-6 shows a general overview of the primary applications of technical illustration. Notice that each area is broken down into more specific areas. These areas can in turn be further divided to more specific illustration tasks, **Figure 1-7**. Illustrations produced for product catalogs, advertising brochures, product manuals, product design evaluation, instructional charts, and informational sheets represent the majority of workload for technical illustrators.

Careers in Technical Illustration

An individual who performs any of the illustration tasks shown in **Figure 1-6** will most likely have the job title of technical illustrator. If you look in the help wanted section of a newspaper, you will see listings for technical illustrators. However, most positions that use a variety of technical illustration skills are *not* advertised under the heading of technical illustrator. There are a number of other job listings with positions that include technical illustration duties. These listings have the following titles:

- Commercial designer
- Design drafter
- Designer
- Drafter
- Drafting illustrator
- Engineering illustrator
- Graphic artist

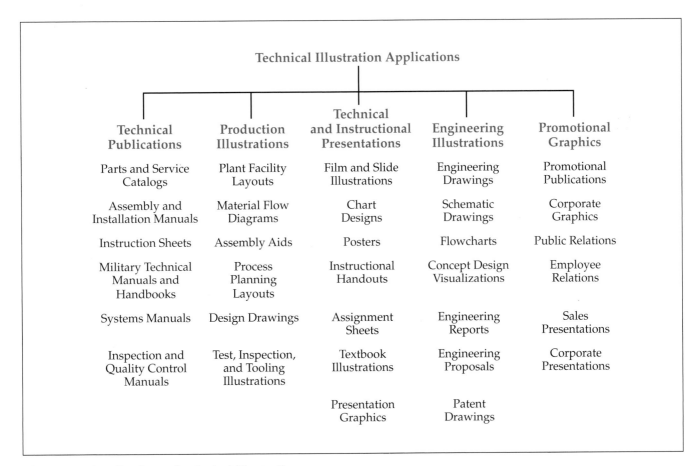

Figure 1-6. Applications of technical illustration.

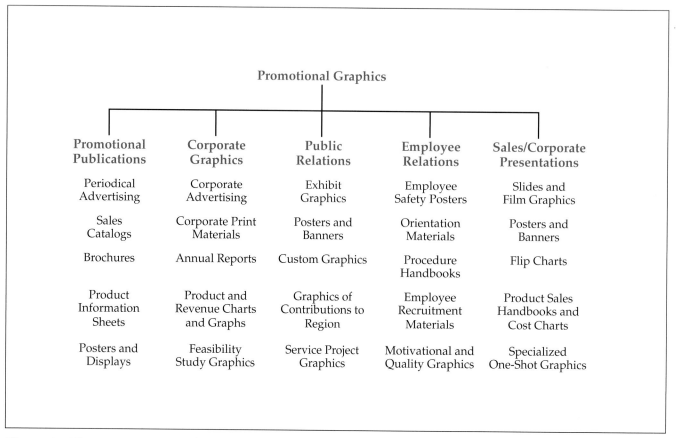

Figure 1-7. Illustration activities in promotional graphics.

➤ Graphic designer
➤ Illustrator
➤ Product illustrator
➤ Production illustrator

Although individuals with these job titles do not perform the same work duties, they all have a common core of skills possessed by a technical illustrator.

Most job listings include a short description of the duties involved. Regardless of the job title, carefully read the description for certain key words identifying technical illustration activities. These key words include *illustrate, shade, render, photographic touch-up, pictorial,* and *3D.* In addition to the specific job skills required, job descriptions often list a variety of personal characteristics the employer would like to see in a candidate. The next section identifies some of the skills needed to be a technical illustrator. It includes a brief overview of personal characteristics often found in job descriptions.

Skills and Characteristics Required in Technical Illustration

Becoming a technical illustrator requires a number of drafting skills. It also requires a variety of supporting skills and abilities. In addition, there are a number of personal characteristics crucial to success as a technical illustrator. These same characteristics are also extremely important to your academic success.

Drafting Skills

An ability to interpret orthographic drawings is vital to success in both learning technical illustration and performing it adequately. A few types of technical illustrations, such as the one in **Figure 1-8**, use simple two-dimensional (2D) views to convey a message. However, most technical illustrations are pictorial.

Pictorials usually start as orthographic drawings, and are then converted into pictorial form for illustration purposes. It is highly unlikely that an illustrator can become successful without the ability to convert the two-dimensional views of an orthographic projection into three-dimensional pictorial illustrations. To do this, an illustrator must be able to visualize objects in three-dimensional format.

Objects drawn by an illustrator within a pictorial view often need to be rotated to show some special aspect of the object. The object may also need to be developed as an exploded-view drawing to show how parts go together. Look at the sample exploded drawing in **Figure 1-9**. The illustrator had to visualize the lawn mower assembly "exploded" into its individual parts to develop this pictorial. This required the ability to envision the assembly in three dimensions, develop the drawing, and rotate the parts so that the entire object could be viewed accurately.

As you study this text, you will develop specific technical illustration skills. These skills must be built on a solid foundation of fundamental drafting skills.

Supporting Skills

Experience with a computer-aided drafting (CAD) system is helpful in learning and practicing technical illustration. An overview of basic computer applications used in technical illustration is presented in this text. If you can develop computer drafting skills, you will be a much more effective technical illustrator. The illustration field has adopted computer technology to a large extent, **Figure 1-10**. For some applications, computers have completely replaced manual illustration techniques.

WELDS AND THEIR SYMBOLS

TYPE OF WELD		SYMBOL	WELD ARROW SIDE	WELD OPPOSITE SIDE	WELD BOTH SIDES	EXAMPLE
FILLET						
G R O O V E W E L D S	SQUARE					
	V					
	U					
	J					

Figure 1-8. An example of a technical illustration using simple two-dimensional graphics.

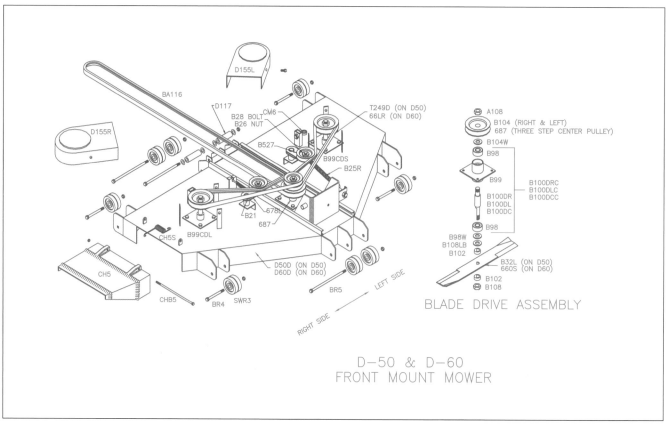

Figure 1-9. An exploded-view drawing shows how individual parts fit together.

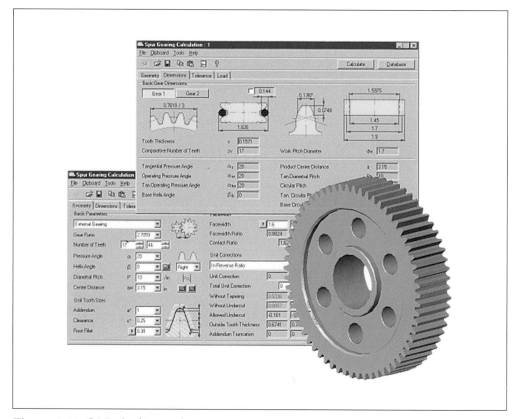

Figure 1-10. CAD drafting software was used to create this 3D model of a gear. The software tools shown permit mechanical calculations based on part dimensions and other manufacturing specifications. (MechSoft.com)

Knowledge of manufacturing processes can be extremely helpful to a technical illustrator. This knowledge can provide insight into how objects are transformed from raw material into a finished shape. Knowing how individual pieces are assembled and how finishes are applied helps in developing realistic shaded pictorials and illustrations. How an object is made determines resulting shapes and surface finishes, and can even determine the color of the object. The more you know about manufacturing processes, the more accurate you can be in developing realistic illustrations.

Basic artistic skills related to color, drawing composition, and balance are extremely important to the illustrator. A technical illustration involves shading, rendering, using special color, and applying surface treatments. Artistic concepts are important to so many aspects of technical illustration that a successful technical illustrator must be very knowledgeable in fundamental artistic principles. See **Figure 1-11**.

Personal Characteristics

A good attitude and good work habits are extremely important to employers. Many studies have been completed documenting why individuals are terminated from employment. Poor personal characteristics and attitude problems are always near the top of the list.

The workforce in American society continues to change from producing goods to providing services. This has created a work environment where work values, habits, and attitudes relate more closely to job success than abilities alone. No single personal characteristic has been identified as more important

Figure 1-11. A technical illustrator needs to have a good understanding of color, composition, and balance to create an illustration such as this for a product catalog. (Ford)

than the others. Generally speaking, individuals who have normal intelligence, average eye-hand coordination, and sound, fundamental skill training in an occupation can be successful *if they also have positive personal characteristics and work competencies.*

The following groups of work competencies have been identified by numerous researchers as essential to job success:

➤ Ambitious

➤ Cooperative, helpful

➤ Adaptable, resourceful

➤ Considerate, courteous, well-mannered

➤ Independent, initiating

➤ Capable of accurate, high-quality work

➤ Careful, alert, perceptive

➤ Pleasant, friendly, cheerful, positive

➤ Responsive, attentive to directions

➤ Emotionally stable, poised

➤ Persevering, patient, enduring, tolerant

➤ Neat, orderly, aware of personal appearance

➤ Dependable, punctual, reliable, responsible

➤ Efficient, good at producing a given quantity of work, able to achieve goals

➤ Dedicated, devoted, honest, loyal, conscientious

These positive traits and work habits are essential for your success both on the job and as a student. Related to the above list are some characteristics that are especially important to have as you study the materials in this text. These characteristics include the following:

➤ A desire to expand your level of technical skills.

➤ An ability to be critical of your own work.

➤ A capacity to maintain a level of self-confidence and self-motivation.

➤ Patience with yourself as you expend the time necessary to develop illustration skills.

➤ Respect for the learning process and an understanding that it will require considerable practice, experimentation, and conscientious effort to achieve your maximum potential.

Summary

Technical illustration is a specialized field of drafting. Graphic materials are prepared for audiences that usually do not have an engineering or technical background. Nearly all technical illustrations use some form of pictorial drawing to provide the audience with a single view that has a three-dimensional appearance. These pictorial drawings may be produced manually or by using computers.

Technical illustrations are used in an extremely wide range of applications. The more prominent applications of technical illustration include promotional graphics and publications, technical publications, production illustrations, technical and instructional presentations, and engineering illustrations. The range of technical illustration applications is so broad that they can be found in virtually every home, school, and job environment.

There are a number of job titles that require duties similar to those of a technical illustrator. These related job titles include commercial designer, design drafter, designer, drafter, drafting illustrator, engineering illustrator, graphic artist, graphic designer, illustrator, product illustrator, and production illustrator. Although these jobs may vary from one company to another, the basic task of developing pictorial drawings for a specific audience is common to all of them.

To become a technical illustrator, you must have a good grasp of fundamental drafting techniques. Additionally, you need to develop specialized technical illustration skills. A number of supporting skills are also needed to help you succeed in the field of technical illustration. These skills include familiarity with computer drafting techniques, an understanding of basic manufacturing materials and processes, artistic skills, and positive personal characteristics.

Review Questions

1. Define *technical illustration*.
2. Why is technical illustration necessary?
3. List the principal applications of technical illustration.
4. Why are technical illustrations often a combination of a pictorial drawing and a written narrative?
5. Give two examples of technical illustrations that require only a pictorial view.
6. Give two examples of technical illustrations that require a written narrative to complement the pictorial view.
7. What types of materials might an illustrator develop that are neither a pictorial view nor a written narrative?
8. Identify at least five job titles that may involve a significant amount of technical illustration tasks.
9. Why must a technical illustrator have a good command of basic drafting skills?
10. What is the value of understanding manufacturing processes for an individual working as a technical illustrator?
11. Identify at least 10 positive personal characteristics that are important to obtaining and keeping employment.
12. List at least three supporting areas of skill, knowledge, or expertise that can make you a more effective technical illustrator.

Activities

1. Search your home or one of your classrooms at school for at least five examples of technical illustrations. Write a short report explaining why illustrated materials were required for each application. Also, identify the primary application of each example.
2. Write a general job description for five of the job titles listed in the *Careers in Technical Illustration* section of this chapter. Use the Internet or a newspaper collection at a local library. You may also use a current copy of the Occupational Outlook Handbook, a reference available at most libraries.
3. Ask to interview at least two employers and find out what they expect of new employees. Ask them what personal characteristics and job survival skills are important for their employees in addition to the knowledge and skills required.

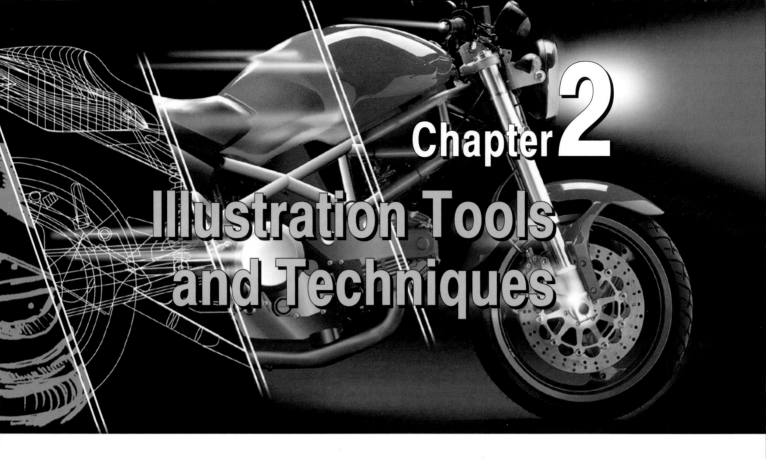

Chapter 2

Illustration Tools and Techniques

Learning Objectives

At the conclusion of this chapter, you will be able to:

☐ Identify common drawing tools.

☐ Describe how to use various drawing tools.

☐ List the major parts of a computer system and their functions.

☐ Identify types of software used for technical illustration.

☐ Explain important safety rules to keep in mind when using drawing tools.

Introduction

Technical illustration tools have evolved over time. Originally, all drawings were done using T-squares, triangles, and protractors. These tools were then incorporated into one unit called a drafting machine. Now, computers take the place of drafting machines and other tools previously used to make drawings, **Figure 2-1**.

Computers are being used for a great deal of illustration today. However, manual techniques are still important. Some technical illustrations are best done using traditional means. Technical illustrators use a variety of tools, including both manual tools and computers. The method used for a particular job depends on which one will produce the desired result in the shortest amount of time and at the least expense. Therefore, it is important to have some knowledge of all the illustration tools available so that you can make an illustration by hand or by computer.

Basic illustration tools, including computer hardware and software, are covered in this chapter. Proper techniques for using the equipment are also discussed.

Measurement Systems

The two most common systems of measurement used with illustration tools and measuring devices are the *US Customary system* and the *SI Metric system*. The US Customary, or English, system is the standard system of linear measurement used in the United States. In this system, distance is measured in inches,

Figure 2-1. Computer-aided drafting programs are widely used in technical illustration. (Autodesk, Inc.)

feet, yards, and miles. Fractions or decimals are used for measurements smaller than a whole number (for example, 12 1/2").

The SI Metric system is the most commonly used measurement system in the world and is recognized as the international standard. In this system, the meter is the basic unit for length. The meter is subdivided into decimeters (1/10 meter), centimeters (1/100 meter), and millimeters (1/1000 meter), **Figure 2-2**. The kilometer (1000 meters) is used for long distances. When using the SI Metric system, decimals are always used for measurements less than a whole number. For example, 12 1/2 mm is properly written as 12.5 mm. If the measurement is smaller than one, then a 0 precedes the decimal point (for example, 0.45 mm).

Different devices using the US Customary and SI Metric systems are discussed throughout this chapter. The most common drawing tools are discussed in the following sections.

Drafting Pencil

The pencil is one of the most basic and important illustration tools. Pencils are often used to create quick sketches that can be examined and easily changed. Pencils are also used for finished drawings or to make drawings that can be used as a pattern for pen-drawn or computer illustrations. Three major types of drafting pencils are the *wooden pencil*, *mechanical pencil*, and *fineline pencil*.

Technical drawing pencils contain special drafting leads. Graphite leads are used for drawing on paper. Sometimes plastic is added to the graphite for strength. These leads come in 18 degrees of hardness, from very hard to very soft, **Figure 2-3**. Softer leads do not require heavy pressure for drawing, but tend to smudge. Hard leads are very good for drawing light lines. For example, a 6H lead might be used for construction lines to lay out a drawing. These lines are light and normally drawn with a hard lead, such as a 5H or 6H. In general, the medium-hardness leads are used for most illustration purposes.

When drawing on polyester film, a special plastic lead is used. Grading systems for these plastic leads may vary by manufacturer. In general, there are 5 or 6 degrees of hardness available for plastic leads.

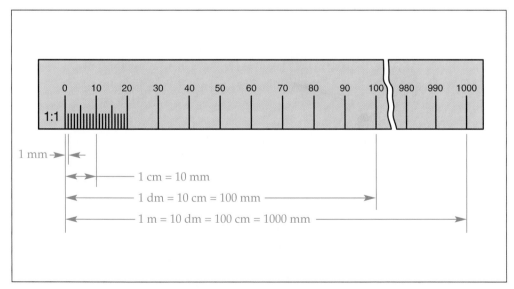

Figure 2-2. A meter and its subdivisions.

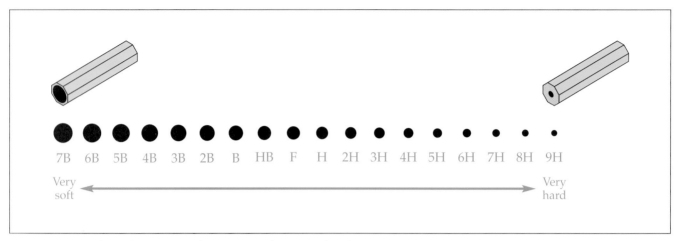

Figure 2-3. Drafting leads range from very soft to very hard.

Wooden Pencils

Wooden pencils are available in various grades of lead and are popular because of their low cost, **Figure 2-4**. However, they do require extra work in sharpening. To sharpen a wooden pencil, you must first remove the wood. Then, you must sharpen, or *point*, the lead.

Use a drafting pencil sharpener or knife to remove the wood from the end opposite the pencil grade. After exposing 3/8″ of lead, point the lead by rotating the pencil on a *sandpaper pad*. The sandpaper pad is 220 grit sandpaper with a wood backing. A *lead pointer* can also be used. Finish the desired point by drawing on scrap paper until the correct line thickness is obtained. Clean excess graphite off the point with a cloth, or by sticking the point into a piece of foam.

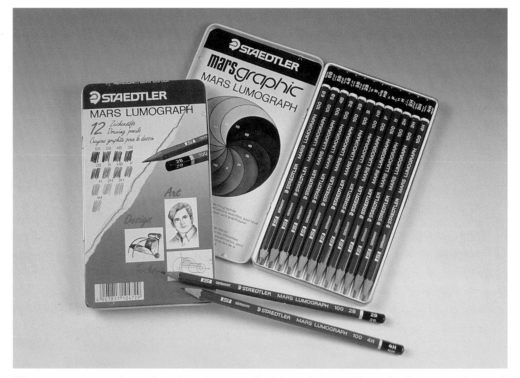

Figure 2-4. A set of wooden drawing pencils. Note the pencil grade shown on the end of the pencils. (Staedtler, Inc.)

When using a wooden pencil with drawing tools, always draw moving your hand away from yourself. Tilt the top of the pencil slightly in the direction of travel. Rotate the pencil slowly as you draw so that the point wears evenly. The side of the pencil lead can be used for shading.

Mechanical Pencil

The mechanical pencil is often called a mechanical *lead holder.* It holds 2 mm diameter leads and can be refilled, **Figure 2-5**. Any lead grade can be used in this pencil. For inserting and extending leads, a release button is pressed on the end of the pencil. This opens a chuck inside the pencil. In order to sharpen the lead, use a lead pointer or sandpaper pad. Expose 3/8″ of lead and sharpen it as you would a wooden pencil. This type of pencil is used in the same way as the wooden pencil.

Fineline Pencils

Fineline pencils are purchased according to the line diameter needed, **Figure 2-6**. The pencil and lead are classified in millimeter sizes. Common sizes are 0.3 mm, 0.5 mm, 0.7 mm, and 0.9 mm. A 0.3 mm or 0.5 mm lead is used for thin lines, such as centerlines and hidden lines. A 0.7 mm or 0.9 mm lead is used for thicker lines, such as visible (object) lines or border lines. The lead is inserted and extended, as it is with a mechanical pencil. A wire comes with the pencil for clearing clogged points.

Fineline pencils are available in three different tip classifications, **Figure 2-7**. The tip is the part of the pencil that presses against the straightedge as you draw. The *fixed tip* works well for most applications. This type of tip does not move. It has an advantage over sliding tips in that it is easier to use with

Figure 2-5. A mechanical pencil holds various grades of lead. (Staedtler, Inc.)

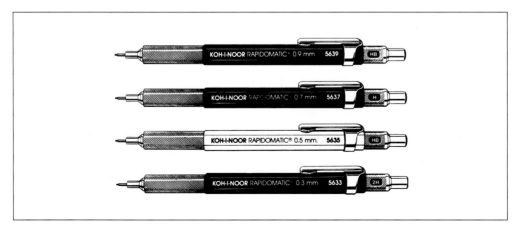

Figure 2-6. Fineline pencils are purchased in different line diameters. (Koh-I-Noor, Inc.)

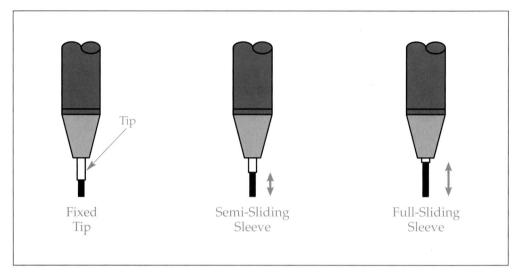

Figure 2-7. Several tip options for fineline pencils are available. Select a pencil that has the appropriate tip for your application.

templates, especially those used for lettering. A sliding sleeve is sometimes more difficult to place in the template because it can slide up above the template.

The *semi-sliding sleeve* exposes lead as you draw. This protects smaller leads, such as 0.3 mm, from breakage. Another advantage is that more lead is available for use, so the lead is advanced less often. The *full-sliding sleeve* retracts completely into the pencil as the lead is used. This can be a disadvantage when using a straightedge because the sleeve can retract and cause the lead to press against the straightedge. This type of pencil is most often used for writing or freehand lettering rather than drafting purposes.

When using a fineline pencil, hold the pencil almost vertically to make sure the correct line diameter is drawn. Pull the pencil toward you when it is being used with a straightedge. You do not need to rotate the pencil as you draw. If you are using a sliding tip pencil, be sure the tip does not "ride up" on the straightedge.

Erasing Pencil Marks

When erasing errant marks on paper, use a rubber or vinyl eraser. An abrasive *rubber eraser* is good for dark lines. The less abrasive *vinyl eraser* or *gum eraser* is good for light lines and smudges. On polyester film, a nonabrasive eraser must be used to avoid removing the rough, "frosted" surface of the film. If this happens, the film will not accept pencil lines. The procedure for erasing marks is as follows:

1. Hold the paper around the error to avoid tearing the paper while erasing. If possible, use an *erasing shield* and choose the hole that best fits the mark, **Figure 2-8**. Make sure that good lines are covered completely with the shield to avoid erasing them.
2. Choose the correct eraser and make sure it is clean. Clean it on scrap paper or sandpaper if necessary.
3. Erase with the least amount of pressure possible. If too much pressure is used, you can rub through the paper or leave a "ghost." A *ghost* is graphite embedded in the paper. An *electric eraser* can be used for large mistakes. This type of eraser has an electric motor that automatically moves the eraser from side to side.

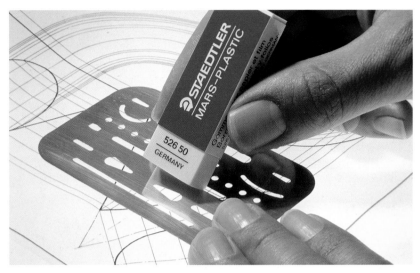

Figure 2-8. A vinyl eraser and erasing shield being used to erase a small area of the drawing. (Staedtler, Inc.)

4. Brush away eraser particles with a ***drafting brush***. Do *not* brush the particles with your bare hand; this can smudge the drawing and deposit oil from your hand on the drawing.

Drawing Media

There are a variety of drawing papers and films available for technical illustration. Drawing papers are also called ***substrates*** or ***drawing media***. With a basic knowledge of these materials, you should be able to decide which one is best for your needs.

Drafting sheets come in both roll and sheet form. Standard sheet sizes are designated in both US Customary and SI Metric units. In the US Customary system, the letters A through E are used to indicate size, **Figure 2-9**. Each sequential letter represents a sheet size twice the area of the sheet represented by the previous letter. For example, an A-size sheet is 8 1/2″ x 11″ or 9″ x 12″. The next largest sheet is B-size. This sheet is 11″ x 17″ or 12″ x 18″, which is twice the area of the A-size sheet. Metric drafting sheet sizes are also shown in **Figure 2-9**.

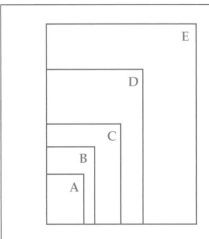

American Standard Sheet Sizes		Metric Sheet Sizes	
A	8 1/2″ x 11″ or 9″ x 12″	**A4**	8.27″ x 11.69″
B	11″ x 17″ or 12″ x 18″	**A3**	11.69″ x 16.54″
C	17″ x 22″ or 18″ x 24″	**A2**	16.54″ x 23.39″
D	22″ x 34″ or 24″ x 36″	**A1**	23.39″ x 33.11″
E	34″ x 44″ or 36″ x 48″	**A0**	33.11″ x 46.81″

Figure 2-9. American Standard and metric drafting sheet sizes.

Drafting sheets usually have a smooth and rough side. The rough side is up for drawing. A package of paper will usually give this information on the label.

Grid Sheets

Grid sheets are normally used when making sketches. There are square grids used when making oblique sketches and isometric grids used when making isometric sketches. See **Figure 2-10**. Grid sheets are also available for perspective drawing.

Translucent Paper

Translucent paper allows light to pass through it. This paper can be used for most copying and tracing methods. The two major categories in this group are *tracing paper* and *vellum*. Tracing paper is normally thin, untreated paper that can be used for pencil and ink drafting. Vellum is treated tracing paper that contains oils and waxes. These additives improve the quality of the paper and make it more durable. Vellum provides the following advantages for technical illustration:

➤ It has a hard surface that is not grooved easily with pencils.

➤ It is good for inking because it is moisture-resistant and will not absorb the ink.

➤ It resists discoloring and deterioration.

➤ It resists moisture-related shrinking and swelling. This makes it more stable than most other paper-based drawing media.

Drafting Film

Drafting film consists of a plastic sheet with one side roughened (dull side) to accept ink and pencil lines. It has all the advantages of vellum with the following additional qualities:

➤ It is moisture-proof. This means drawings are safer and do not change size with a change in humidity.

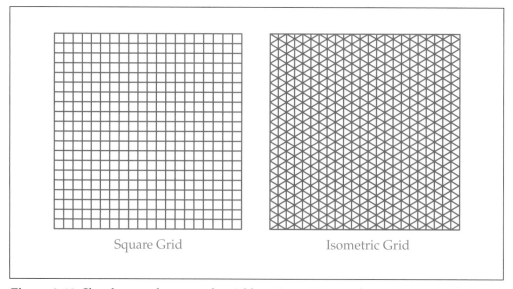

Square Grid   Isometric Grid

Figure 2-10. Sketches can be created quickly using square and isometric grid sheets.

➤ Inked errors can be removed with special erasers and ink solvent without damaging the sheet.

Drafting film is popular for inking because inked errors can be corrected. Erasing on drafting film should be done with a vinyl eraser, which will not mar or smooth the film surface.

Illustration Board

Illustration board is heavy paper specially designed to accept wet materials (such as ink, paint, or watercolor) without buckling. *Hot-pressed* illustration board has a smooth finish and is not very absorbent. *Cold-pressed* illustration board has a textured surface and is more absorbent.

Thinner illustration boards are used for chalk, charcoal, lead/graphite, and other dry media. Thicker boards are typically used to resist bowing when a liquid medium is to be applied to the drawing surface. In addition, some illustration boards have a foam core to provide extra strength.

Measurement Tools

Measurement tools include both scales and dividers. A *scale* is a device used for measuring distances on drawings. Scales come in a variety of shapes and are usually around 12″ (300 mm) long.

Scales can be used to draw an object full size, reduce it, or enlarge it, **Figure 2-11**. This is often called "drawing to scale." The proper scale to use depends on the type of drawing needed and the size for the finished drawing. For example, a skyscraper and a watch can both be drawn on the same size paper using different scales.

The term "scale" also refers to the formula used for reducing or enlarging actual measurements. The scale is almost always indicated on the drawing. When a scale is given, the first number is the drawing size and the second is the object size. For example, 1/4″ = 1″ means that 1/4″ on the drawing equals 1″ on the actual object.

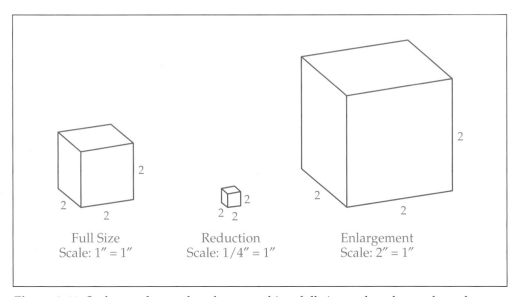

Full Size
Scale: 1″ = 1″

Reduction
Scale: 1/4″ = 1″

Enlargement
Scale: 2″ = 1″

Figure 2-11. Scales can be used to draw an object full size, reduced, or enlarged.

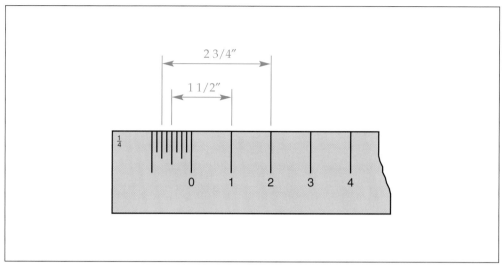

Figure 2-12. A mechanical engineer's scale is an open-divided scale. Whole inches are measured inside the scale. Fractions of an inch are measured at the end of the scale.

Mechanical Engineer's Scale

A *mechanical engineer's scale* uses inch measurements to make drawings 1/16 size, 1/8 size, 1/4 size, 1/2 size, or full size, **Figure 2-12**. For example, using the 1/4″ = 1″ (1/4 size) scale, 1/4″ on your drawing represents 1″ on the object.

A mechanical engineer's scale is known as an *open-divided scale*. The scale is marked in whole inches, beginning with zero. This gives it a very uncluttered appearance. At the end of the scale is one inch divided into fractional increments. This part of the scale is used for making any measurement that falls between whole inches, such as 1 1/2″. The divisions within the one inch are typical fractions of an inch, such as eighths or sixteenths. The number of divisions depends on the amount of room available in an inch using the particular reduction scale.

To measure, place the correct scale over the line to be measured and align it to the nearest whole inch along the scale. Any remainder is found at the end of the scale in fractions of an inch.

Architect's Scale

An *architect's scale* is most often used for drawing buildings. Very large objects require a large reduction in size. Therefore, marks on the scale are used to represent feet. For example, an architect's scale marked 1/4 means that 1/4″ = 1′-0″.

The architect's scale is an open-divided scale. It looks similar to the mechanical engineer's scale. However, the divisions at the end of this scale represent portions of a foot (inches). There are typically 12 divisions, each representing one inch. However, there may be more or less divisions, such as 6 or 24, depending upon how much room is available.

To measure, find the nearest whole number of feet first, **Figure 2-13**. Find the remaining number of inches at the end of the scale. Measurements are usually shown in feet and inches, even if the measurement is less than one foot. For example, measurements might be 7′-5″, 0′-3″, or 5′-0″.

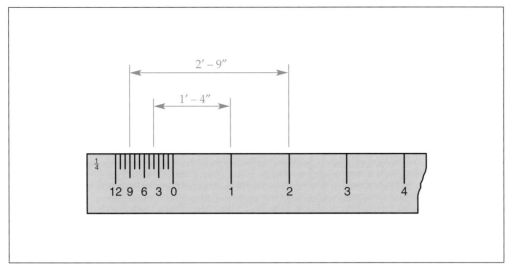

Figure 2-13. When using an architect's scale, numbers along the scale represent feet and divisions at the end of the scale represent inches.

Civil Engineer's Scale

A *civil engineer's scale* is a fully-divided decimal scale often used for drawing highways and maps. It includes scales that divide an inch into parts that are multiples of 10, such as 10, 30, 50, or 60. These divisions can be used to represent common distances, such as feet, yards, or miles. For example, the 40 scale is used for 1″ = 40 miles. In this case, the smallest division equals 1 mile.

The 50 scale is useful for measuring in decimal fractions of an inch, **Figure 2-14**. The smallest division on this scale then equals .02″. When decimal inches are used, inch marks are omitted, as well as the zero for dimensions of less than one. For example, .25 and 1.30 are both in the correct format for decimal inch measurements.

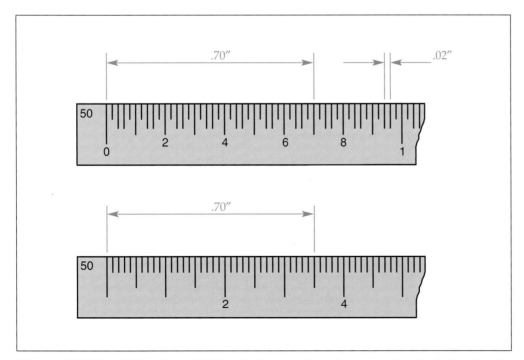

Figure 2-14. The civil engineer's 50 scale is convenient for measuring in decimal inches. Two types of scales are shown.

Metric Scale

A *metric scale* is used to make drawings in metric units, such as millimeters. Scales are shown on the drawing as ratios, such as 1:1 for full size, **Figure 2-15**. A 1:2 scale means that 1 mm on the drawing equals 2 mm on the object being drawn. A 2:1 scale means that 2 mm on the drawing equals 1 mm on the object.

A single metric scale can be used for various reductions and enlargements. For example, a 1:2 scale can be used for 1:2, 1:20, 1:200, and 1:2000 scales. Multiply the measurement on the 1:2 scale by 10 for each zero added. When used for a 1:20 scale, a 50 mm measurement on the 1:2 scale is 500 mm. In order to change from a larger to smaller ratio, divide the measurement by 10. A 100 mm measurement on a 1:20 scale equals 10 mm if used for a 1:2 scale.

Before using a metric scale, be sure you know what the numbers and divisions represent. On smaller ratio scales, such as 1:3 or 1:5, the numbers usually stand for centimeters. For larger ratio scales, such as 1:500 or 1:2500, the numbers usually stand for meters.

Combination Scale

A *combination scale* is a single instrument that contains many of the most commonly used scales. Many illustrators prefer this scale because it serves several purposes. A typical combination scale has a mechanical engineer's scale, an architect's scale, and a civil engineer's scale, and may have both US Customary and SI Metric measurements. For example, a combination scale might have a decimal inch and fractional inch scale, a metric 1:100 scale, a civil engineer's 50 scale, and a mechanical engineer's 1/4 size and 1/2 size scales.

Dividers

Dividers have two adjustable legs and a steel point at the end of each leg. See **Figure 2-16**. This tool is often used for layout. For example, suppose a line needs to be divided into 8 mm intervals. Set the dividers at 8 mm and mark off the distances on the line.

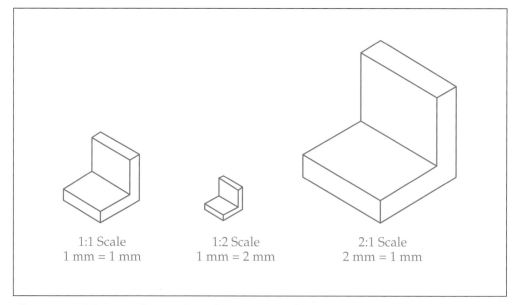

Figure 2-15. Metric scales used on drawings are shown as ratios.

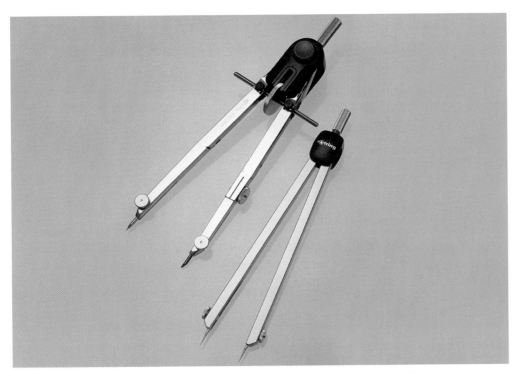

Figure 2-16. Dividers (bottom) appear very similar to a compass (top). (Koh-I-Noor, Inc.)

Dividers can also be useful for transferring distances. Instead of constantly measuring a line that needs to be drawn more than once, draw one line. Then, use the dividers to transfer this distance to other lines. When setting dividers, take measurements from the drawing and not directly off the scale. This prevents the scale from being scarred by the divider points.

Tools for Drawing Straight Lines and Angles

A combination of tools is normally used when drawing straight lines and angles. The most basic combination is the T-square, triangle, and protractor. The parallel straightedge and drafting machine are also used to draw straight lines and angles. These tools are described in the following sections.

T-square

The **T-square** consists of a head attached at 90° to a blade. The blade is used for drawing horizontal lines and as a guide for other instruments. To use the T-square, pull the head against the working edge of the board with your fingers toward the head-end of the blade. Hold the T-square firmly in this position when drawing horizontal lines. (In general, the head should be opposite your drawing hand. The head is always along the left or right edge of the drawing board; *not* along the top or bottom edge of the board.) For vertical or inclined lines, hold a triangle against the top edge of the blade.

To slide the T-square, first lift the blade by pushing down on the head. This will keep the blade from smudging lines. Then, slide the T-square to the new position and place the blade down on the drawing.

Figure 2-17. A T-square and 45° triangle are shown. The head of the T-square is placed flush against the edge of the drawing surface. In this photo, the illustrator is positioning the triangle as needed. (Staedtler, Inc.)

Triangle

A *triangle* is used for drawing vertical and inclined (slanted) lines, **Figure 2-17**. Triangles are identified by height and angle. The two most useful triangles are the 45° triangle and the 30°-60° triangle. Both triangles are available in various heights, such as 8″ or 10″.

To use a triangle for vertical lines, hold it flush against the top of the T-square with your fingers. Be sure that the T-square head is square against the edge of the board. Place the inclined side toward your drawing hand. Your hand then rests on the triangle when drawing and cannot smudge the paper. Draw the line with slight pressure against the triangle. Use this same procedure for inclined lines also, except place the straight side of the triangle toward your drawing hand.

Any inclined line can be drawn at 30°, 45°, 60°, and 90° with the 45° and 30°-60° triangle. It is also possible to draw other angles by stacking these two triangles in different configurations on the T-square. Adjustable triangles are also available for drawing lines at any given angle.

Protractor

A *protractor* is used for measuring and drawing inclined lines and angles, **Figure 2-18**. Protractors are usually made of clear plastic and have a semicircular shape. When using a protractor, align the baseline of the protractor with a reference line already drawn. Set the center point at the vertex of the angle. Now mark the correct degree setting using the edge of the protractor, starting at zero on either the left or right side. Use a triangle as a straightedge to connect the marks.

To measure an angle with a protractor, align the baseline with one leg of the angle and place the center point on the vertex. Then, read the degree measurement where the other leg meets the edge of the protractor.

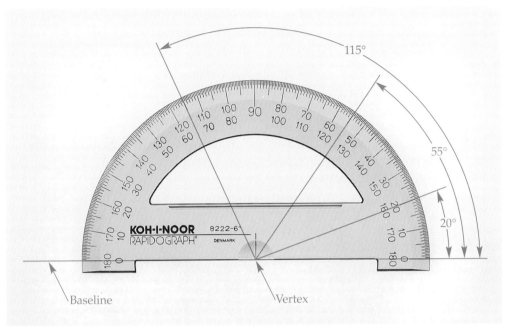

Figure 2-18. Protractors are used to lay out angles. (Koh-I-Noor, Inc.)

There are special protractors available for pictorial drawing. One of these, the isometric protractor, is discussed in Chapter 5 of this text.

Parallel Straightedge

A *parallel straightedge* is attached to the drawing surface by wires and pulleys. The straightedge remains horizontal when the board is raised and lowered, **Figure 2-19**. A parallel straightedge functions in much the same way as the T-square for horizontal lines. Triangles and protractors are used with the

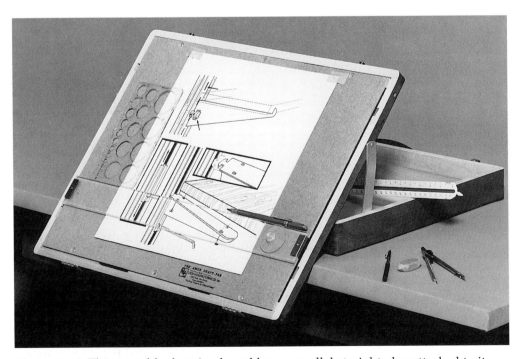

Figure 2-19. This portable drawing board has a parallel straightedge attached to it. (Olson Manufacturing and Distribution, Inc.)

parallel straightedge to create inclined lines in the same way as with the T-square. A parallel straightedge is especially useful for large drawings because it does not flex when moved (as compared to a T-square).

Drafting Machine

A *drafting machine* combines the functions of the T-square, triangle, protractor, and scale. It is attached to the drawing surface. A horizontal and a vertical scale/straightedge are attached to a protractor head that can be set to draw at any angle. A drafting machine can be moved freely to any place on the board.

There are two types of drafting machines: elbow-type, **Figure 2-20,** and track-type. The track-type machine is preferred for larger drawings. Some drafting machines have electronic components. These may include digital read-outs for angular and linear measurements.

Tools for Drawing Circles and Curves

A compass or circle template is used for drawing circles and common curves. For curves that do not have a common center, an irregular curve, or "French curve," is used.

Compass

A *compass* is a tool for drawing circles and arcs, **Figure 2-21.** The most common type is the bow compass. Often, both a small and large compass are needed to draw a variety of circles up to 6″ in diameter. Pencil and pen attachments are available for compasses.

The leg of the compass that holds the pin is placed in the center of the circle to be drawn. The pin should have a shoulder to reduce the size of the hole

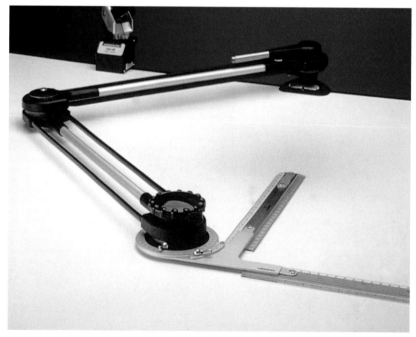

Figure 2-20. An elbow-type drafting machine combines the functions of the T-square, triangle, protractor, and scale. (Vemco Corp.)

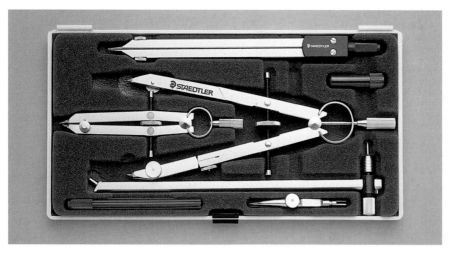

Figure 2-21. This drafting set comes with two bow compasses (middle) and a beam attachment (directly below the compasses) for drawing larger circles. (Staedtler, Inc.)

created in the paper. The other leg holds a wedge-shaped lead. The lead should be one or two grades softer than the pencil lead being used on the drawing. With the legs together, the pin point should be slightly longer than the lead so that the length will be even when the pin is inserted in the paper, **Figure 2-22**.

To use a compass, first mark the radius on the drawing. Never measure the radius by placing the compass on the scale. Insert the pin in the center point of the circle and move the other leg to the desired radius. Tilt the compass in the direction of travel and rotate it between your thumb and index finger to draw the circle. If necessary, rotate the compass more than one revolution to obtain the correct line density. It may be helpful to practice on scrap paper first.

If required, sharpen the compass lead on a sandpaper pad to a wedge shape, as shown in **Figure 2-22**. The beveled (slanted) surface should be on the outside when the lead is inserted into the compass.

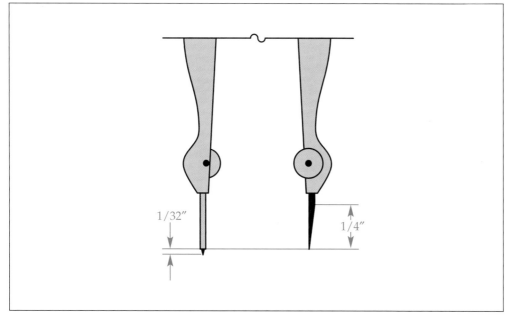

Figure 2-22. When using a compass, the pin point should be slightly longer than the pencil point. The length of the two legs will be even when the point is inserted in the paper.

Irregular Curve

An *irregular curve,* also known as a "French curve," consists of a variety of curves that can be used when a common arc is not satisfactory. To use this tool, first mark points along the curve to be drawn. Find a part of the irregular curve that aligns with at least three points and draw a line lightly between the first few points. Align the curve with the next few points and draw a light line between these points. Continue this procedure with other points on the curve, **Figure 2-23.** When this is completed, go back and darken the line using the irregular curve again. Be sure the finished curve has a smooth appearance.

A *spline* is a flexible type of irregular curve that is often used for long curves. It is made of a plastic material that holds its shape when bent. In order to use it, first bend it to the desired curve. Then, trace along the curve in a manner similar to that used with the irregular curve.

Templates

A *template* is used with a pencil or pen for drawing shapes and symbols. There are templates for drawing circles, ellipses, squares, symbols, and just about every other standard shape. As shown in **Figure 2-24,** a template is used by placing it on the drawing and tracing the desired symbol.

When using a template for inking, the bottom side must be raised slightly to avoid having ink flow under the template. "Build up" the bottom of the template with drafting tape. There are also templates designed for inking, as well as *ink risers* (plastic disks) that can be attached beneath the template.

An *isometric ellipse template* can be used in technical illustration. It is used to draw circles on an isometric drawing (a type of pictorial drawing). Circles on an isometric drawing are not drawn as true circles, but as isometric ellipses, **Figure 2-25.** In order to use the template, centerlines on the drawing are aligned with marks on the template. Detailed instructions in the use of isometric templates are presented in Chapter 5.

Another frequently used template in technical illustration is the *isometric hexagon template.* This template is used for drawing the heads of hexagon-shaped bolts and hexagon-shaped nuts.

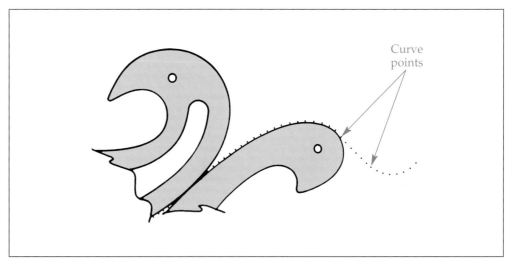

Figure 2-23. When making a curve with the irregular curve, first mark points along the curve as shown. Then draw the curve by aligning the tool with at least three points. When drawing the next part of the curve, be sure to overlap the last part drawn.

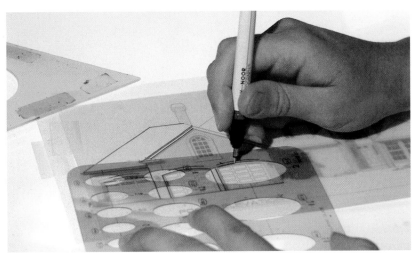

Figure 2-24. An ellipse template used to draw a curve on an architectural drawing.

Figure 2-25. This isometric ellipse template is being aligned with the centerlines on the right side of the cube in order to draw an isometric circle.

Inking Tools

Inking produces a better line density than that achieved using pencil. This is important for technical illustrations that will be reproduced by photographic methods and require a high-quality original. The two types of pens that are available for inking are technical pens and adjustable ruling pens.

Technical Pen

The *technical pen,* like the fineline pencil, comes in a variety of point sizes according to the line width needed, **Figure 2-26.** The pen can be used with a straightedge, as a compass attachment, with templates, or in freehand drawing. The pen has an ink cartridge, so frequent refilling is not necessary.

Most technical pens work in a similar manner. Ink flows down from the cartridge to the point, or *nib.* The nib contains a wire connected to a valve that

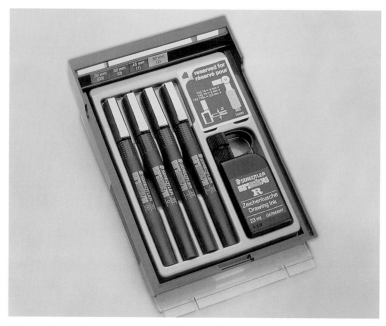

Figure 2-26. Technical pens, like fineline pencils, are designed for a specific line width. (Staedtler, Inc.)

allows ink to flow to the sheet when the wire is pushed up. As ink is used, air enters the cartridge from around the nib to replace the ink. If air cannot enter the cartridge, a vacuum is created and the ink flow stops.

Pens come in US Customary line widths and metric widths. A standard set might begin with 0000 (0.13 mm) for thin lines and progress to 7 (2.00 mm) for thick lines.

To use the pen, hold it horizontally and shake it gently to make sure the valve is free and clicking. Be sure to shake the pen over scrap paper, not over your drawing. Alternate between shaking and drawing until the ink flow starts. Hold the pen vertically and keep it moving when drawing. Do *not* apply pressure on the point as you would with a pencil. Recap the pen when finished to keep the ink from drying in the nib. To fill or clean the pen, do the following:

1. Remove and fill the cartridge 3/4 full if needed, **Figure 2-27.** Slant the cartridge when filling so bubbles do not form.
2. Take the point section apart and wash and dry each part.

Figure 2-27. When filling an ink cartridge, it should be held at a slant. (Staedtler, Inc.)

3. Attach the cartridge to the pen body before reassembling.
4. Screw the nib on and gently tighten.
5. Alternate shaking the pen horizontally and drawing until a constant flow of ink begins.
6. If the point is clogged even after washing, disassemble and soak the point section in ink solvent or place in an ultrasonic cleaner.
7. After cleaning the pen, store it with the nib up.

Adjustable Ruling Pen

An *adjustable ruling pen* consists of two nibs that can be adjusted for a variety of line widths. Though not used much anymore, adjustable ruling pens are still available. Ruling pens can be used with a straightedge or as a compass attachment, but they are not typically used freehand.

To use a ruling pen, fill the pen over scrap paper up to 1/4″ between the nibs. On scrap paper, draw lines and adjust the nibs with the thumbwheel until the line width is correct. Place the stiff nib (opposite the thumbwheel) lightly against the straightedge and draw the line. If you hesitate at the beginning or end of the line, it may thicken. Try to complete all the inking of one line width before changing the width setting. Refill the pen as needed. Clean the pen by passing paper through the nibs. When drawing, some ink left on the pen will help the ink flow. Use an ink solvent for thorough cleaning.

Correcting Errors

Deleting inked lines on paper can be done by erasing, scraping, or opaquing. When erasing, use an eraser shield with a soft eraser and rub gently to remove the error. Use an art knife when scraping and tilt the blade. Scrape gently. White opaque fluid can be used if a translucent sheet is not needed for copying.

Inked errors are corrected on drafting film with a vinyl eraser and deletion fluid. Special erasers with ink solvent are also available. Do not use a highly abrasive eraser on film. This will smooth the rough surface and make it less receptive to ink.

Rendering Tools

Rendering is adding shading and color to a drawing in order to make it appear more realistic. A variety of tools can be used to apply color. A common way to apply color to a drawing is to use colored markers and pencils. (Shading and rendering techniques are discussed in greater detail in Chapters 9 and 10.)

Markers dry quickly, do not require special paper (because they do not buckle the surface), and are easy to use. They are available in a variety of colors, including various shades of gray. Many different tip shapes are also available, **Figure 2-28**. Markers leave a transparent color on the paper. Colors can be blended to achieve different effects.

Colored pencils have similar advantages to markers and are available in hundreds of colors. However, they tend to be semi-opaque when applied to paper. Color can be applied with the point of the pencil or, if a smooth tone is needed, with the side of the lead. Pencil colors can be blended in order to achieve new colors.

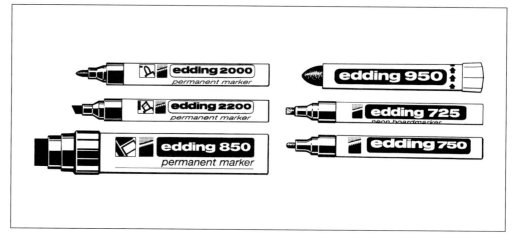

Figure 2-28. Markers are available with various tip shapes. (Koh-I-Noor, Inc.)

Colored pencils sometimes create a "wax bloom" on paper surfaces after a few weeks. This is a result of the wax in the color rising to the surface. It appears as a fading of the color. This is only a problem when there is a lot of color applied to the paper. The wax bloom can be lightly removed with a cloth. Spraying a fixative on the drawing will help prevent bloom from occurring.

Both colored pencils and markers should be used sparingly at first to make sure you do not add too much color. Interesting effects can be created by applying colored pencil lightly over the marker so that they blend to create a different color.

Colored chalks (pastels), charcoal, and soft lead pencils are often used for smudge shading. See **Figure 2-29.** *Smudge shading* involves applying the material to the paper and then rubbing to evenly blend it. This is done with a cotton pad, swab, or blending stump. A *blending stump* has a felt pad on the end specifically for blending. A spray fixative may be used after the drawing is completed to prevent further smudging.

Figure 2-29. These chalks, also known as pastels, work very well for smudge shading. (Koh-I-Noor, Inc.)

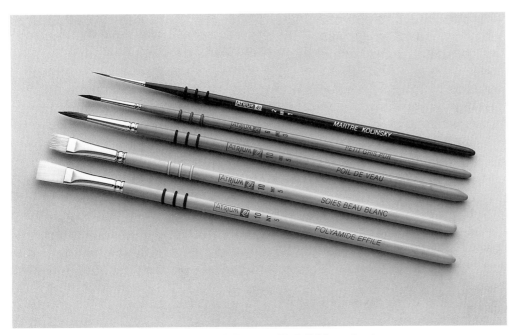

Figure 2-30. Several brush styles and shapes. (Koh-I-Noor, Inc.)

Watercolors work well for shading surfaces of an object, as well as coloring shadows. However, using watercolors requires more preparation than the other techniques. A wash is prepared by mixing the watercolor with water and applying it with a brush in very light coats. The wash is transparent, so you are able to see the surface beneath it. A cotton pad or cloth is used to blot the applied wash in areas where excess color is applied.

Opaque paints are used to shade illustrations when it is necessary to obtain a specific color on the illustration. The color is opaque, so it does not blend with any colors beneath it.

There are a variety of brush styles and shapes used for applying watercolors and paints to a drawing. See **Figure 2-30.** Typical shapes include round brushes and flat brushes. In addition to various shapes, different bristles are available for working with specific kinds of paints. Generally, brushes used with solvent-based paints should not be used with watercolors and vice versa.

The *airbrush* is another tool used for applying shading and shadows in a variety of colors, **Figure 2-31.** Usually, the area to be airbrushed is masked so that ink is applied uniformly in the designated area. Practice on a scrap piece of paper to make sure that the amount of ink being sprayed is correct. When using the airbrush, normally start the spray off the area to be sprayed and spray past the end point. The airbrush should be continually moving in a smooth motion. Leaving the airbrush pointed in one spot will make that area darker. Be sure to follow directions for safe use of the airbrush, as well as cleanup procedures. Airbrush techniques are discussed in greater detail in Chapter 12.

Transfer Sheets

Various images, symbols, and shading screens can be applied to drawings using *transfer sheets*. These are commercially prepared sheets containing decorative images, shading patterns, or letters of the alphabet. Two common types are *dry transfer sheets* and *appliqué sheets.*

Figure 2-31. This airbrush kit comes with the airbrush (bottom left), an air hose for connection to a compressor, and bottles that attach to the airbrush for holding ink. (Badger Air-Brush Co.)

Images on dry transfer sheets are applied by rubbing the top surface with a burnishing tool to transfer the image from the bottom to the surface of the substrate. A *burnishing tool* has a smooth steel or plastic ball on the tip used to rub down the image, **Figure 2-32.** In place of a burnishing tool, a pencil with a rounded point can be used. A cover sheet should be used over completed dry transfer work since the image can be removed.

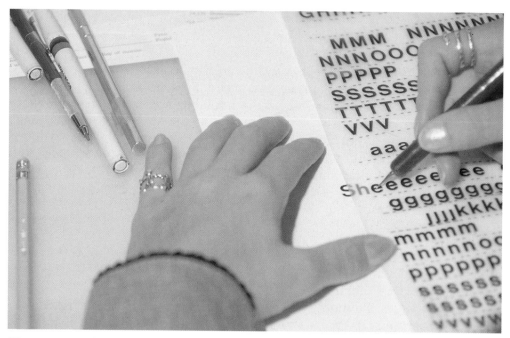

Figure 2-32. A burnishing tool is used to transfer symbols from dry transfer sheets to a substrate.

Appliqué sheets are similar to dry transfer sheets, except that the images are on the top of the sheet and have an adhesive backing. The images are cut and removed from the sheet and applied to the substrate. No burnishing is required.

Drawing Equipment Safety

Use common sense when working with drafting equipment. Doing so will prevent accidents and prolong the life of your tools. Some important safety guidelines to practice with drafting instruments and equipment are as follows:

➤ Carefully pass tools and equipment from one person to another; do *not* throw tools.

➤ Use tools with points, such as dividers, only as directed. Store them in their proper containers when done.

➤ Be careful when using pencils or other drawing instruments with sharp points.

➤ Protect skin and eyes if cleaning solvents are used. Wash exposed skin thoroughly after use.

➤ Make sure there is adequate ventilation when using solvent-based or petroleum-based paints. When using an airbrush or spray fixatives that produce harmful vapors, wear safety glasses and ventilate the work area properly. Wear a face shield or respirator if necessary.

➤ Be knowledgeable of all safety precautions in the laboratory or workplace, as well as procedures to follow in case of an accident.

Computer Hardware

Computers are widely used today to create technical illustrations. This section serves as an introduction to the hardware components making up a computer and their basic functions.

A computer stores, retrieves, and processes data. **Input** is the way that data enters the system, and is generated by using a device such as a keyboard. **Processing** is the way that functions and operations are performed on data, and takes place in the central processing unit (CPU). Output devices include monitors, disk drives, and printers. Information is stored or managed using hard drives, optical drives, disk drives, tape drives, and portable storage media, such as compact discs and Zip disks.

Input Devices

An *input device* is hardware used to get information into the computer. The most common input device for a computer is the keyboard. The keyboard is used to supply alphanumeric data to the computer as input.

A *mouse* is another commonly used input device. When the mouse moves across a flat surface, there is a corresponding movement of a cursor on the display. Buttons on the mouse are used for special functions, such as making selections from a menu. The mechanical mouse has a ball or wheel underneath that rolls against the work surface. A *trackball* is like an "upside-down" mouse. The unit does not move, but you move the ball to control the cursor.

Figure 2-33. A scanner is an input device. It is used to scan images and convert them to electronic form. (Fuji Photo Film USA, Inc.)

A *graphics tablet* is an input device that consists of a flat surface with a wire grid underneath. A *stylus* or *puck* is used to control cursor movement and initiate commands from the tablet. When the stylus or puck moves over the surface of the tablet, electric signals are input into the computer. The computer then moves the cursor. A *digitizer* is essentially a very large graphics tablet that can be used to initiate commands and convert drawings into digital form.

Drawings and other existing images are more commonly converted into digital form using a scanner. A *scanner* is an input device that works much like a copy machine. See **Figure 2-33.** The difference is that the reflected light is turned into electronic data, instead of being transferred to paper. The scanned image is saved as a file. It can then be imported into a software program for modification.

Central Processing Unit

The main part of the computer is called the *central processing unit (CPU).* This is where "work," or processing, is performed on data. The CPU is actually made of small silicon chips known as *microprocessors* or *microchips.* Electrical signals, representing data, flow through the chip in a coded form called *binary code.* This is a code in which all characters are represented as a string of zeros and ones, and is the basic system used by a computer to process and store digital data.

Memory

Memory is where information is stored for use by the computer. All data that enters the computer is placed in *random access memory (RAM).* Information stored in RAM is temporary and is erased when the computer is turned off.

One of the first items stored in RAM is an *operating system.* The operating system "tells" the computer what to do. As you run other programs, additional data is placed in RAM and processed. For example, as a CAD drawing is being made, it is being placed in RAM.

Read only memory (ROM) is installed when the computer is assembled. This memory permanently stores instructions necessary to perform tasks. For example, the first instructions the computer receives when turned on are stored in ROM. Like RAM, this memory can be read by the computer, but it cannot be changed and is not erased when the computer is turned off.

Output Devices

An *output device* is a computer hardware device used to display or produce something generated within the CPU. There are several different types of output devices used with computers. These include video displays and hard copy devices. These devices are covered in the following sections.

Video displays

The most common output device is the video display, or *monitor.* An image is created by a series of "dots" on screen. These dots are called picture elements, or *pixels*. The display resolution of a monitor is measured in pixels. Higher resolution is achieved by having more pixels visible on screen. For example, a monitor with a display resolution of 1024 × 768 means that the monitor is capable of displaying 1024 pixels horizontally and 768 pixels vertically. Higher resolution monitors may display resolutions of 1600 × 1280 or greater.

Hard copy devices

Output devices used to produce hard copy are called *hard copy devices*. The most common hard copy device is a printer. Plotters are also classified as hard copy devices. The most common types of printers are inkjet and laser printers.

Hard copy devices can also be classified as vector or raster. Vector devices, such as pen plotters, create images using lines. Raster output devices create images with dots. Vector and raster output devices are discussed in greater detail in Chapter 14.

An *inkjet printer* forms an image by shooting droplets of ink onto the page. These dots form the image. Inkjet printers are capable of producing high-quality images.

A *laser printer* works on principles similar to those used by a copying machine, **Figure 2-34.** First, a light-sensitive drum or belt is given a positive electrical charge. Then, the image is scanned onto the drum using a laser. This makes this area negatively charged. Positively charged toner powder is attracted to the image area on the drum. The drum is then rolled against the paper, transferring the toner, and heat fuses the toner to the paper. Laser printers are comparable to inkjet printers in printing quality.

A *thermal printer* uses heat to form an image on paper. A special type of thermal printer called a thermal transfer printer uses an inked ribbon for color output. A printhead heats the ribbon, which transfers ink to the image area.

A *plotter* is used for outputting drawings where high-quality line work is needed. Hard copies of technical drawings are often created with a plotter. Pen plotters use one or more pens to produce the image. Inkjet imaging systems are also used.

Figure 2-34. Laser printers are used to produce high-quality copy. (Hewlett-Packard Co.)

A *flatbed plotter* holds the paper stationary while the pens move over the surface horizontally (along the X axis) and vertically (along the Y axis). See **Figure 2-35.** A *drum plotter* uses a cylinder to move the paper forward and back, **Figure 2-36.** The pens move left to right. This combination of motion creates the image. Another type of plotter is the ***grip-wheel plotter,*** or ***microgrip plotter.*** Instead of a drum, small wheels grip the paper at each side. These wheels turn to move the paper.

Figure 2-35. A flatbed plotter. (Graphtec)

Figure 2-36. A drum plotter. (Hewlett-Packard Co.)

Data Storage

Computers process large amounts of data, and it is necessary to have a way of storing this information electronically. *Storage devices* use both magnetic and optical means to store data to disk or tape. Most of these devices are *input/output devices.* This means that they can provide data to the computer (input) and also store data from the computer (output).

Magnetic storage

Magnetic storage devices include hard drives, disks, and tapes. A computer hard drive is the most common storage device. Common portable magnetic storage devices are Zip disks and diskettes. Both types of disks are 3 1/2" in size. However, most diskettes have a storage capacity of only 1.44MB (megabytes) and are no longer widely used. Zip disks are available in the 100MB and 250MB storage format.

Data stored on a disk is placed on a specific track and sector on the disk surface. *Tracks* are concentric, magnetic circles on the disk surface. *Sectors* are divisions of a track. A sector is like a slice of pie. See **Figure 2-37.**

Data is stored on and retrieved from a disk with the *read/write head* of the disk drive. As the disk spins at a constant rate, the head moves over the disk's surface and either places (writes) data on the disk or retrieves (reads) data. Diskettes have a write-protect tab that can be closed to stop anything from being added, or "written," to the disk.

Precautions should be taken when using portable storage disks. They should not be dropped, bent, or placed near a magnetic field or heat source. When writing on the label, a felt-tip pen should be used so the disk surface is not dented. The plastic disk, which can be seen through the access hole, should never be touched.

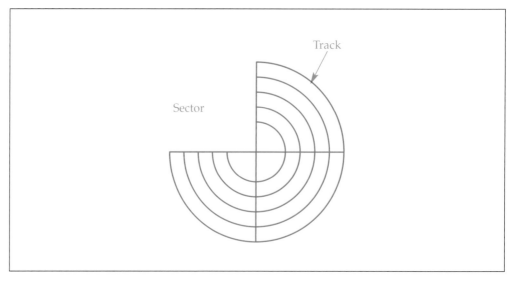

Figure 2-37. A disk stores data on specific tracks and sectors.

When a large amount of magnetic storage is needed, the **hard drive** of the computer is used. Hard drives are capable of providing more than 100GB (gigabytes) of space. The hard drive consists of metal platters where the data is stored. These platters are in a sealed case with the read/write head, so no contaminants can get in.

Sometimes a hard drive might "crash." This means that some sort of error has occurred so that the data on the hard drive cannot be accessed, or the computer is not functioning as it should. If the computer crashes, you may need to restart your computer, and your current work may be lost. Therefore, any files you are working on should be saved often. In addition, it is important to "back up" files to prevent work from being lost.

Removable hard drives, or cartridge drives, are an alternative to hard drives for storing data. A removable cartridge is used in the same way a diskette is used in a 3 1/2" disk drive. However, the removable cartridge can store several hundred megabytes of data. The magnetic-type cartridge is similar to a regular computer hard drive in how it functions.

A **magneto-optical (MO) drive** is similar to a disk drive, but it uses different methods to read and write data. Electromagnetism is used to write to MO discs, but the data is read with a laser. Optical discs and drives are discussed next.

Optical storage

Optical storage devices are a highly popular form of storage media. The most common optical storage device is the **optical disc**. It is used in many different applications because it can store large amounts of data and offers long-term durability. When compared to magnetic media, optical discs have a much longer "shelf life." Data is written to and read from optical discs using a laser. The most common types of optical discs are CD-ROMs, CD-Rs, CD-RWs, DVDs, and MO discs.

CD-ROM stands for *compact disc read-only memory*. This means that it contains data that can be read on a CD-ROM drive, but new data cannot be written. A CD-ROM has a metal base with a plastic coating. The metal has tiny pits that contain digital information. This information is read with a laser. Up to 700MB of information can be stored on a typical CD-ROM. See **Figure 2-38**. Many CD-ROMs contain entire software programs. They are used to store

Figure 2-38. CD-ROMs are typically used for storing photographic images. (Eastman Kodak Co.)

various kinds of data from simple image files to fully illustrated, audible presentations.

Two other types of optical discs similar to CD-ROMs are CD-Rs and CD-RWs. As previously discussed, CD-ROM drives allow data to be read, but not written. By comparison, CD-Rs and CD-RWs are *writable*. **CD-R** stands for *compact disc-recordable*. With this type of disc, data can be written, read, and erased. However, sectors of the CD that have been written to cannot be recorded on again. This is not the case with CD-RWs. **CD-RW** stands for *compact disc-rewritable*. With this type of disc, data can be written, read, and erased, and previously recorded sectors can be written to again. In many cases, CD-RW drives are replacing CD-R drives on computers.

DVDs are similar to the different types of CDs, but they can store much more data. **DVD** stands for *digital versatile disc* or *digital video disc*. DVDs have very large storage capacities and may hold up to 17GB of material. DVDs are most commonly used in the film industry, but they can also store other types of data. They are available in writable and rewritable formats. A major advantage of DVDs is that DVD drives can be used to read CDs.

Another type of optical disc similar in function to a CD is the magneto-optical (MO) disc. **Magneto-optical discs** combine optical and magnetic technology to store and provide data. Information is read from the disc using a laser. However, there is also a write head in the MO drive that writes information to the disc magnetically. MO discs typically store from 100MB to 500MB of information. One advantage of MO discs is that stored data can be accessed nearly as fast as data on a hard drive.

Computer Software

Without the necessary software and computer programs, hardware is of no use. *Software* is a term used to describe the specific instructions for directing the operation of the computer. The two primary forms of software that are necessary to operate the computer are systems and applications software.

Systems software consists of a collection of programs that allow the parts of the computer to work together as a single unit. Specifically, this software coordinates the computer, peripheral devices, and software so they all work together correctly. Systems software includes the operating system of the computer and the utilities. Utilities are programs that carry out useful functions such as formatting disks, copying disks, deleting files, and viewing folder directories.

Applications software consists of programs that perform specific tasks. The most common application programs are word processors, database programs, spreadsheets, graphics software, and integrated software. *Suites,* which combine several applications in a single package, are also commonly available.

Graphics Software

Graphics software consists of one or more programs used to create images. Graphics software programs are widely used in technical illustration. Products created with these programs include charts, photo enhanced images, and technical drawings.

There are a wide variety of graphics software packages available today. For example, presentation graphics programs are used to create appealing images in business reports, **Figure 2-39.** CAD software is used for creating precise technical drawings, such as machines, buildings, and shapes for mechanical parts.

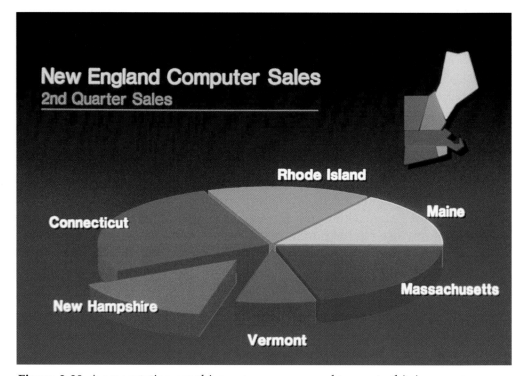

Figure 2-39. A presentation graphics program was used to create this image. (Polaroid Corp.)

See **Figure 2-40.** Paint software is typically used for generating freehand images similar to what artists create, **Figure 2-41.**

Most graphics software programs can be divided into two categories— paint programs and draw programs. These categories are defined by how the images are created. *Draw programs* create images using vector graphics. *Vector graphics* are images created using objects such as lines, circles, and arcs. The elements making up the image are individual entities. For this reason, vector graphics are also called *object-oriented graphics*. When an object such as a line is drawn, the computer "sees" it as a line, not a series of dots. Creating drawings

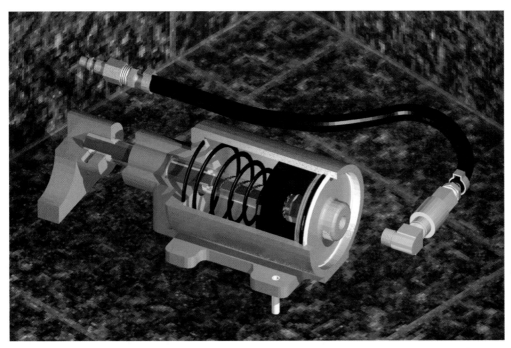

Figure 2-40. A CAD software program was used to draw this three-dimensional model. (Autodesk, Inc.)

Figure 2-41. A paint program was used to create this image. (National Computer Graphics Association)

in this manner is similar to using traditional technical drawing techniques and tools, such as the T-square and compass. A CAD program is an example of a draw program. Vector graphics usually produce images with very high quality. Using a CAD program to create drawings is discussed in Chapter 4.

Paint programs create images using bitmap graphics. *Bitmap graphics* are images made up of patterns of dots (pixels). Bitmap graphics are also known as *raster images*. Paint programs are often used to generate freehand art much in the same way an artist creates a picture. An advantage of bitmaps is that individual pixels can be altered. With vector graphics, the entire entity has to be changed. Paint programs and image editors are discussed in Chapter 10.

Health and Safety at the Computer Workstation

As with any tool, there are some precautions that will make computer work more comfortable and safe. A well-designed workstation can reduce fatigue and increase productivity, **Figure 2-42.**

The computer keyboard should be placed at a height so that your upper and lower arms form approximately a 90° angle. This should be about 24" to 28" from the floor. At this height, the upper arms are at rest and do not strain your back.

Use a wrist rest when working on the keyboard for extended periods of time. A supportive rest will help to prevent carpal tunnel syndrome. If you are working with a mouse, a wrist rest might also be helpful.

The computer monitor should be placed so that the top is no higher than eye level. It should also be about 18" from your eyes. Tilting the screen back 10° to 20° will make the screen easier to see, as long as glare is not increased. Closing blinds or using an anti-glare filter can help prevent glare problems.

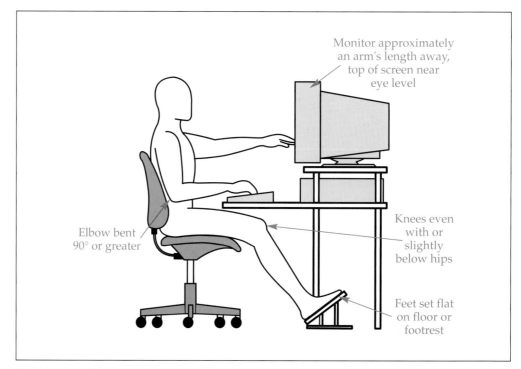

Figure 2-42. You can be more comfortable, productive, and safe if your computer workstation is well designed and adjusted for your size.

Chairs are another consideration for comfort and health. The height should be adjusted so feet are flat on the floor or a footrest. In addition, the backrest should fit comfortably at the small of the back for good back support.

Summary

A variety of tools used for technical illustration have been introduced in this chapter, including drawing tools used by hand and computer-based tools. It is important for the illustrator to be familiar with all drawing options, and not limited to one method of creating a drawing. A variety of image creation options are usually available in the workplace, and different options can create different end results.

In order to manually complete a technical illustration, you first need a technical drawing pencil or pen, a substrate to draw on, and a means of accurately drawing lines and shapes on the substrate. For straight lines, a T-square, triangle, and protractor might be used. A drafting machine, which takes the place of these three pieces of equipment, might also be used. A scale is needed to verify the correct length of lines, and tools such as a compass, circle template, and irregular curve are needed to draw circles, arcs, and curves.

If color or shading is to be added to a drawing, other tools will be needed. Examples are markers, colored pencils, chalks, watercolors, and opaque paints. An airbrush can also be used for this purpose. Practice will be needed to gain proficiency with any of these tools.

As computers and software have become more powerful and less expensive, the use of electronic methods for technical illustration has risen dramatically. A knowledge of graphics software is critical if you wish to create technical illustrations with a computer. Safety is also an important factor with computers. By setting up the workstation using the guidelines provided in this chapter, you will be able to be more comfortable and safe while working.

Review Questions

1. Describe the differences in the three types of technical drawing pencils.
2. How is an erasing shield used?
3. What are three important characteristics of a drawing substrate?
4. What are the two meanings of the term *scale*?
5. The mechanical engineer's scale and architect's scale look very similar in that they are both open-divided and have similar numbering along the scale. What, then, is the distinguishing characteristic, besides the fact that some may be labeled?
6. Explain two ways that dividers can assist you in making a drawing.
7. Which manual drawing tools are incorporated into a drafting machine?
8. List two differences in the way a person uses a technical pen as opposed to an adjustable ruling pen.
9. How is *rendering* defined in technical illustration?
10. Describe one important safety rule in technical illustration.
11. List three different input devices used by a computer.
12. In regard to computer memory, what is the difference between RAM and ROM?
13. Why does a computer need applications software?
14. Explain the difference between *draw programs* and *paint programs*.
15. How can you determine the proper height for a computer keyboard?

Activities

1. Interview an illustrator who has worked in the drafting field for the past 20 years. Ask about the major changes that have taken place in regard to tools, drawing techniques, and computer-based methods.
2. Visit a drafting firm. Determine what types of scales and other illustration tools are used and identify their purpose. Share this information with the class.
3. Conduct a search for sites on the Internet using "Technical Illustration." Locate sites that sell technical illustrations and/or the tools used in their production. List at least four sites and identify the types of content available.
4. Visit at least five different computer workstations in your school and determine whether they are configured to reduce fatigue and maximize productivity. Report on what you find.

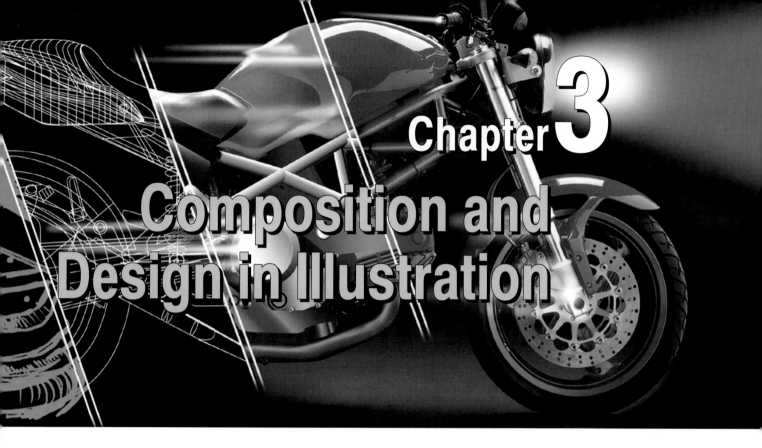

Chapter 3

Composition and Design in Illustration

Learning Objectives

At the conclusion of this chapter, you will be able to:

☐ Describe the design problem-solving process.

☐ List and describe the basic principles of design composition.

☐ Explain specific techniques used to establish proportion, balance, and unity in a drawing.

☐ Identify the design applications of lines, planes, three-dimensional forms, and surface characteristics within an illustration.

☐ Explain how to plan, organize, and develop the design for a technical illustration.

Introduction

The design of an illustration requires planning, organizing, and developing visual elements. A technical illustration is always designed to communicate a purpose, message, mood, or visual goal. Illustrations are a visual medium, so the principles and elements of visual design must be applied in an organized way.

To be effective, a technical illustration must be pleasing to the viewer. People tend to understand and respond more favorably to designs that are pleasing to the eye. If the image is negative or unattractive, or if there is visual detraction, an undesirable impression may be created. This can hinder the goal of the illustration.

In order to achieve visual attractiveness, a number of design and composition principles must be applied. You need to follow a certain design procedure. You need to follow proper composition guidelines. You must also use the basic elements of lines, planes, forms, and surface characteristics. All of these principles will allow you to create an effective technical illustration.

Applying the Design Problem-Solving Process

The design problem-solving process can be applied to a wide range of situations. To be an effective illustrator in areas where you do not have experience, you need to understand the problem-solving process. The following six steps compose this process:

1. Statement of the problem.
2. Analysis of design parameters, criteria, requirements, and project limitations.
3. Research.
4. Possible solutions.
5. Final solution.
6. Evaluation.

These six steps are described in the next sections.

Statement of the Problem

In one clear sentence, the basic goal of the illustration should be indicated. See **Figure 3-1**. If you make a general statement, you may get a range of possible results. However, if you state the problem in specific terms, you are more likely to meet your objective. Therefore, be as specific as possible. For example, the statement "To show a drill press" presents too many illustration possibilities. A statement such as "To identify the component parts and assembly of a drill press" is more likely to lead to an appropriate illustration.

Analysis of Design Parameters, Criteria, Requirements, and Project Limitations

The next step is to identify the parameters, requirements, and criteria for the design. See **Figure 3-2**. These include such things as the object or activity to be shown, customer demands, color and style parameters, legal requirements,

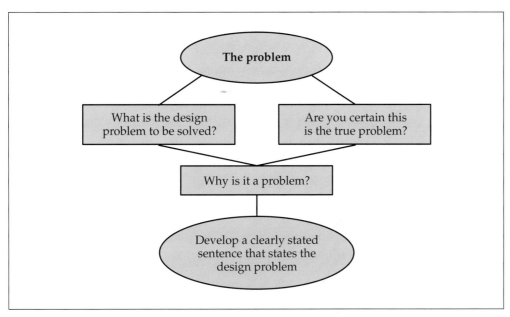

Figure 3-1. The design problem-solving process begins with a statement of the problem.

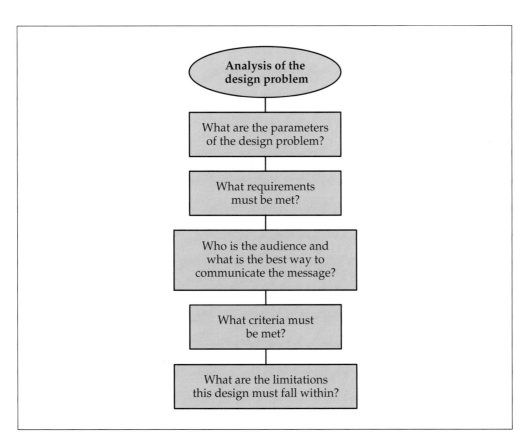

Figure 3-2. The parameters, requirements, criteria, and limitations of the design project are analyzed after stating the problem.

safety requirements, and the audience. Then, define the project limitations and list all factors that restrict your solution. These might include the time deadline or production schedule, size restrictions, material or construction limits, finish limitations, project cost, and weight. However, do not create unnecessary restrictions that may artificially limit the solution.

Research

Research should include examining the types of solutions developed by competitors and how others in your own organization have solved the related problem or similar problems. See **Figure 3-3**. Research should also include considering standard size data, laws or regulations applicable to the project, preliminary planning and development that may have been completed, and marketing research information.

Research can be recorded as sketches, pictures of printed material, copyright search data, or ideas from related products. Write a brief description of anything found in your research that cannot be shown pictorially. Sources should always be documented for future reference.

Possible Solutions

When producing possible solutions, divide the project into different parts and activities. Then, create ideas for solving each part. See **Figure 3-4**. It is easier to solve a large number of small problems than to solve one large problem. Get as many ideas, partial solutions, or variations of design concepts as possible. Be sure that all the potential solutions and ideas are recorded. Try to develop a minimum of four to five thumbnails of tentative solutions to your

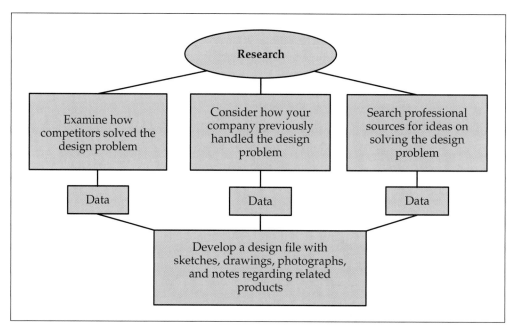

Figure 3-3. Research should explore the various ways the design problem has been addressed by competitors and those within the company.

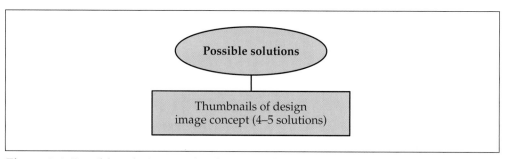

Figure 3-4. Possible solutions to the design problem are developed as thumbnails.

design problem. A *thumbnail* is a rough draft sketch, approximately 3″ × 5″, drawn to record a visual thought or design concept.

Final Solution

Evaluate your tentative solutions and select the most appropriate one, or combine the best features from several tentative solutions. Sketch the final solution in the form of an illustration rough. See **Figure 3-5**. An *illustration rough* (or *rough*) is normally a shaded sketch approximating the full size of the finished illustration. Objects in the rough are sketched in proportion. Text is normally represented by blocked areas.

When the rough is finished, check to see that sound design principles and elements are used. Add any appropriate design notes and write a descriptive paragraph listing the illustration size, type of illustration materials or media, color requirements, and type fonts. Also, describe any photographic, production, or printing information needed.

Evaluation

Evaluate how well the final solution meets the design criteria. If it fails to meet some criteria, identify what is needed to meet these criteria. These changes are incorporated into the creation of an *illustration design comprehensive*. The comprehensive is the full-scale illustration with all colors, textures, text, and other features needed to develop the final illustration. **Figure 3-6** illustrates how a design is evaluated to create a comprehensive.

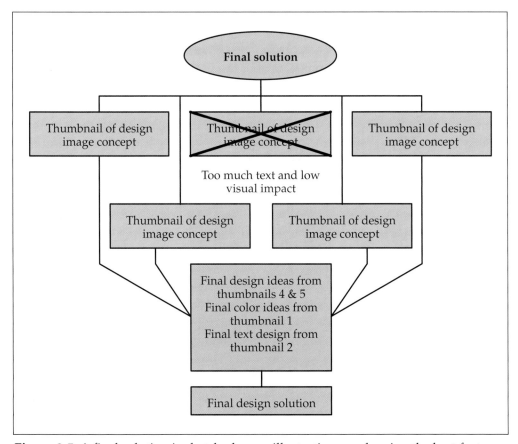

Figure 3-5. A final solution is sketched as an illustration rough using the best features from several thumbnail sketches.

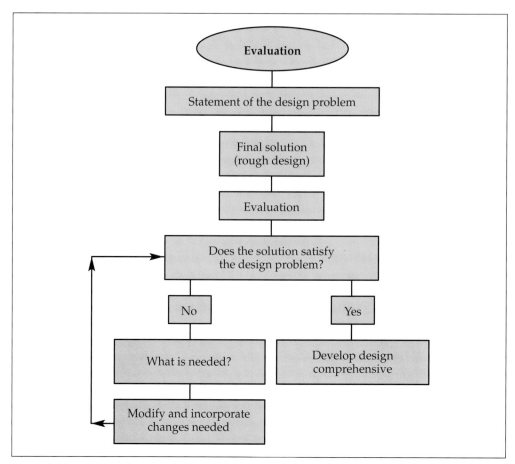

Figure 3-6. The design comprehensive is developed after evaluation of the rough design.

Composition and Design Principles

Composition describes the process of using design principles to arrange individual elements into a completed product. While this section serves as an introduction to design principles, there are no absolute rules of design. Instead, there are general guidelines to follow. Select the appropriate composition and illustration design concepts that apply to the project.

Some common design principles are listed below. Each of these is explained in the next sections. As you plan thumbnails and roughs, integrate these fundamentals to make the illustration as effective as possible.

➤ Define and work with a picture area.

➤ Scale objects properly.

➤ Control the perception of depth.

➤ Develop a focal point for the viewer.

➤ Maintain consistent values and establish contrast.

➤ Generate proportion, balance, unity, and variety.

➤ Maintain clarity within the illustration.

Defining a Picture Area

The *picture area* is the space within the borders of an illustration or graphic. The size of the picture area determines the size of objects, text, and elements. For example, if an illustration needs to fit on an A-size sheet (8.5″ × 11″), then the objects in the illustration must be scaled to fit within that area.

The size of the sheet is not necessarily the size of the picture area. For example, if you are going to use an A-size sheet for a parts manual, the white border around the printed and bound sheet must be subtracted from the sheet size to determine the picture area. For a technical manual using A-size sheets in a vertical format, you will normally deduct at least 1″ from the binding side (gutter) and 1/2″ from the outer edge (thumb). The outer, top, and bottom edges are normally trimmed after binding so all pages have the same edge alignment. This trimming reduces the available picture area even more. Therefore, there is usually a maximum of 7″ × 9 1/2″ for the picture area on an A-size vertical format sheet. See **Figure 3-7**.

Scaling Objects

In most cases, the drawing scale is affected by the sheet orientation as much as by the sheet size. A vertical A-size sheet format only allows 7″ of picture area width to work with. A horizontal format allows approximately 9″. This extra 2″ may make considerable difference in the scale used. For example, if a vertical format is required and the illustration is of a wide object, then you must make the object smaller by reducing the scale. If company standards allow, you can also rotate the view 90° counterclockwise on the page. Rotating the view allows the scale to be larger for a wide object. The difference between using a horizontal and vertical format for the same drawing is shown in **Figure 3-8**. As shown in the figure, the object is displayed at a larger scale in the horizontal format.

What the audience is supposed to notice within the picture area affects what you actually draw in the illustration. For example, if one specific component of a large assembly is being explained, that component must be scaled large enough to clearly show details. The surrounding components then must be drawn at the same scale. This may mean that only a partial view of the assembly can be shown in order to fit the entire illustration within the picture area.

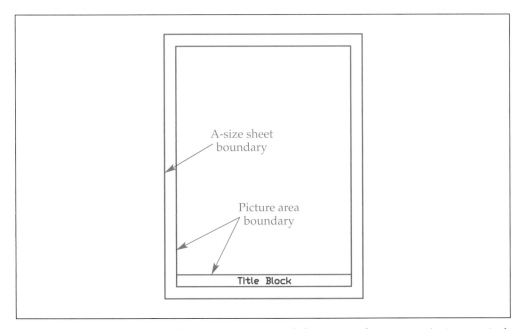

Figure 3-7. A comparison of the picture area and the outer edges on an A-size vertical sheet.

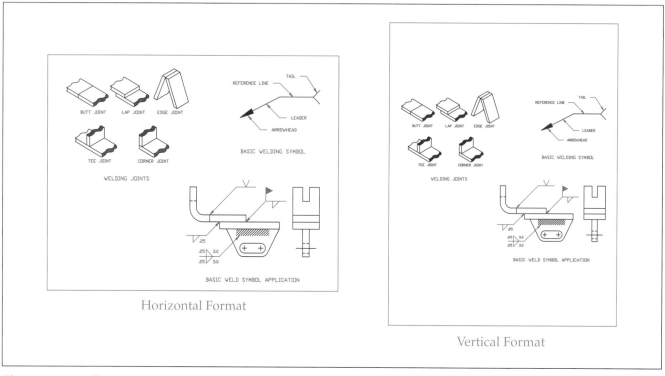

Horizontal Format

Vertical Format

Figure 3-8. An illustration in horizontal and vertical format. The drawing can be displayed at a larger scale using the horizontal format.

The illustration in **Figure 3-9** shows an example of cropping an image to fit a portion of the view into a picture area. *Cropping* a view or scene is cutting off one or more edges to remove an area or feature that is not needed. If the primary object is complete and surrounding objects are incomplete, this tells the viewer that the incomplete objects are not as important as the complete objects.

Controlling Depth Perception

Generating depth perception is a common way to make an illustration appear realistic. There are some applications where the illustration is best presented as a perspective drawing. A *perspective drawing* appears realistic,

Figure 3-9. An example of cropping. Software tools are used here to remove the right portion of the image while retaining the front view of the car. (Honda)

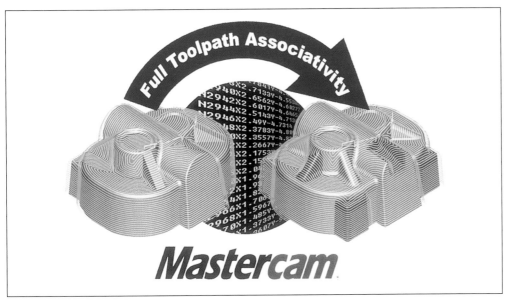

Figure 3-10. Object overlap is used in this illustration to emphasize the solid models and create the perception of depth. (CNC Software, Inc.)

with receding lines that converge at a vanishing point. When working with a perspective illustration, you need to make sure that the viewer will focus on the most important object in the illustration. There are several ways to do this.

The size of an object can determine the most important features in an illustration. When compared to other objects within a scene, the larger an object, the more important it appears. Therefore, an object of importance should be drawn in the foreground and appear larger than objects in the background.

Objects that are closer to the viewer not only appear larger, they also overlap and hide features of objects farther away. This principle is called *object overlap.* This is how objects appear in nature. An illustration can be confusing if this principle is not used to show which objects are near and which are distant.

The illustration in **Figure 3-10** shows how object overlap within a picture area creates the feeling of closeness. The larger scale used for the main focal objects seems reasonable to the viewer. More importantly, object overlap establishes the larger objects as the most important.

Varying size or scale in an illustration generates depth perception. As you look down a row of telephone poles, notice that the poles in the distance appear shorter than those closer to you. Objects behind the main object in the illustration should be drawn smaller to create a perception of distance. Also, parts of the objects located behind the main object should be hidden from view. This overlapping also helps create depth perception in an illustration.

Aerial perspective

Utilizing aerial perspective in a drawing is another way to create depth perception. *Aerial perspective* is viewing an object or scene from above the normal elevation. Normal elevation is about five and one half feet from the ground or floor. This is approximately how high your eyes are off the ground. The drawing views in **Figure 3-11** show how a viewing elevation located above the drawing produces an aerial perspective. Notice in the figure how overlapping helps create depth in each representation. Also, notice how the objects in each drawing are the same size and located very close together. However, **Figures 3-11B** and **3-11C** are aerial perspective views. The viewing angle has

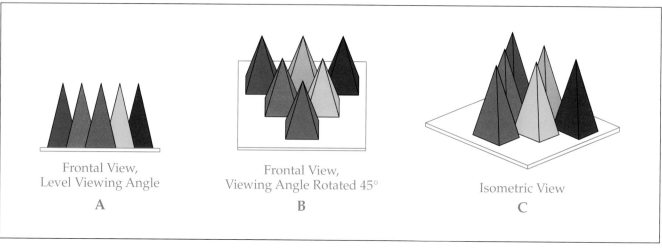

Figure 3-11. Three different views of an arrangement of pyramids. A—A view with no elevation provides very little perception of depth. B—Rotating the viewing angle to an aerial perspective provides a better perception of depth. C—An isometric view creates a greater perspective and provides the best representation of the objects.

been rotated to create depth perception. The view in **Figure 3-11C** is an isometric view and the most realistic representation. Isometric views and drawing techniques are discussed in Chapter 5.

Notice in **Figure 3-11** that the objects are not scaled differently to generate depth perception. If the second and third rows of pyramids were created at a reduced scale, there would not be enough size difference to be seen. Therefore, distant objects should not be scaled when they are close to the main object.

Atmospheric effects

Atmospheric effects can be used to generate depth perception in drawings. An **atmospheric effect** is caused when the air naturally reflects light in different directions. Atmospheric effects are always present, but you may only think about them on hazy or foggy days. Technical illustrators may use a specific atmospheric effect to create an optical illusion that gives a sense of depth.

To understand atmospheric effects, simply look out the window at trees, telephone poles, cars, or houses. Notice the detail of individual features. You can often see very small, intricate details on the objects nearest you. However, on a similar object farther away, you may only be able to see the large features. The smaller details may be blurred and indistinct. The large features on a similar object even farther away may be blurred. Look at one of these objects at an extreme distance and only the overall shape and color can be identified.

Color also tends to change with distance due to atmospheric effects. What is a deep, pure color close up appears lighter farther away.

When creating an illustration, blur objects in the distance. Also, change the shade of color used on distant objects. These two techniques together help add depth to the illustration. See **Figure 3-12**.

Moving the Viewer to a Focal Point

A **focal point** is any important aspect or feature on an illustration upon which focus is intended. See **Figure 3-13**. There are many ways to move the viewer to a focal point. A typical technique involves developing lines, voids, and a perspective effect. This very subtle technique can be accomplished by

Copyright Torbjorn Lagerwall

Figure 3-12. Color gradients and shading lines help establish a sense of distance in this illustration. (Macromedia FreeHand)

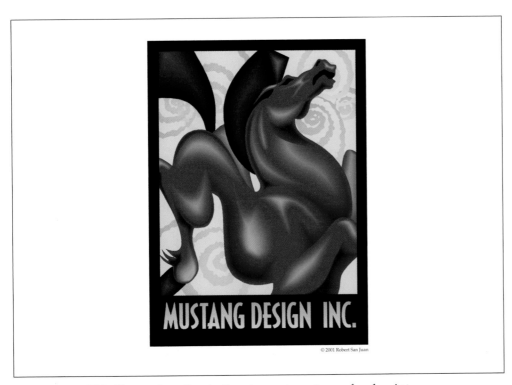

© 2001 Robert San Juan

Figure 3-13. This illustration directs the viewer to a strong focal point. (Macromedia FreeHand)

using lines or groups of lines to point the viewer in the desired direction. Normally, using lines alone is not sufficient to shift the viewer's focal point. By grouping objects in patterns and using some perspective drawing techniques, you can create subtle pointing directions to lead the viewer's eyes.

Controlling Values and Establishing Contrast

A *value* is the relative lightness or darkness of a color, shade, or tint. Maintain consistency of values within an illustration as much as possible. Consistent values appear realistic because lighting is naturally consistent throughout a scene. Whether the lighting intensity is low, moderate, or high, try to maintain consistent values.

Contrast is achieved when the variation of elements in a design is used to draw attention or provide meaning. Adding contrast while maintaining consistency is an important aspect of design. The proper amount of contrast can create a striking visual effect. For example, white space and different styles of type are often used to create contrast.

Use care when applying contrast to ensure that it does not detract from the design. For example, when selecting colors, avoid extremely dark colors or shading in areas with light colors. The overall illustration should have some variation in values, but not too much.

Establishing Proportion and Balance

Proportion and balance must be considered together, rather than as separate design principles. Achieving proportion without maintaining balance, or vice versa, will seriously detract from the illustration. *Proportion* is the size relationship of one part of an object to the size of the whole object. *Balance* is the sense of visual equilibrium in appearance.

Balance may be classified as formal or informal. *Formal balance* exists if one side of an illustration is a mirror image of the other side. *Informal balance* exists when opposing sides are not identical but appear to contain approximately the same number of lines, mass, or balance of colors.

There are no rules on proportion and balance that apply to all design situations. However, there are several guidelines that can be used in most instances. First, proportion does not include just the overall shape or mass of objects on the illustration. Proportion relates to combinations of small and large groupings of lines, areas, space, and entities.

Generally, proportional relationships of 1:1.618, 1:3, or 2:3 are considered more pleasing to the eye than the absolutely symmetrical formal balance found in a proportional ratio of 1:1. You will find that applying different proportional relationships will allow you to create appropriate arrangements depending on the nature of the design. For example, when working with illustration designs that deal primarily with rectangular shapes, you may consider using the golden section technique as the basis for establishing proportional guidelines for the shapes.

The *golden section* is a traditional design concept that is used to set proportion. The golden section creates a ratio of 1:1.618. The process for creating a rectangle using this ratio is shown in **Figure 3-14**.

Creating a golden section

Refer to **Figure 3-14** as you go through this example. First, draw a 1″ cube, **Figure 3-14A**. Then, locate the midpoint of one side. Using the radius value measured from the midpoint on the base to the top right corner of the cube, scribe an arc. Extend the baseline of the cube to meet the arc. Finally, draw the remaining lines. The rectangle has a height of 1″ and a length of 1.618″, so it has a ratio of 1:1.618.

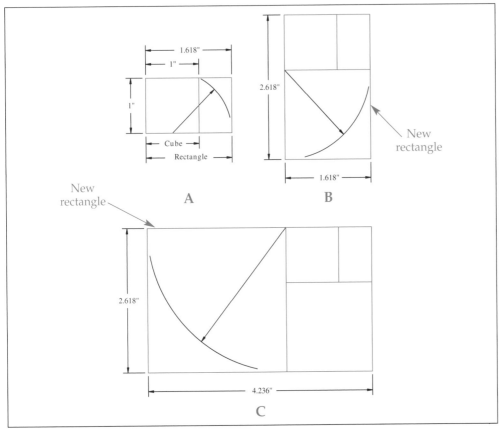

Figure 3-14. The golden section technique can be used to create rectangular shapes with a proportional relationship of 1:1.618. A—Starting with a 1″ cube, an arc is drawn using the length from the base midpoint to a corner as the radius. The base is then extended to meet the arc to form the base of the first rectangle. B—A second rectangle is drawn using the length of the longest side (1.618″) as the radius for an arc. The 1″ side of the existing rectangle is extended to meet the base of the arc, forming a side of the new rectangle. C—A third rectangle is drawn in the same manner.

To create another "golden rectangle" from the existing one, use the length of the longest side (1.618″) as the radius value and the lower left corner as the center. Scribe an arc as shown in **Figure 3-14B**. Then, extend the short side of the existing rectangle through the center to meet the bottom of the arc. This is the base of the new rectangle. Draw the remaining sides. The new rectangle measures 1.618″ × 2.618″. This measurement is also equal to a ratio of 1:1.618.

The rectangle in **Figure 3-14C** uses the rectangle in **Figure 3-14B** as the basis. The rectangle is drawn in the same way as the second rectangle, using the length of the longest side as the radius value and a corner as the center. The third rectangle measures 2.618″ × 4.236″. This measurement is also equal to a ratio of 1:1.618.

Establishing Unity and Variety

Establishing unity and establishing variety may sound like opposite goals. In reality, a good illustration will have both unity and variety at the same time. **Unity** is achieved when many different elements of an illustration are combined into an organized layout that creates a pleasing whole. Unity is needed for a graphic design to look appropriate and fit within the context of its application.

If you look at a design and something about it distracts you enough to ask the question "What's wrong with this design?," you are likely looking at an illustration without unity. The following are some examples of a lack of unity:

- Text styles inappropriate for the nature of the project are used.
- Color looks artificial in relation to the surfaces or objects to which it is applied.
- Portions of the graphics do not match the rest of the illustration.

Two layouts are shown in **Figure 3-15**. The layout in **Figure 3-15A** lacks unity because the text, graphics, and white space do not blend together. The graphics look like they have been added quickly as an afterthought. The lines of text at the bottom have a lack of emphasis and make the layout look disorganized.

The layout in **Figure 3-15B** has better unity. The graphic and text elements are pulled together by the surrounding white space. The graphic images have an orderly appearance and the bulleted lines of text help create a more pleasing layout.

While trying to make the various components of the illustration appear unified, you also want to create enough variety so the layout is not boring or monotonous. Variety is needed to create a level of intrigue in the overall presentation. See **Figure 3-16**. Variety can be achieved by using informal balance,

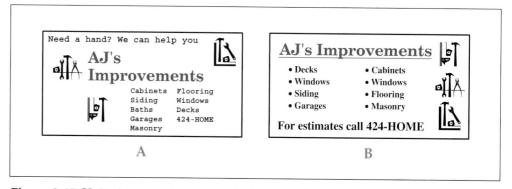

Figure 3-15. Unity is a very important design consideration. A—This layout lacks unity. There is little connection between the graphic images and text, and the text at the bottom has little emphasis. B—This layout has unity. The design elements are connected to create a pleasing layout.

Figure 3-16. The water and surrounding objects help add variety to this advertisement illustration. (Mitsubishi Electronics)

object size variations, and pattern breaks. A *pattern break* is created when an object overlaps a repeating pattern to disrupt the monotonous effect.

Maintaining Clarity

Clarity is a measure of how easily an illustration is understood. Clarity is a design principle that usually relates to text content. Using text that is legible, appropriately sized, and consistent with the style of the illustration improves clarity.

Clarity also relates to the words chosen for the message within the project. It is important that words be appropriate for the audience and spelled correctly, and that correct grammar and punctuation are used. Above all, the text must clearly convey the message of the project.

There are several things to keep in mind when developing a text message in an illustration. First, reduce the text content as much as possible without negatively affecting the illustration. In other words, provide all the information needed using as few words as possible. Second, try to frame text with white space. Use a suitable amount of white space around headings, titles, and other section designations. If you are using desktop publishing software, you may want to change the kerning, leading, gutter widths, or text fonts.

Kerning is a technique that reduces the amount of space between individual letters. By closing up the space between letters in a title or heading, you can give the printed letters more unity while increasing the white space surrounding the letters. *Leading* (pronounced *led-ing*) is the space between lines of type. Reducing the leading between related text elements can project the letters and words as a single unit or entity. Increasing the leading produces a more open, less congested appearance to large blocks of text. Keep the changes in leading minor. A small change in leading makes a significant difference in the text appearance. *Gutters* are the vertical spaces between columns of text. If you have a lot of text, wider gutters and a larger type size may make reading easier. However, this may also force you to consider more than one page or sheet if there is a great deal of text.

Using different styles of text fonts and changing the size of text are two ways to improve clarity. For example, some type fonts are more difficult to read than others. Serif font styles are generally more useful in straight text applications. Conversely, sans serif fonts are primarily used for visual display applications of text such as key words and illustration labels. Other specialized fonts are available for a variety of ornamental or decorative purposes. Text can also be printed in **boldface** style. A boldface character is printed approximately twice its normal thickness.

The following sentences are examples where special fonts and treatments are used. *This sentence is written with a font style called President, which may be appropriate for a formal invitation design, but not for a newspaper or book.* **This sentence is printed in the same size, but with a boldface sans serif font called Helvetica.** By comparison, this sentence is printed in the same size Helvetica font without boldface lettering.

Too many different fonts can be disruptive to the visual effect of the illustration. Generally, you should use no more than two or three different font styles on the same illustration. If you are developing a series of related illustrations, use the same font styles on the whole series to aid in unity and clarity.

The color of text can also help with clarity. There should be a high level of contrast between the text color and the color of the surface on which it lies. Dark lettering on a dark background can make the message difficult to read.

Light-colored text on a light background can also blend together to reduce legibility. Generally, use light text on a dark background or dark text on a light background for clarity.

Design Elements

Drawings used in technical illustration are developed using combinations of lines, planes, surfaces, and three-dimensional forms. These are basic design elements. Individually, each of these elements is generally limited in what it can accomplish. However, an illustrator can blend these entities and take advantage of their own special applications to achieve a visual goal. These design elements are discussed in the following sections.

Lines

A line represents the path of a point as the point moves through space. The most important characteristics of a line are its direction and the path it defines. The directions and paths of lines are used to describe the outlines or shapes of objects. A series of three or more lines with connecting endpoints defines a boundary that can be seen as a plane or surface. A series of interlaced lines can also be used to represent a three-dimensional form similar to a model made of wire. By developing a line along a curved path, rounded, warped, or irregularly shaped objects can be shown.

Planes and Surfaces

A *plane* is created by three or more connecting straight lines or a closed arc. In technical illustration, planes generally refer to the frontal, horizontal, and profile principal planes. They can also be used for inclined or skewed planes where images are projected.

The vast majority of technical illustration work is based upon developing pictorial views, rather than orthographic views. The importance of planes in technical illustration is primarily in their two-dimensional reference applications as you develop a pictorial drawing. For example, an isometric front face of an object uses the frontal plane as a reference to obtain the height and width measurements. A profile or horizontal reference plane provides the depth measurements.

Surfaces are used in technical illustration to represent the exterior boundaries of an object. Surfaces can be flat, regularly curved, or irregularly curved, or they can represent a compound curve. A compound curve arcs in two directions at the same time. Lines may be added to surface features to help define an edge.

In **Figure 3-17**, a portion of a spoked hub is shown illustrated with fillets and rounds. The elliptically shaped arcs are not real lines on the object. Rather, they are used to symbolize the surface characteristics where the fillets and rounds are located.

Surfaces are drawn and shaded to make the illustration appear realistic. Refer to the objects shown in **Figure 3-18**. In **Figure 3-18A**, a simple object is drawn as an isometric drawing. Although a viewer can visualize the relationship of one surface to another, the image of the actual surfaces has to be generated in the mind. For a viewer without technical training in drafting, this type

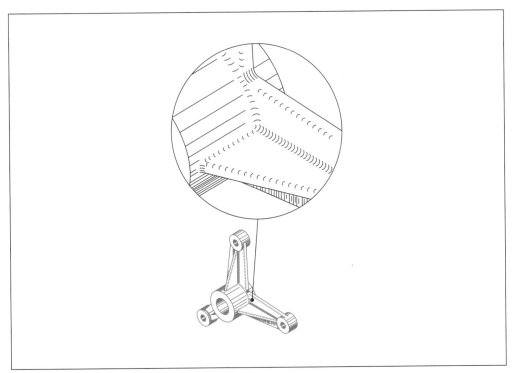

Figure 3-17. In this removed view, curved lines are used to show the surfaces created by fillets and rounds.

of drawing may not communicate the nature of the object. In **Figure 3-18B**, the same object is shown with stipple dots added. *Stipple shading* is applying a pattern of small dots in an image to create shadows or surface textures. The shading in **Figure 3-18B** allows all audiences to visualize the individual surfaces.

As you develop pictorial drawings using lines to define surface areas, it is often necessary to enhance the surfaces with a surface treatment. This helps the viewer distinguish one surface from another.

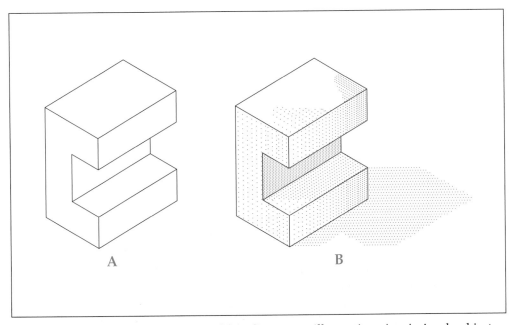

Figure 3-18. Shading surfaces can add realism to an illustration. A—A simple object shown in isometric view. B—Shading with stipple dots adds realism.

Three-Dimensional Forms

Pictorial drawings allow illustrators to show the height, width, and depth of an object in a single view. Pictorial drawing is covered in detail in Chapters 5 through 7. Developing an accurate three-dimensional (3D) illustration as a perspective drawing is often more difficult than developing a 3D model with a computer-aided drafting program.

As you evaluate design concepts used with lines, surfaces, 3D forms, and solid models, you will need to understand how they relate to one another. For example, suppose you have a box of toothpicks to use in illustrating an object. By arranging the toothpicks in an orthographic shape of an object on a flat sheet of paper, you have applied lines to the illustration. If you rearrange the toothpicks on the flat sheet of paper to represent an isometric view of the object, you have applied lines to symbolize surfaces. If you glue the toothpicks together in the shape of the object, you have used lines and surfaces to generate a 3D representation. The 3D representation can be further enhanced by covering the skeleton framework, or wireframe, with paper to further define the surfaces. This is similar to rendering a computer-generated 3D surface model.

The final step toward realism is to fill the inside of the paper skin with molten wax and let it harden. Now, when you pick up the object, you will be holding a 3D solid model that has substance and weight. When using a computer modeling program, you can even change the object by adding or removing portions. See **Figure 3-19**. This type of modeling can be used to simulate a variety of production machining processes.

Surface Characteristics

Whether developed manually or by using a computer, an illustration showing an object, a group of objects, or an assembly can appear artificial. This problem can be typically solved by adding some type of texture, color, or other surface treatment to enhance realism. You will need to use a variety of techniques to make

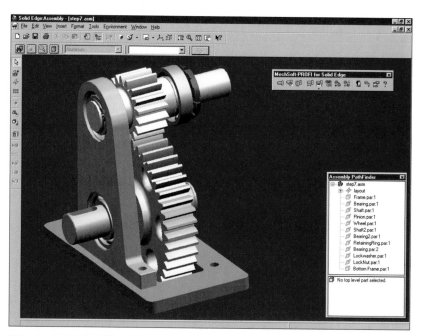

Figure 3-19. An illustration of a 3D solid model created in a CAD software program. Various modeling operations were used to create the mechanical gears and hole features. (MechSoft.com)

an audience perceive individual surface features, such as the surface boundaries, color, texture, and any other characteristics that should appear realistic.

Most illustration work is printed on paper as the final product. Paper-reproduced drawings have a tendency to appear flat, or two-dimensional. Therefore, you need to develop the image with surface treatments that will help create the appearance of being three-dimensional.

Applying various illustration techniques to produce realism is one of the major distinctions between a drafter and a technical illustrator. Where a drafter may produce an orthographic or pictorial drawing, a technical illustrator tries to make the object "come to life" for the viewer.

Applying surface characteristics is the primary way to produce realism. For example, you may need to use certain colors to show a specific type of material. A range of silver colors can be used to simulate aluminum, stainless steel, or other reflective metals. Several shades of orange can produce a look of copper or brass. Many computer rendering programs have predefined materials with colors already assigned. Most allow you to create new materials as well.

The texture of a surface is also an important characteristic to illustrate accurately. Sand-cast objects made of aluminum or cast iron have a surface texture similar to coarse sandpaper. This texture can be simulated using stipple shading. As discussed earlier, stipple shading is used to create shadows or surface textures. After shading is applied, the patterns of tiny speckles are seen as a whole pattern of dots. This makes the surface appear textured. The larger the dots in the stipple pattern, the rougher the surface will appear to the eye.

Airbrush shading is another technique used to create realistic surface characteristics. An example of a very realistic drawing with airbrush shading applied is shown in **Figure 3-20**. By varying the direction and the density of the shading, the shapes and contours of the car and its internal parts become quite realistic. This object drawn as a simple outline of the pictorial shape would be much less recognizable. Adding shading to create metallic reflections and transparent features gives the object instant recognition.

Figure 3-20. Numerous shading effects define the various engine parts and other details in this highly realistic airbrushed illustration. (David Kimble)

Summary

A high-quality illustration should be both visually appealing and appropriate in relation to the purpose of the design. The design problem-solving process is used to make sure that the illustration meets the visual requirements of the application. The design problem-solving process as applied to technical illustration consists of six basic steps:

1. Statement of the problem.
2. Analysis of design parameters, criteria, requirements, and project limitations.
3. Research.
4. Possible solutions.
5. Final solution.
6. Evaluation.

There are a number of basic design principles and concepts used in composition. Composition is the process of applying design principles and elements to develop a pleasing illustration. The primary principles of composition include the following:

- Defining and working with a picture area.
- Controlling depth perception.
- Developing a focal point for the viewer.
- Maintaining consistent values and establishing contrast.
- Generating proportion, balance, unity, and variety.
- Maintaining clarity.

Each of these principles is important. An illustrator normally applies most or all of these principles to develop the finished illustration.

Illustrations use lines, planes, surfaces, and three-dimensional shapes to explain visual concepts to an audience. Although the shape, direction, and boldness of lines may cause an emotional response, lines alone usually cannot communicate a visual message. Lines may be combined with other elements in a pictorial drawing to create surfaces that appear three-dimensional. Surface shading may also be applied to help simulate a 3D effect.

Above all else, a technical illustrator needs to develop illustrations that combine strong composition principles and effective pictorial drawing techniques so they are easy to understand. By using the problem-solving process to plan and develop illustrations, you will be a much more effective technical illustrator.

1. List and briefly describe the steps in the design problem-solving process.
2. How will performing the problem-solving process make you a better technical illustrator?
3. What is composition as it relates to technical illustration?
4. List the basic design principles used in composition.
5. Why is defining the picture area one of the very first activities performed when developing an illustration?
6. When would you omit some features and background portions of an object on an illustration?
7. Identify and describe at least three different techniques used to create a perception of depth in an illustration.
8. Define *focal point*.
9. Explain why some contrast should be added to the value of colors in an illustration. What happens if too much contrast is added?
10. What is the difference between proportion and balance?
11. What is the purpose of the golden section technique?
12. How can an illustrator have both unity and variety in an illustration?
13. Identify three simple techniques used to establish clarity in an illustration.
14. Define *kerning* and *leading* and describe the purpose of each.
15. What are *gutters*?
16. Explain how the selection of color for text affects clarity in a design.
17. What is the difference between an object drawn manually in isometric view and a solid model?
18. Identify three ways to make surface features on a drawing appear more realistic.

1. Using the design problem-solving process, complete the following project. Assume you are a technical illustrator who has been assigned to develop an illustrated cover sheet for the menu at a popular full-service restaurant called "The Hangar." This restaurant is located at a major city airport. Be creative, but do not include any text information beyond the name of the restaurant, the address, the phone number, and the owner's name. The menu cover should have a tasteful, visually appealing design. Final printed copies of the cover sheet will be inserted into an 8 1/2″ × 11″ transparent pocket in the front of a plastic folder holding other sheets with the menu items available. However, you are only designing the cover sheet. Use all six steps of the design problem-solving process to complete the project.
2. Develop two thumbnail sketches of graphic images for a safety awareness illustration. The designs should illustrate the use of proportion, balance, and unity. Each image should contain a safety warning with a graphic sketch and a minimum of words.
3. Collect three examples of technical illustrations. Identify how the illustrator used lines, planes, three-dimensional forms, and surface characteristics to communicate the message of each illustration.
4. Write a brief essay describing how an individual plans, organizes, and develops the design used in a technical illustration.

Advertisement illustrations are designed to represent the product in an attractive manner. In this example, the background shading complements the vehicle and generates appeal. (Ford)

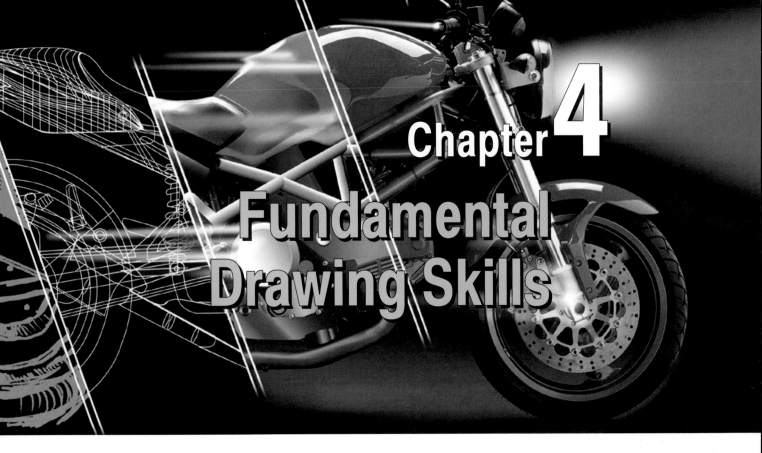

Chapter 4

Fundamental Drawing Skills

Learning Objectives

At the conclusion of this chapter, you will be able to:

- ☐ Identify and describe the lines in the alphabet of lines.
- ☐ Explain the fundamentals of orthographic projection, including first-angle projection and third-angle projection.
- ☐ Identify common lettering typefaces and measurement guidelines for type.
- ☐ Identify and describe the use of various lettering aids.
- ☐ Sketch multiview and pictorial drawings.
- ☐ Explain the basic functions and techniques used to create drawings with a CAD system.

Introduction

Certain basic drawing skills are needed for technical illustration. These skills include using the proper drawing technique, as well as having a good understanding of drafting conventions. This chapter covers basic drawing techniques such as orthographic projection and sketching. In addition, computer-aided techniques are introduced. Whether you are creating technical illustrations using traditional or computer-aided techniques, it is important to understand basic drafting conventions. This will allow you to easily convey your ideas to others.

Alphabet of Lines

Lines of various thicknesses and shapes are used to produce technical drawings. The *alphabet of lines* is an industry standard developed by the American Society of Mechanical Engineers (ASME) that classifies the different types of lines used in drawings. See **Figure 4-1.** The alphabet of lines is used by drafters and illustrators the world over. In general, lines are classified by line thickness. As shown in **Figure 4-1**, lines are typically drawn thin or thick, and the way they appear on a drawing indicates specific information.

The drafting standard for line conventions and lettering developed by the American National Standards Institute (ANSI) also classifies lines as thick and thin. According to the ANSI Y14.2M—1992 standard, there should be an

Line	Appearance
Construction	
Visible/Object	
Hidden	
Center	
Dimension	
Extension	
Phantom	
Long Break	
Short Break	
Cutting-plane/ Viewing-plane	
Section	
Chain	

Figure 4-1. Line conventions used in the alphabet of lines. Lines are classified as thin and thick. The ANSI Y14.2M—1992 standard recommends an approximate ratio of 2:1 for thick and thin lines. Thin lines are 0.3 mm or 0.5 mm thick; thick lines are 0.7 mm or 0.9 mm thick.

approximate 2:1 ratio of line width for thick and thin lines when drawn by hand. If a 0.9 mm line is used for thick lines, a 0.5 mm line should be used for thin lines. If a 0.7 mm line is used for thick lines, a 0.3 mm line can be used for thin lines. When working with CAD drawings, a single line width is acceptable for all lines.

On technical drawings, thick lines are drawn very dark (dense) with a soft lead. If the lead is too soft, the lines on your paper will smudge. If the lead is too hard, the lines will be gray and fuzzy. A 0.7 mm or 0.9 mm mechanical pencil can be used for thick lines. Thick lines are used as visible (object) lines, cutting-plane lines, and some break lines, **Figure 4-2.**

Thin, dark lines are also drawn with a soft lead. A 0.3 mm or 0.5 mm mechanical pencil can be used for thin lines on most drawings. Thin lines include hidden lines, centerlines, extension lines, and dimension lines.

Construction Lines and Guidelines

Very light lines are used for construction lines and guidelines. *Construction lines* are used to initially lay out a drawing. *Guidelines* are used to make sure that freehand lettering is uniform in size. Both types are drawn so thin they are barely visible. The lines should be light enough not to show on copies. They are made with a 0.3 mm pencil or a sharpened drafting pencil with a hard lead. If these lines are drawn lightly, they will not need to be erased when the drawing is completed.

Visible Lines

Visible lines are also called *object lines*. These lines represent the visible edges of an object. They are drawn thick and dark. A 0.7 mm or 0.9 mm drafting pencil with a soft lead is used to draw visible lines.

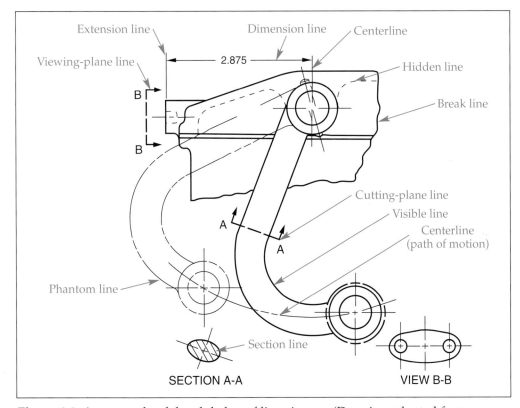

Figure 4-2. An example of the alphabet of lines in use. (Drawing adapted from ASME Y14.2M)

Hidden Lines

Hidden lines are used to represent edges that are hidden from view. They are drawn by placing 3 mm dashes about 1.5 mm apart. This size and spacing can vary slightly depending on the size of the drawing. Hidden lines are drawn as thin, dark lines.

Centerlines

Centerlines are drawn as thin lines. They are used to show the center of circles and arcs. Centerlines consist of two long dashes and a short dash separated by a 1.5 mm gap. The short dash is about 3 mm long and is placed at the center point of the circle or arc. The length of the long dash varies, depending upon the size of the object, but should extend just beyond the last concentric circle.

Dimension Lines, Extension Lines, and Leaders

Dimension lines, extension lines, and leaders are used for dimensioning and notes on the drawing. They are all drawn as thin lines. *Extension lines* are used to extend dimensions from the objects in the drawing. A short space (1.5 mm) is placed between the object and extension line. The extension line extends to about 3 mm beyond the last dimension line.

A *dimension line* is typically placed between extension lines and has an arrowhead on each end. The actual dimension is placed in a break along the line.

A *leader* consists of an arrowhead and angled line connected to a shoulder. The shoulder points to a note or dimension. The angle of the leader is typically 30°, 45°, or 60°.

Cutting-Plane and Viewing-Plane Lines

Cutting-plane lines and viewing-plane lines are used to indicate that another view of the object is drawn. A *cutting-plane line* indicates where an imaginary cut is made on the object. The "cut" surface is shown in a section view. A *viewing-plane line* indicates the area for which a separate view is drawn. For both types of lines, the arrows indicate the direction the viewer is looking at the section or view.

Both cutting-plane and viewing-plane lines can be drawn in either of the ways shown in the alphabet of lines in **Figure 4-1.** When using alternating long dashes and pairs of short dashes, the short dashes are about 3 mm long, and the spaces next to dashes are 1.5 mm.

Section Lines

Section lines represent a surface that has been cut by a cutting-plane line. For general applications, these lines are drawn thin and equally spaced at 45°. However, if other lines on the object are 45°, some other angle should be chosen, such as 30° or 60°. Spacing for section lines varies, depending on the size of the drawing. Normally, about 1.5 mm spacing is used.

Break Lines

Break lines indicate that part of the object has been removed. A short break is indicated by a thick (0.7 mm or 0.9 mm) freehand line. A long break line is drawn as a thin (0.3 mm or 0.5 mm) line with freehand "zigzags" at equal intervals.

Phantom Lines

Phantom lines are used to indicate alternate positions for moving parts on the object or to indicate repeated details. For example, a phantom line can be used to show the two positions of a switch. An example of an object with repeated details is a spring. The beginning and end of the spring can be drawn with two phantom lines connecting them. This implies that the same details are carried throughout. Phantom lines are drawn thin and are made up of long dashes connected by two short (3 mm) dashes. Spacing is approximately 1.5 mm.

Chain Lines

Chain lines are placed next to surfaces that are to have some treatment, such as hardening. They are drawn as thick lines made up of alternating long and short dashes.

Orthographic Projection Fundamentals

Orthographic drawing is a standard form of drawing used in drafting. This type of drawing is made with the alphabet of lines. *Orthographic* means *right-angle drawing*. **Orthographic projection** is the process of representing several different views of a three-dimensional object to provide a complete description. Orthographic drawings are also known as **multiview drawings**. When products are designed to be manufactured, multiview drawings are typically used. Each view shows one side of the object in two dimensions. Since only two dimensions are shown on each view, more than one view is usually needed to fully describe the object.

Projecting Views

Orthographic projection involves projecting each side, or view, of an object onto an imaginary plane known as a **projection plane.** Using this technique, each view of an object has a specific location on a drawing in relation to the other views.

Orthographic projection is easily visualized by thinking of an imaginary glass box placed around the object, **Figure 4-3.** Each side of the box becomes one view of the object. A *projector* transfers each point on the object to the side of the glass box. These projectors are at right angles, or perpendicular, to the glass box.

Each side of the box is a projection plane. When all the views are projected onto the sides of the glass box, it can be unfolded, **Figure 4-4.** Notice that the views are aligned with each other. There are six **principal views,** or **normal views,** of an object, just as there are six sides of the glass box. The top and bottom views are also known as **horizontal views.** The side views are called **profile views.** The front view is called the **frontal view.**

Choosing the Front View

Any of the principal views of an object can be used as the front view. However, take the time to choose the one that is most appropriate. The front view should be the view that best shows the shape or contour of the object while having the fewest hidden lines.

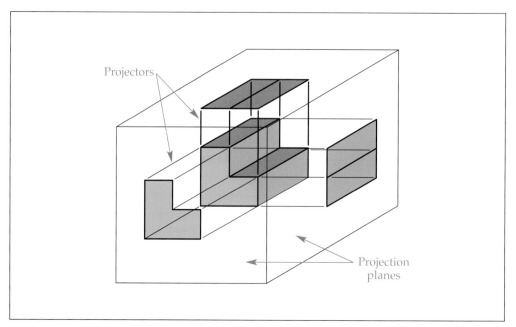

Figure 4-3. Three views of the object are shown projected onto the projection planes.

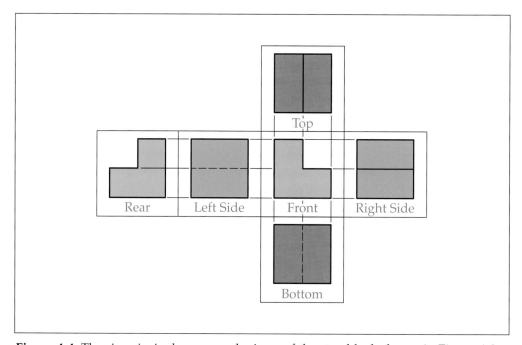

Figure 4-4. The six principal, or normal, views of the step block shown in Figure 4-3.

Determining the Number of Views Needed

Once you decide on the front view, then you can decide on the number of views needed. The front view is always shown on a multiview drawing. After drawing the front view, you should try to draw the fewest views possible, while still fully describing the object. The views chosen should best describe the object with the least number of hidden lines.

Three views of an object are commonly used in a multiview drawing. These views are typically the front, right side, and top views, **Figure 4-5.** However, sometimes more than three views are needed. For some objects, one or two views is satisfactory. See **Figure 4-6.**

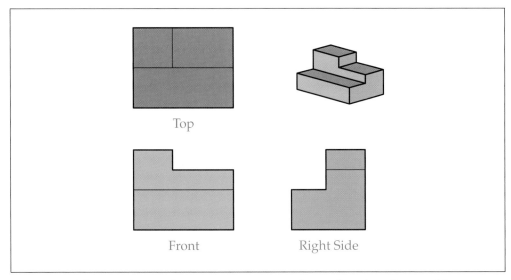

Figure 4-5. Three orthographic views are commonly used to describe an object.

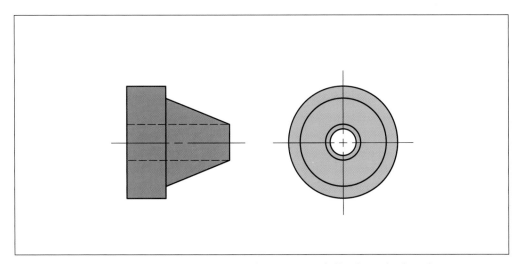

Figure 4-6. Many circular objects, such as this one, are fully described with two views.

Understanding Third-Angle and First-Angle Projection

There are two types of projection used in orthographic drawings: third-angle projection and first-angle projection. Third-angle projection is used in the United States. However, first-angle projection is used in most European countries.

In *third-angle projection*, the sides of the object are projected to the sides of the imaginary glass box and then toward the viewer. Referring to **Figure 4-4**, this can be visualized as unfolding the sides of the box with the projected views forward. In other words, when looking at the front of the box, the sides of the box are unfolded toward the viewer. The views then appear in their natural configurations when revolved to the frontal viewing plane. In *first-angle projection*, the sides of the object are projected to the sides of the imaginary glass box and then away from the viewer. This can be visualized as unfolding the sides of the box with the projected views toward the "back" of the box. In other words, the views are projected to the rear and the frontal viewing plane is "behind" the object. This, in turn, reverses the configurations of the views from their orientations in third-angle projection. When looking at the frontal viewing plane, the top view is below the front view, the bottom view is above the front view, the right view is to the left, and the left view is to the right.

As shown in **Figure 4-7,** different symbols are used to indicate whether an object has been drawn using first-angle or third-angle projection. If no symbol appears on the drawing, it is assumed that the drawing has been prepared using third-angle projection.

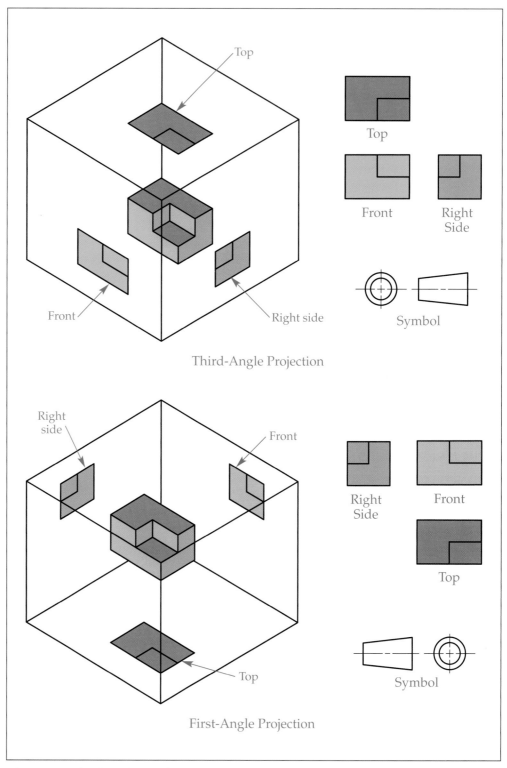

Figure 4-7. Third-angle and first-angle projections of the front, top, and right-side views of an object. Note the differences between the views. The symbols for first-angle and third-angle projection are also shown.

Projecting Features and Surfaces

In orthographic projection, a feature is projected in its true size and shape if it is parallel to the projection plane. If the feature is not parallel to the projection plane, it is foreshortened. *Foreshortening* means that the feature has been drawn smaller than the true size and shape. For example, if a circle representing a hole is at an angle to the projection plane, it appears as an ellipse in the orthographic view. See **Figure 4-8.** The angle of the ellipse is dependent upon the angle from which the circle is viewed in a specific projection plane. For example, if the viewing angle is 45°, then a 45° ellipse is drawn.

Surfaces that are parallel to the projection plane are called *normal surfaces.* They are projected as true size. *Inclined surfaces* are perpendicular to two planes of projection, but inclined to all others. These surfaces appear as lines where they are perpendicular to the projection plane and as foreshortened surfaces in other views, **Figure 4-9.**

Skewed surfaces, or *oblique surfaces,* are not perpendicular or parallel to any plane of projection, **Figure 4-10.** An oblique surface appears foreshortened in all views.

Producing the Drawing

Often, a sketch of a multiview drawing is made before the final drawing. Grid paper helps in creating the sketch. After the sketch is done, you must

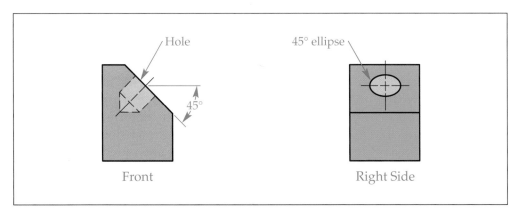

Figure 4-8. A circle that is not parallel to the plane of projection is seen as an ellipse because it is viewed at an angle to the line of sight.

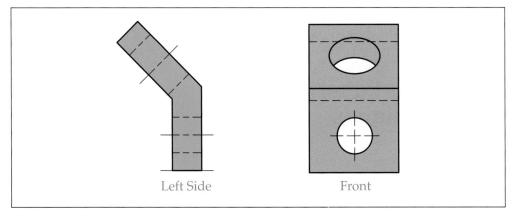

Figure 4-9. The upper inclined surface is seen as a line in the left-side view and as a foreshortened surface in the front view.

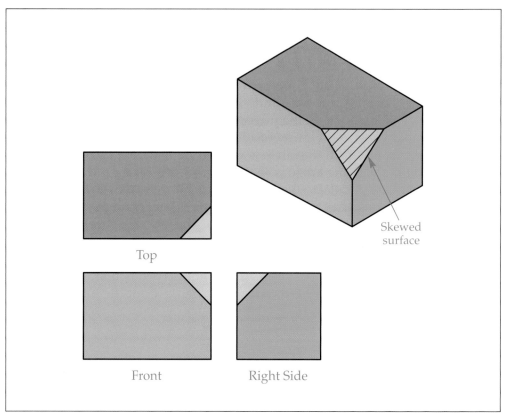

Figure 4-10. A skewed surface is foreshortened in all views.

determine the scale and position of the drawing. Use the following steps to create the drawing:

1. Center the drawing in the drawing space provided. In a three-view drawing, the horizontal dimensions of the front and right side views (plus space between the views) are added to determine horizontal centering. For vertical centering, the vertical dimensions of the front and top views (plus space between the views) are added together.
2. Starting with the front view, "block in" the views with construction lines using the overall width, height, and depth of the object. Make sure the views are aligned and in correct position. Some features can be projected between views by extending construction lines.
3. Lay out the rest of the drawing. To save time, transfer distances from one view to another using dividers.
4. Finish the drawing by darkening the lines. Start with circles and arcs first (if any). Then, connect straight line segments to the ends of the arcs. Finally, darken the remaining lines.

Illustration Lettering Guidelines and Techniques

Various guidelines are used for lettering, or placing text on, technical illustrations. Lettering for illustrations can be done manually, with lettering aids, or by using a CAD program. Regardless of how the lettering is done, it is important to understand typefaces, how type is measured, and the terminology used with typefaces.

Typeface Terminology and Classifications

Special terms are used to describe typefaces and typography applications. Capital letters are known as *uppercase* letters. Small letters are known as *lowercase* letters. In publications, large letter type used to attract attention is called a *headline* or *display type*. Small letter type is called *text* or *body copy*. All the characters of a single typeface style are known as a **font**.

There are four major categories of typefaces. These are Roman, or serif, sans serif, script, and Old English, **Figure 4-11.** Within each of the typeface categories, there are hundreds or even thousands of specific typefaces from which to choose. For example, Times Roman and English Times both belong to the typeface classification of serif type, but each is a different typeface.

Roman

Roman or **serif** typefaces have thin horizontal and vertical strokes at the ends of the letters. These thin strokes are called **serifs.** It is thought that serifs were originally copies of chisel marks used to place letters on stone buildings. This type is often used in publications such as books and magazines for text or body copy. It also works well for headlines.

A special type of serif typeface is called *square serif.* This is similar to normal serif type, but the serifs are blocks. The square serif typeface captures your attention, but it is not as easy to read in large blocks of text. It is primarily used in headlines and advertisements.

Sans serif

The French word *sans* means "without." Thus, **sans serif** typefaces do not have serifs. The clean design of sans serif type is very legible. It is used for a variety of publishing applications and works well for headlines or text. Most lettering done on technical drawings uses sans serif type.

Roman (Serif)	ABCDEFGHIJKLMNOPQRSTUVWXYZ abcdefghijklmnopqrstuvwxyz
Sans Serif	ABCDEFGHIJKLMNOPQRSTUVWXYZ abcdefghijklmnopqrstuvwxyz
Script	*ABCDEFGHIJKLMNOPQRSTUVWXYZ* *abcdefghijklmnopqrstuvwxyz*
Old English	𝔄𝔅ℭ𝔇𝔈𝔉𝔊𝔥𝔦𝔧𝔨𝔩𝔪𝔫𝔬𝔭𝔮𝔯𝔰𝔱𝔲𝔳𝔴𝔵𝔶𝔷 abcdefghijklmnopqrstuvwxyz

Figure 4-11. Four common typeface categories.

Script

Script typefaces resemble cursive handwriting, giving the characters a warm, personal look. Some script faces have joining letters and some do not. A script typeface is almost never set in all uppercase letters. Script is popular for announcements and advertisements.

Old English

Old English is a special script typeface classification. It is a version of the style that was used by scribes in the Middle Ages to write books. The letters have extra strokes for ornamentation. Old English type should never be used in all uppercase because it is difficult to read. This style provides a formal appearance to the printed product. It is used for invitations, announcements, and advertisements.

Novelty

Novelty typefaces include special typefaces that do not fit in the other four major categories. These are unusual and decorative typefaces. Large companies often have a novelty style developed to represent the company. Advertising is the major area of use for novelty typefaces.

Type Measurement Guidelines

Once a typeface is selected, a size for the type is determined. There are several standard ways to measure type sizes. In some cases, the height of type may be measured in inches or millimeters. However, typeface measurements are most often made in points and picas. These are standard units of measure used in printing. A *point* is equal to 0.0138″. A *pica* is equal to 0.166″. There are approximately 72 points in one inch, and 12 points are equal to one pica. Six picas are approximately equal to one inch. A *line gauge* is used to measure type, **Figure 4-12.**

The point size of a typeface is measured from the top of the highest ascender to the bottom of the lowest descender, **Figure 4-13.** For example, the

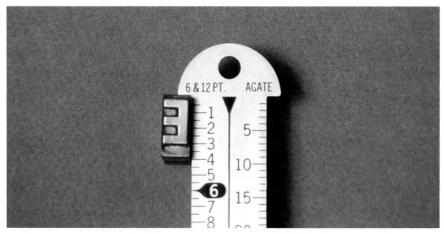

Figure 4-12. A line gauge is used for measuring type and line lengths in points and picas. When measuring type sizes, 12 points are equal to one pica and six picas are approximately equal to one inch. In this photo, the line gauge scale being used is divided into increments of six and 12 points. The reading indicates that the block of type is 48 point.

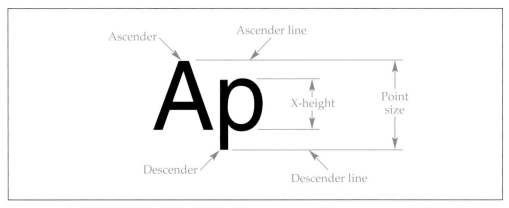

Figure 4-13. Terminology associated with type size.

highest ascender may be the top of a *B*, while the lowest descender may be the bottom of a *y*. Therefore, the point size is measured from the top of the *B* to the bottom of the *y*.

The numbers on the side of a line gauge represent picas. Conversion to points can be easily made by multiplying by 12. Picas are often used for line lengths, while points are used for type height. For example, a newspaper might use 12-pica columns. The other side of the line gauge allows measurement in inches or millimeters. Some gauges are also divided in agates. One *agate* is approximately 5 1/2 points.

In addition to type size, you must specify the amount of space between lines. In the publishing industry, the total line spacing is known as *leading.* Leading includes the point size of the type and the space between lines. However, the term *line spacing* often refers to just the space between the lines, and discounts the type size. Be sure of what is actually being referred to when using the terms "leading" and "line spacing." For example, "ten on twelve" leading typically means that 10-point type is set with two points of space between the lines.

Leading, type size, and typeface style are usually specified together. For example, 10/12 Helvetica means to use the Helvetica typeface with 10-point size and 12-point leading. If no extra space is used between lines, the type is said to be *set solid.* An example of this would be 10/10 Helvetica. Setting a typeface solid is not usually done because the ascenders and descenders of two adjacent lines come into contact with one another.

Freehand Lettering

Placing text characters such as letters and numbers on a technical drawing is called *lettering.* Lettering requires practice, especially if it is to be done free-hand. A 0.5 mm pencil with a soft lead, such as an F grade lead, is normally used. This section discusses manual lettering techniques. When using a CAD program, the lettering process is performed with text commands. Basic CAD drawing functions are discussed later in this chapter.

A single-stroke, sans serif typeface called Gothic is used on most technical drawings, **Figure 4-14.** This style is easy to draw and read. It is called a single-stroke face because each character is made with single strokes that connect to one another. The characters vary in width. On many drawings, all uppercase letters are used.

As shown in **Figure 4-14,** Gothic lettering can be vertical or inclined. Vertical characters, as the name implies, have their upright parts straight up, or vertical.

ABCDEFGHIJKLMNOPQRSTUVWXYZ
abcdefghijklmnopqrstuvwxyz   0123456789

ABCDEFGHIJKLMNOPQRSTUVWXYZ
abcdefghijklmnopqrstuvwxyz   0123456789

Figure 4-14. The single-stroke Gothic typeface is commonly used in lettering. Characters can be vertical or inclined.

Inclined characters are tilted at 68°. Either style can be used, but they should not be mixed.

Characters are usually drawn 3 mm (1/8″) high. However, larger characters may be used on larger drawings. Lettering should be kept in proportion to other parts of the drawing. It may also be necessary to increase the line thickness of the characters as the drawing size increases.

Spacing is very important when lettering by hand, **Figure 4-15.** *Letter spacing* is the spacing between letters in a word. This should be done accurately by eye so the space between letters appears equal. *Word spacing* is the spacing between words. The spacing between words should be equal to the height of the letters. *Line spacing* is the space between lines. This is normally equal to one-half the letter height or the full letter height.

When lettering by hand, guidelines are used to help with spacing, **Figure 4-16.** As previously discussed, guidelines are very light construction lines. Horizontal guidelines are used for letter height and line spacing. Vertical guidelines are used to keep letters straight and are drawn randomly. Inclined guidelines are used to keep inclined letters at 68°. Guidelines can be drawn with a scale and common drafting tools or with a lettering guide.

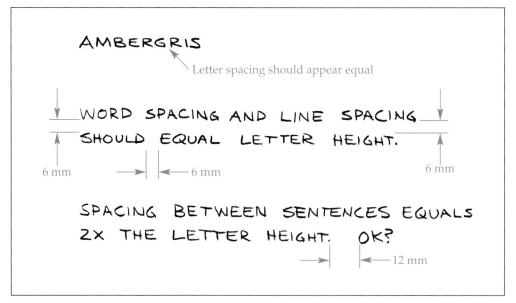

Figure 4-15. Letter spacing, word spacing, and line spacing principles used in lettering.

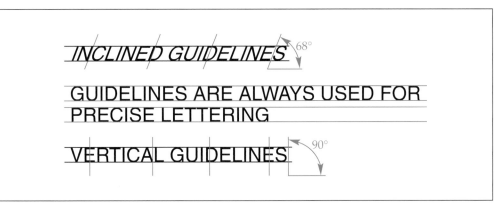

Figure 4-16. Guidelines are useful when lettering by hand.

Lettering Tools and Devices

Although lettering can be done freehand on drawings, there are drawing tools available that allow you to use other techniques. These include lettering templates, templates and scribers, and lettering machines.

A *lettering template* is a plastic strip with precisely cut characters along its length. The side of the template rests against a straightedge to ensure alignment of the letters. See **Figure 4-17A.** Lettering templates are often used when inking a drawing and are available with openings of different widths corresponding to the pen diameter being used.

Another lettering tool consists of a special template and scriber used with a technical pen, **Figure 4-17B.** This device is commonly known as a Leroy lettering guide. The scriber holds a technical pen. A tracing pen on the scriber traces the letter on the template while the technical pen duplicates the letter on the paper. The scriber can be set for vertical or inclined lettering.

Lettering can also be produced using a lettering machine. A *lettering machine* is used to create strips of lettering on clear, adhesive-backed plastic. After the lettering is created, it is simply removed from the strip and pressed in place on the drawing. A variety of fonts are usually available with a lettering machine.

A   B

Figure 4-17. Templates assist an illustrator when lettering a drawing. Note the vertical orientation of the technical pen in both figures. A—Lettering template. B—Template and scriber in use. (Staedtler, Inc.)

Fundamental Sketching Techniques

Sketching is a quick way to communicate ideas visually. You can use a sketch to store your mental image on paper. You can also share sketches with others to develop different ideas. When you know how to sketch skillfully, your ideas for products can be developed and shared more accurately and quickly.

In industry, sketches are used in the product planning stage. By drawing sketches first, changes can be made before a large amount of time and money is invested in finished drawings. For example, a floor plan for a home might be sketched by an architect and shared with the client. The client and architect then discuss changes and revised sketches are made prior to the finished drawing. Eventually, a complete set of plans for the house can be produced.

Materials Used in Sketching

Only a few drawing tools are needed for sketching. In fact, many sketches are done with any pencil and paper that is available. However, there are materials that can assist you in making good sketches. These materials are covered in the next sections.

Pencils

Use a drafting pencil with a soft lead or a writing pencil for sketching. For thin lines, use a sharpened point. For thick lines, use a well-worn conical point. Any pencil eraser or drafting eraser is good for erasing. Be sure that the eraser, regardless of the type, is clean and free of contaminants (such as pen or marker ink).

Paper

Plain paper or grid paper can be used for sketches. However, sketching is usually easier with grid paper, producing better results. The grids most often used are square and isometric grids, as discussed in Chapter 2. Various grid sizes are available. If you use plain, translucent paper for sketching, a grid can be placed behind the paper as a guide and removed for reuse. When sketching, the paper is usually not fixed to a surface so it can be moved to the most comfortable drawing position.

Other aids

For transferring distances on a sketch, the edge of a second sheet of paper can be used. Place the edge so that it just touches the endpoints of the distance. Then, mark the distance on the paper. Finally, move the paper to where the distance is being transferred and mark it on the sketch.

Sketching Procedure

Sketches can be made in a variety of ways. Regardless of the means chosen to sketch an object, the following items should be considered:

1. Decide how many sketches are needed to completely describe the object and determine how much time will be spent on each one. Sketching ideas (visual brainstorming) usually takes less time than a final sketch.

2. Hold the pencil in a writing position. As an alternative, the pencil can be held so you are sketching more with the edge of the pencil.
3. If the object or drawing is accessible, study it. Notice the proportions of parts. Begin with height, width, and depth. What is the approximate angle of lines to be drawn in relation to an imaginary horizontal or vertical line?
4. Relax and have confidence. Sketch the basic shapes of the object using construction lines. Make changes as needed until the proportion is correct. Use very light lines so you can draw new lines for corrections without erasing the first lines.
5. Darken the correct lines.
6. Add details after the general proportions are established.

Establishing Proportion

Proportion is a design concept that involves drawing each part of an object so it is the correct size in relation to the entire object. If a square is drawn correctly, the width and height will be the same, or "in proportion." In order to draw an object in proportion, study it first and determine the size relationship of various parts. For example, a rectangle might be twice as long as it is high. Start by looking at the height, width, and depth of the object. Sketch so the proportion appears visually correct. Always think about proportion before sketching.

Sketching Lines and Shapes

Lines and shapes are the basic elements of a technical drawing. Shapes have two dimensions—width and height. Sketching lines and shapes properly helps the appearance of the drawing, and also improves sketching speed. In fact, sketches are usually just a group of lines and shapes.

Straight lines

When drawing a straight line, locate the beginning and ending points of the line with dots. Draw short construction lines between the dots to connect them, **Figure 4-18.** Then, darken the visible line. Short lines can be drawn in one stroke.

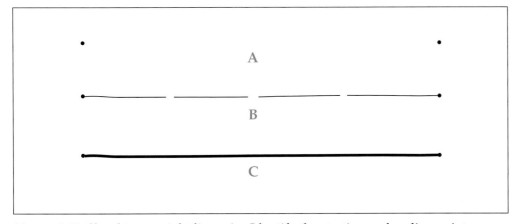

Figure 4-18. Sketching straight lines. A—Identify the starting and ending points. B—Draw short construction lines. C—Darken the visible line.

Irregular curves

An *irregular curve* is a curved line with no single center, **Figure 4-19.** A string tossed on the ground forms an irregular curve. To sketch an irregular curve, first make light dots on the curve. Connect the dots smoothly with short construction lines. Finally, darken the curve.

Angles

Angles are created by drawing two lines that intersect at a point. A right angle is 90°. Acute angles are less than 90°. Obtuse angles are greater than 90°. See **Figure 4-20.** To sketch an acute or obtuse angle, first draw a 90° angle as a guide. Then, to form the acute angle, draw a line inside the 90° angle. To sketch an obtuse angle, draw the correct angle outside the right angle.

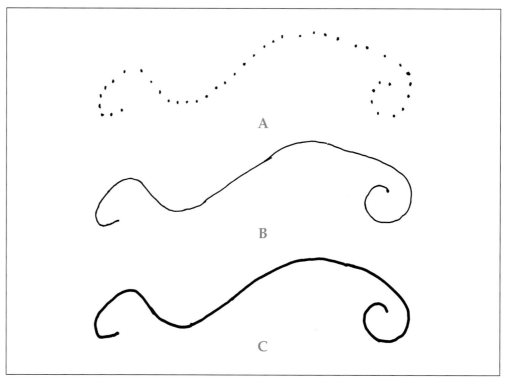

Figure 4-19. Sketching an irregular curve. A—Identify points along the curve. B—Draw short construction lines to connect the dots. C—Darken the curve.

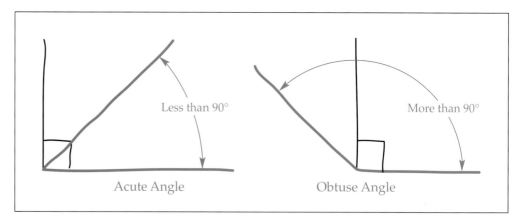

Figure 4-20. Begin with a right angle when drawing an acute or obtuse angle.

Squares and rectangles

To sketch a square or rectangle, first draw a set of axes with construction lines. Then, lay out the width and height, **Figure 4-21A.** If plain paper is being used for the sketch, use the edge of a second sheet of paper to get an approximate measurement. Lightly sketch the sides. Finally, darken the lines.

Circles

A circle is a 360° curved line drawn about a center point, **Figure 4-21B.** To sketch a circle, first draw a horizontal and vertical axis for the circle. Next, draw two axes at 45° to the ones already drawn. (Smaller circles may require only the horizontal and vertical axes.) Lay out the radius of the circle from the center point on each axis. Sketch the circle lightly, turning the paper as necessary. Finally, darken the lines.

Ellipses

An ellipse is an elongated circle. Unlike a circle, the arcs that form an ellipse have two different radii. These values are measured on a *major axis* and a *minor axis*. Start by drawing the axes, **Figure 4-21C.** Next, sketch a rectangle with a length equal to the major axis and a width equal to the minor axis. Then, sketch construction lines diagonally across the rectangle. From the intersection of the axes, mark a point about two-thirds the distance out on each diagonal. These will be points on the ellipse. Sketch the ellipse, making sure the arcs meet smoothly.

Arcs

Arcs are lines that curve less than 360° around a center point. Many times, arcs are used to round a corner, **Figure 4-22.** To round a 90° corner with an arc,

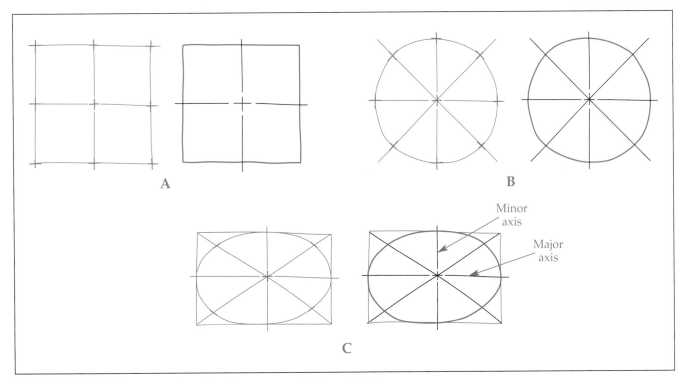

Figure 4-21. Procedures used to sketch a square, circle, and ellipse.

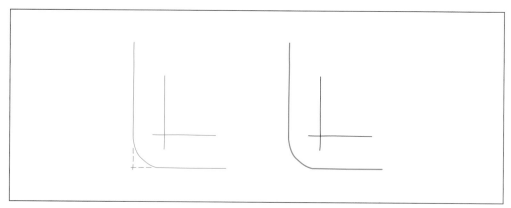

Figure 4-22. Sketching a rounded corner with an arc.

first draw a construction line parallel to one of the two lines forming the 90° angle. The distance the construction line lies from the 90° lines should be equal to the radius of the arc. Repeat the procedure on the other line, making sure the two construction lines intersect. Lay out the radius from the intersection of the two construction lines and sketch in the arc.

Creating Pictorial Sketches

Pictorial drawings show height, width, and depth in a single view, **Figure 4-23.** The three major types of pictorial drawings are axonometric, oblique, and perspective. The most common pictorial sketch is a type of axonometric drawing called an *isometric drawing*. An isometric drawing uses isometric lines to define its shape. Isometric lines are drawn vertically (to represent vertical lines on the object), or at 30° from horizontal to represent the width and depth of an object. Refer to **Figure 4-23.** Isometric drawings and other types of axonometric drawings are discussed in greater detail in Chapter 5.

Special graph paper called isometric grid paper can be used to help create an isometric sketch. Isometric grid paper is similar to normal graph paper, but the angled lines correspond to isometric lines.

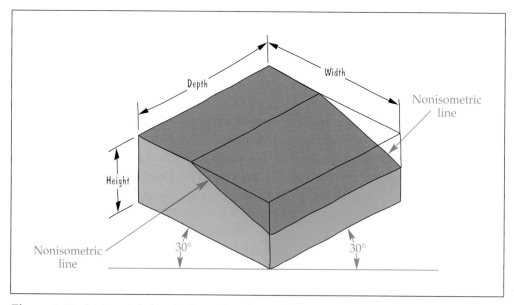

Figure 4-23. A pictorial sketch shows height, width, and depth. This isometric sketch consists of isometric and nonisometric lines.

To start an isometric sketch, draw a box using the maximum dimensions of the object. Draw other isometric lines as necessary for details.

Not all lines in an isometric drawing are drawn at 30° angles or vertical. Lines that are drawn at an angle other than 30° or 90° from horizontal are called *nonisometric lines*. Referring to **Figure 4-23**, these lines are drawn by first locating points on isometric lines and connecting the points. Darken the sketch when all lines are located.

Circles appear as ellipses in isometric drawings, **Figure 4-24.** To draw an isometric circle, start by drawing an "isometric square" with the sides equal to the circle diameter. Then, draw centerlines that intersect the midpoint of each side of the box. The midpoints are tangent to the isometric circle. Next, sketch the ellipse using construction lines. Finally, darken the lines when the ellipse appears correct.

In **Figure 4-24**, notice the different orientations of the isometric circles on each face of the isometric box. When adding isometric circles, the resulting ellipses must be oriented correctly on each isometric plane. The object in **Figure 4-24** shows the correct orientation of ellipses in an isometric view. Constructing isometric circles properly is discussed in greater detail in Chapter 5.

Creating Multiview Sketches

For production applications, it is often necessary to provide detail by showing different views of an object. A multiview drawing is used for this purpose, and it is typically dimensioned to show actual sizes of features on the object. Multiview sketches often precede finished multiview drawings. The sketching techniques presented in this chapter work well for multiview sketches.

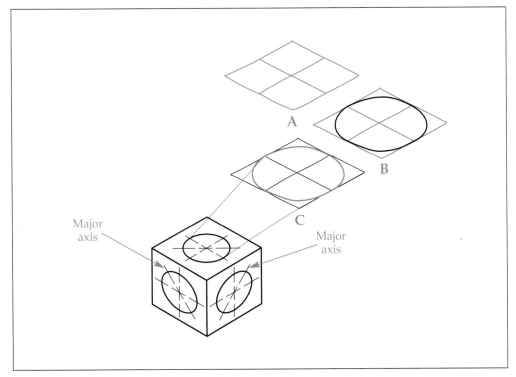

Figure 4-24. Isometric circles are drawn as ellipses. The ellipses shown are oriented correctly on each face of the isometric object. A—First, draw an "isometric square" and centerlines that intersect the midpoint of each side. B—Sketch the ellipse using construction lines. C—Darken the lines forming the ellipse.

Basic Computer-Aided Drafting Functions

Many of the manual drawing tasks discussed in this chapter have been simplified by tools provided in computer-aided drafting (CAD) programs. There are several basic features and functions common to most CAD programs. These include the following:

- Coordinate systems
- Unit format settings
- Command functions
- Drawing aids
- Drawing entities
- Geometric construction tools
- Object property controls
- Editing tools
- Display controls

The following sections introduce these common features. In addition, the general steps used to create drawings on most CAD systems are discussed.

Coordinate Systems

Coordinate systems are used by CAD programs to precisely place or locate objects. The Cartesian coordinate system is the most common. In the *Cartesian coordinate system,* objects are located using three coordinates. The horizontal dimension is called the *X coordinate.* The vertical dimension is called the *Y coordinate.* The *Z coordinate* "projects out" of the computer screen and is used to specify the third dimension for a three-dimensional object. For two-dimensional drafting, only two coordinates are used—the X coordinate and the Y coordinate.

There are four quadrants in a two-dimensional Cartesian coordinate system, **Figure 4-25.** Coordinates in this system are located in relation to the origin (X=0,Y=0). Coordinates in the upper-right quadrant have a positive X value and a positive Y value (+X,+Y). Coordinates in the upper-left quadrant have a negative X value and a positive Y value (–X,+Y). Coordinates in the lower-left quadrant have a negative X value and a negative Y value (–X,–Y). Finally, coordinates in the lower-right quadrant have a positive X value and a negative Y value (+X,–Y). For most 2D drawings, the upper-right quadrant is used so that only positive coordinate values are used for point location.

There are three ways to locate coordinates in the Cartesian coordinate system. Points can be entered using absolute coordinates, relative coordinates, and polar coordinates.

Absolute coordinates

Absolute coordinates specify all X, Y, and Z coordinates in relation to the origin (0,0). For example, the absolute coordinate (4,2) means the point is located 4 units horizontally and 2 units vertically from the origin, as shown at Point A in **Figure 4-25.**

Relative coordinates

Relative coordinates specify all coordinates "relative" to the last point drawn. Relative coordinates are also called *delta coordinates.* For example, the

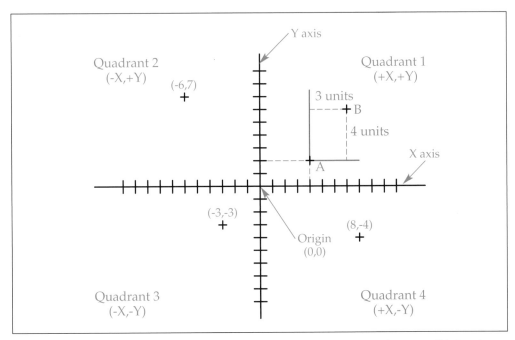

Figure 4-25. The Cartesian coordinate system. The absolute coordinates of Point A are (4,2). Using Point A as a base point for relative coordinates, the relative coordinates of Point B are (3,4).

relative coordinates for Point B in **Figure 4-25** are specified as (3,4) relative to Point A.

Polar coordinates

Polar coordinates specify locations as a specified distance and angle from the coordinate system origin or a fixed point, **Figure 4-26.** Polar coordinates are also called *radial coordinates.* Polar coordinates are typically used for round objects or for placing an object, such as a line, at a specific angle.

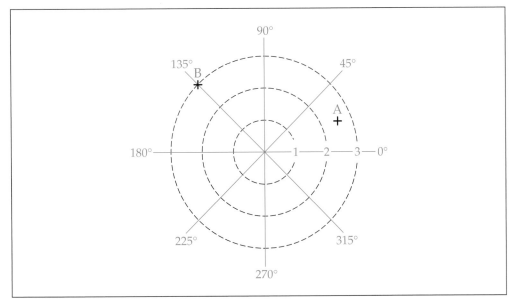

Figure 4-26. Polar coordinates are specified by their distance and angle from the origin or a fixed point. The polar coordinates for Point A relative to the origin are (2.5,22.5). For Point B, the polar coordinates relative to the origin are (3,135).

Think of the coordinate system origin as being in the center of a series of concentric circles. When using polar coordinates, the radius of a given circle is the distance measurement. A point is located on the given circle with an angle from 0° to 360°. A polar coordinate is specified as (*distance,angle*). Relative polar coordinates can also be used. This allows the distance and angle to be specified relative to the last point drawn.

Unit Format Settings

Units on a drawing specify the type of measurement system being used for distances, dimensions, and other parameters. For example, units may be specified in inches and feet or in metric format. Most CAD systems have settings that control the unit format for a drawing. Typically, parts of an inch may be specified in fraction or decimal form. Dimensions applied to the drawing are then displayed in the chosen format.

In addition, the level of precision can be specified for the measurement format. For example, assume a decimal inch format is specified with 0.0000 precision. This format would result in dimensions such as 35.4565.

Command Functions

Commands are your instructions to the computer. There are a variety of commands available for CAD functions such as drawing, editing, dimensioning, and placing text. For example, a simple box might be drawn with the **Line** drawing command. The **Erase** editing command may be used to delete any unwanted lines from the box. The box can then be dimensioned and text added using the appropriate commands.

Each CAD program has its own terminology associated with the various commands. If you are using a CAD system to develop technical drawings, you will need to refer to the software help system or the user's manual for the specific terminology. Therefore, in this chapter, generic command names are used. The CAD system you are using may have slightly different terminology.

Commands can be input in several ways, depending upon the CAD system. These input methods include typing commands at the keyboard, making a selection from a toolbar or screen menu, picking a screen button, or making a selection from a digitizer tablet. Often, a combination of techniques is used in making a drawing.

Commands have a certain syntax. *Syntax* is the "grammar" of a computer language. The typical syntax for a CAD system is verb, noun, modifier, and location. For example, **Draw** (verb) might be chosen, followed by **Circle** (noun). Then, the **3 Point Method** (modifier) might be selected to draw the circle by identifying three points on the circle (location). The circle is then drawn through these three points. Some CAD software allows the user to modify the command order. For instance, a command sequence may involve placing the noun before the verb.

Drawing Aids

Drawing aids are features of CAD programs that assist in the drawing process. In most cases, the desired drawing aids should be set prior to beginning a drawing. However, a user still has the opportunity to modify the settings while working on a drawing. The most common drawing aids are grid, snap, and object snap.

A *grid* is a configuration of dots or lines on the screen arranged in rows and columns. The grid can be used much like graph paper. However, it does not print or plot with the drawing. The user can specify how far apart the dots or lines are and even rotate the grid for pictorial drawing. The grid is a useful tool for determining locations when drawing objects.

Snap is a function that confines cursor movement to an invisible grid. The cursor jumps from snap point to snap point when snap is activated. Thus, the cursor is always precisely on the snap grid. Snap is a very useful drawing aid since it is sometimes difficult to be sure that the cursor is directly on a grid dot. Snap can be set to the same increment units used by the visible grid on screen, but it is often set to a smaller unit increment.

Object snap allows the cursor to "attach" to a specific location on an object, such as the endpoint, midpoint, or center. For example, a line can be drawn from the exact center of a circle with object snap. In order to perform this function, the circle is first drawn. Then, the object snap mode **Centerpoint** is selected and the circle is picked. The end of the line automatically snaps to the center of the circle. A second point for the line can then be entered.

Drawing Entities

Entities, or *objects,* are the elements that make up a drawing. Objects in a CAD program include points, lines, shapes, dimensions, and text. A variety of commands are used to place objects on a drawing. Object locations can be specified by entering the defining coordinates of the object. The number of coordinates required and the type of coordinates depend on the type of object being drawn. For example, a point is located with a single coordinate, such as (2,4). However, a line is defined by its endpoints, such as (2,4) and (3,6).

Geometric Construction Tools

In manual drafting, geometric constructions are created with basic drafting tools—pencils, a compass, and dividers. In a CAD system, the tools of the software replace manual drafting tools. The following sections discuss common CAD software tools and commands used to create geometric constructions. Keep in mind that the exact command names may differ based on the CAD system used.

Creating parallel lines

Drawing a line parallel to an existing line can be achieved by using the **Offset** or **Parallel** command. This is a convenient command to use when creating multiview or pictorial drawings. The parallel line can be drawn through a selected point or at a desired distance from the first line.

In some CAD systems, parallel lines can be created using a **Double Line** or **Multiline** command. By using these commands, the user saves a couple of steps, thus increasing productivity. The user specifies the distance between the lines and identifies the beginning and ending points.

Establishing tangency

The **Tangent** command is used to draw new objects tangent to an existing object. A *tangent point* is a point on an arc or curve that touches another entity at a single point. For example, to draw a line tangent to an existing circle, the

Tangent command is entered and the first endpoint of the line is located. The object to which the line is to be tangent (the circle) is then selected. Next, the second endpoint of the line is located. The **Tangent** command can be used to create circles tangent to lines, circles tangent to other circles, and arcs tangent to lines.

Creating polygons

A *polygon* is a closed shape composed of lines. Examples of polygons are triangles and rectangles. A *regular polygon* has equal sides and angles. Most CAD systems have a **Polygon** command that provides a quick way to create regular polygons. After the command is entered, the number of sides for the polygon is specified. A radius is also specified. Then, the polygon is drawn.

There are two ways to draw a polygon. The polygon may be inscribed or circumscribed, **Figure 4-27.** An *inscribed* polygon has each corner touching an imaginary circle, with the polygon inside the circle. The radius defines the polygon by measuring across corners of the shape. A *circumscribed* polygon surrounds an imaginary circle with each side of the polygon tangent to the circle. The radius defines the polygon by measuring across its flat sides. In some CAD programs, a polygon is drawn by specifying "corners" or "flats" rather than "inscribed" or "circumscribed."

Object Property Controls

Objects created in a CAD program have common properties and characteristics that may be set or changed depending on the requirements of the drawing. *Object properties* include characteristics such as the object's layer, line type, line width, and color. Most CAD systems have various tools that let you manage these properties.

Layers are settings that are similar to different sheets of paper aligned and overlaid on top of each other in a drawing project. Objects are typically grouped on layers to show different details of a drawing in relation to each other. For example, in an architectural drawing, layers corresponding to different plan drawings ensure that building details are in proper alignment with one another. A floor plan can be drawn on one layer, the electrical plan on a second layer, the plumbing plan on a third layer, and so on. By plotting layers 1 and 2 together, the floor plan is drawn with the electrical details. Layers 1 and 3 provide plumbing details on the floor plan. This technique allows the floor

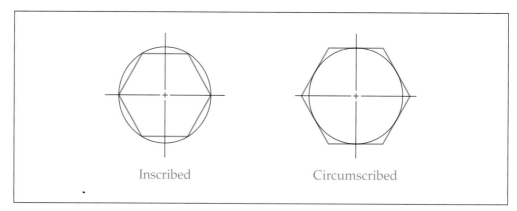

Inscribed   Circumscribed

Figure 4-27. Regular polygons are drawn inside a circle (inscribed) or outside a circle (circumscribed) depending on whether the distance across the corners or flats of the polygon is known.

plan to be used for several drawings without redrawing it. In the same way, different layers can be assigned to individual objects on the same plan drawing depending on the needs of the project. Layers can be plotted separately or together as needed.

Layers are typically managed using commands, toolbars, and other controls. Often, different colors are assigned to different layers to distinguish them. Using layers, you might choose to have a different layer with a special color for each of the lines in the alphabet of lines.

In a CAD system, lines in the alphabet of lines are referred to as *linetypes*. For example, in a multiview drawing, a solid, or continuous, linetype is used for visible lines. A dashed linetype is used for hidden lines, and a centerline linetype is used for the centers of circles and arcs.

Linetypes and colors can be assigned to layers, or they can be assigned to individual objects. Custom linetypes and colors can be created on most CAD systems. As with layers, linetypes and colors can be modified with relative ease.

Editing Tools

Editing a CAD drawing involves changing an object that has been previously drawn. Common editing functions include erasing, trimming, moving, rotating, copying, and mirroring. Some CAD systems may have more editing functions. Editing commands allow a user to put the power of a computer to work and enhance productivity.

One of the most common editing commands is **Erase**. (Some CAD systems call this command **Delete**.) The **Erase** command removes selected objects from the drawing.

In some instances, only part of an object needs to be removed. The **Trim** or **Break** command is used for this purpose. The **Trim** command removes part of an object at a specified cutting line. This is similar to using a pair of scissors to cut off an unneeded portion of an object. The **Break** command removes the part of an object between two selected break points.

The **Move** command is used to relocate an object to a different location on the drawing. The **Rotate** command is similar to the **Move** command. However, the object moves at an angle around a selected base point. The base point is often on the object, but it can be a point in space as well.

The **Copy** command makes a duplicate of an object. The second object can then be placed elsewhere on the drawing, **Figure 4-28.** This command is the

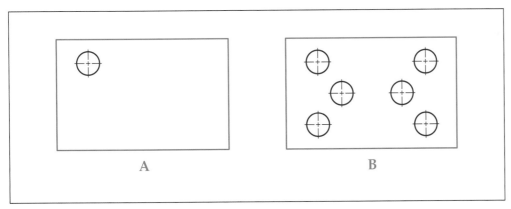

Figure 4-28. Using the **Copy** command to make copies of previously drawn objects. A—The original circle with centerlines. B—The object is copied to several new locations.

basis of the CAD philosophy "never draw the same thing twice." Rather than drawing an object that has been already created again, a user can simply use the **Copy** command to carry out this task.

The **Offset** command, discussed previously, is similar to the **Copy** command. It can be used to construct a copy parallel to the original. The **Offset** command can be used when a series of copies is to be made, with all of the copies at the same distance from the original objects. For example, if an outline of a floor plan has been drawn, the **Offset** command can be used to create the thickness of the walls by offsetting the original lines 6".

The **Mirror** command is also similar to the **Copy** command. The **Mirror** command copies an object as a mirror image about a selected mirror line, **Figure 4-29**. The **Mirror** command is very helpful when both halves of an object are symmetrical. You can draw half of the object and then use the **Mirror** command to create the other half. Once again, the user increases productivity with this command.

Display Controls

Display controls are tools used to change the way a drawing appears on screen. While the specific commands vary with CAD systems, there are some basic commands that are common to most. These commands include **Zoom**, **Pan**, and **Viewport**. Display controls are also called *view controls*.

The **Zoom** command is used to enlarge or reduce a portion of the drawing. In some programs, this is known as *windowing*. Zooming is necessary when trying to see small details on large drawings. *Zooming in* is enlarging the view. This makes small details appear larger. *Zooming out* is reducing the view. When working on a "zoomed-in" view, you must zoom out to see the entire drawing.

The **Pan** command is used to move the drawing on the screen. This is like moving a sheet of paper on your desk. For example, you may zoom in and find that the part of the drawing you need to work on is partially off the screen. The **Pan** command can be used to move that part of the drawing to the center of the screen. Panning does not enlarge or reduce the view.

A *viewport* is a "window" showing one view of an object. Several viewports can be created with the **Viewport** command and then shown at the same time, each displaying a different view of the drawing. For example, you may wish to show both orthographic and pictorial views of an object simultaneously, **Figure 4-30**.

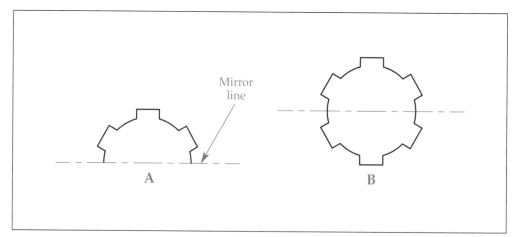

Figure 4-29. Using the **Mirror** command. A—The original object. B—The final object after mirroring.

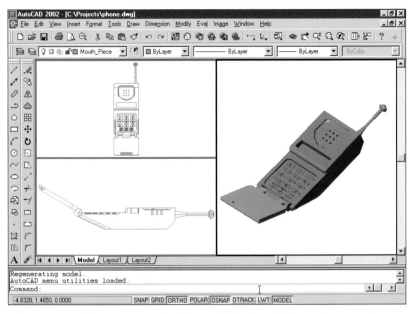

Figure 4-30. Viewports make this model of a wireless phone easier to visualize and draw. In this illustration, the same model is shown in each viewport. The upper-left viewport shows a plan view. The lower-left viewport shows a right-side view. The right viewport shows the model with hidden lines removed. (Anthony J. Panozzo)

Creating a CAD Drawing

The specific steps used to create a CAD drawing vary according to the CAD software and the type of drawing. However, the following is a general sequence that can be followed to create a drawing on a computer.

1. Make a sketch of what is to be drawn. This step helps in visualizing the drawing and reduces mistakes. Sketching and planning are keys to any good drawing.
2. Start the CAD program.
3. Set up parameters for the drawing as needed. These include such things as the unit format, drawing sheet size, layers, and color settings. Properly setting these items early on will save time and increase productivity for the project.
4. Set up a grid to help you in creating the drawing. Set up snap as well.
5. Draw objects using absolute, relative, or polar coordinates. Use editing commands to correct errors and to manipulate the objects in the drawing. Display control commands, such as **Zoom** and **Pan**, can be used to help draw detailed portions of the drawing.
6. Save the drawing every 10–15 minutes throughout the drawing process. Many programs provide an option to automatically save the drawing periodically.
7. Dimension the drawing if needed.
8. Place text, such as notes, on the drawing.
9. Save the drawing at the end of the drawing session.
10. Check the drawing for completeness and accuracy.
11. Plot or print the drawing.
12. Review the plot to verify that the output meets your needs. Edit the drawing as needed until it is correct.

Summary

Basic drawing skills for technical illustration have been introduced in this chapter. Whether sketching or drawing with manual instruments, you need to understand orthographic projection and pictorial drawing, as well as lettering tools and techniques. You also need to be familiar with the basic concepts involved in using a CAD system.

Much like the alphabet is used for making words and sentences, the alphabet of lines is used for making drawings. This basic language of technical drawing needs to be understood by any technical illustrator, whether using traditional means or a computer.

Even though many technical illustrations are pictorial drawings, the drawings are normally based on orthographic drawings. If the multiview drawing is not interpreted correctly, there will be difficulty making a pictorial drawing of the same object. The basic fundamentals of orthographic projection were covered in this chapter, as well as the general steps for making multiview drawings.

Lettering can be done freehand, with the help of special instruments, or by using text commands in a CAD system. Most freehand lettering is done using a single-stroke Gothic typeface, but some drawings require a greater degree of sophistication. In this case, it is important to understand typefaces so the appropriate type can be selected for the project. Using different lettering techniques and tools, such as guidelines and lettering templates, is important to overall drawing productivity.

Many drawings begin as sketches, and these sketches are used to communicate basic ideas in visual form. Therefore, it is important to be able to sketch accurately. Basic sketching techniques were covered for both pictorial and orthographic drawings in this chapter.

Today's technical illustrator must have some background with computers and CAD. This chapter covered the basic terminology associated with CAD programs as well as the procedures for producing a CAD drawing.

1. Three of the most common lines in the alphabet of lines are the visible line, hidden line, and centerline. Explain the purpose of each type of line.
2. What is the purpose of the projection plane in creating a multiview drawing?
3. What are two criteria in choosing a front view for a multiview drawing?
4. From a drawing standpoint, what is the advantage of aligning views in a multiview drawing?
5. If a feature in a multiview drawing appears foreshortened, what do you know about this feature?
6. Describe the basic differences between typefaces in relation to the four primary typeface classifications.
7. What does 12/24 Times Roman mean, assuming Times Roman is a typeface?
8. What is the difference between horizontal and vertical guidelines used for lettering?
9. Why is it important, even when drafting tools are available, to be able to sketch accurately?
10. Briefly explain how to sketch a circle.
11. Using the Cartesian coordinate system and absolute coordinates, a line is drawn from (2,2) to (4,2). Is the line vertical, horizontal, or angled?
12. Which of the coordinate entry types—absolute, relative, or polar—would be best to use when drawing a detailed wheel of a car?
13. Give an example of a command sequence using typical CAD syntax.
14. What is the advantage of using several viewports when creating a CAD drawing?
15. Give an example of why using layers can be important when creating a CAD drawing. Do not use an example from the book.

1. Using a sans serif typeface such as single-stroke Gothic, letter the alphabet and the numerals 0–9 by freehand. Start by drawing light guidelines. Letter the characters in uppercase using a 3 mm (1/8″) height and 3 mm line spacing. Letter a second set of alphanumeric characters using a 6 mm (1/4″) height and 3 mm line spacing.
2. Make a freehand sketch of the illustration shown. Use lines, circles, and arcs. The outside edges of the drawing should match the edges of your drawing paper. Begin with construction lines and darken with visible lines when done with the sketch. Place your name in the blank using Gothic lettering.

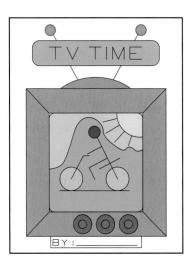

For Problems 3–5, use the isometric views shown to sketch multiview drawings containing the top, front, and side views. Each drawing should fully describe the object shown. Use your own dimensions and hidden lines where necessary. Keep the height, width, and depth in proportion. Align the views correctly. Use square grid sheets to assist you in the sketching process. When done with each sketch, make a final drawing.

3.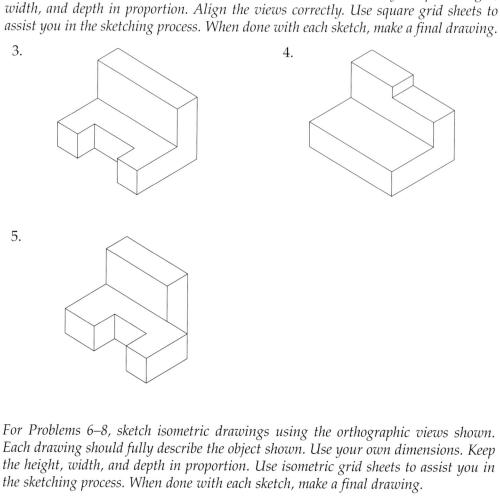

4.

5.

For Problems 6–8, sketch isometric drawings using the orthographic views shown. Each drawing should fully describe the object shown. Use your own dimensions. Keep the height, width, and depth in proportion. Use isometric grid sheets to assist you in the sketching process. When done with each sketch, make a final drawing.

6.   7.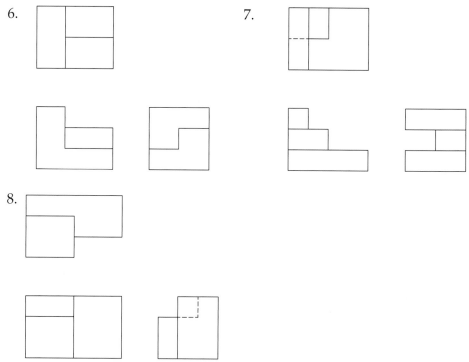

8.

9. Make an isometric sketch of the object shown using the metric dimensions provided. Use isometric grid sheets to assist you in the sketching process. When done with the sketch, make a final drawing. Do not dimension.

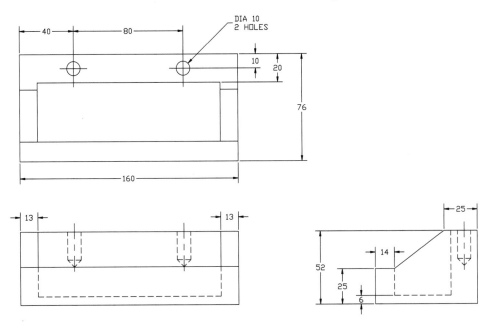

10. Make a multiview sketch of the object shown using the metric dimensions provided. Use hidden lines where necessary and align the views correctly. Use square grid sheets to assist you in the sketching process. When done with the sketch, make a final drawing. Do not dimension.

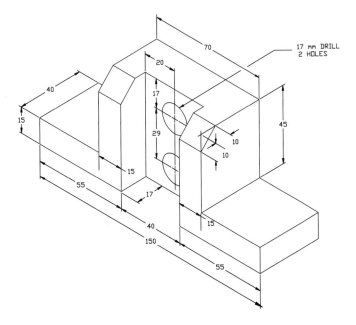

11. Using discarded magazines and newspapers, locate examples of the major typefaces discussed in this chapter. Unusual typefaces can be typically found in advertisements. Cut out the examples and mount them on sheets of paper. Identify the sources and describe in several sentences why you think the particular typefaces were used.

Section 2

Pictorial Generation

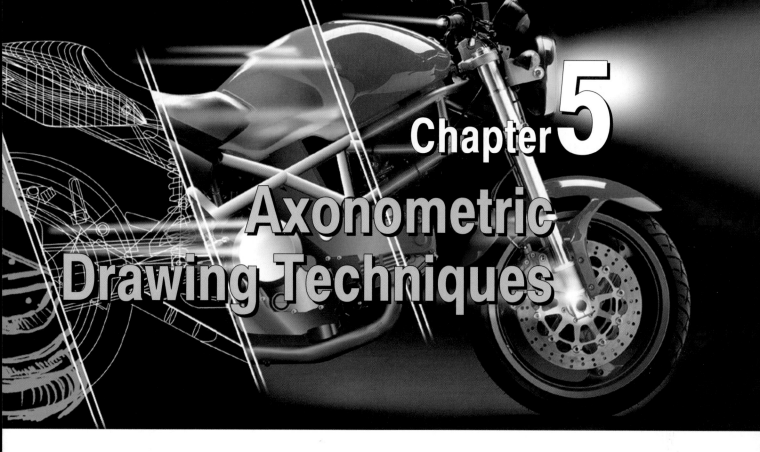

Chapter 5
Axonometric Drawing Techniques

Learning Objectives

At the conclusion of this chapter, you will be able to:

☐ Identify and explain the different types of pictorial and axonometric drawings.

☐ Plan, lay out, and draw isometric views of objects containing normal, inclined, skewed, irregularly curved, and circular surfaces.

☐ Use isometric ellipse templates, isometric protractors, and angle ellipse templates as drawing aids.

☐ Identify and explain the primary applications of computer-generated axonometric drawings.

☐ Develop drawings using the isometric, dimetric, and trimetric methods of pictorial drawing.

Introduction

A pictorial drawing describes all three dimensions of an object in a single view. Technical illustration uses pictorial drawings almost exclusively. Therefore, it is very important for you to learn the principles of pictorial drawing. As discussed in Chapter 4, the three major types of pictorial drawings are axonometric, oblique, and perspective. This chapter looks at the principles, procedures, and drawing aids used in creating axonometric drawings.

Pictorial Drawing Overview

Each of the three types of pictorial drawing uses a single three-dimensional view. The axonometric, oblique, and perspective drawing classifications can be further divided into specific types, as shown in **Figure 5-1**.

Axonometric Drawings

The three types of axonometric drawings are isometric, dimetric, and trimetric drawings. These are illustrated in **Figure 5-2**. Isometric drawings were introduced in Chapter 4. The term *isometric* means equal measure. The term *dimetric* refers to two measures. The term *trimetric* refers to three measures. Each of these drawing styles has specialized applications in the field of drafting, design, and illustration.

Referring to **Figure 5-2**, the major differences between these drawings are the scales used to measure the horizontal and vertical axes and the angles produced by the intersections of the three axes. An *isometric drawing* has equal angles (120°) between each of the three width, height, and depth axes. The scales used to measure each axis are also equal. A *dimetric drawing* uses two different scales to measure the axes. Two of the axes use the same scale, while a third axis is measured using a different scale. In addition, the intersections of the three axes produce different angles. Two of the angles are equal. A *trimetric*

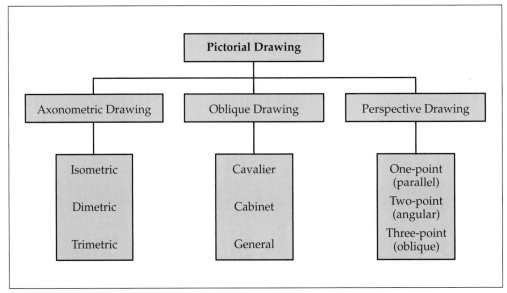

Figure 5-1. There are three general types of pictorial drawings. Each general classification can be further divided into specific types.

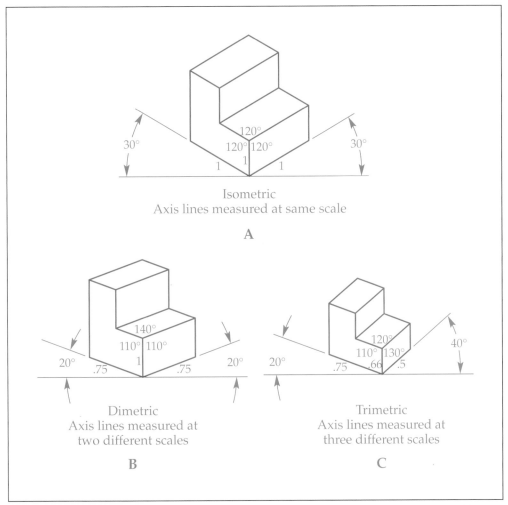

Figure 5-2. The three types of axonometric drawings. Note the angles produced by the intersections of the axes. A—An isometric drawing uses the same scale to measure each of the width, height, and depth axes. B—A dimetric drawing uses the same scale to measure two of the axes and a different scale to measure the third axis. C—A trimetric drawing uses different scales to measure the three axes.

drawing uses three different scales to measure the width, height, and depth axes. The intersections of the three axes produce three different angles.

Oblique Drawings

Oblique drawings are similar to axonometric drawings. There are three types: cavalier, cabinet, and general. See **Figure 5-3**. An *oblique drawing* is a pictorial drawing in which the front view is parallel to the projection plane and the top and side views are viewed at an oblique angle. The width and height axes are drawn at full scale, or true size. The primary difference between each type of oblique drawing is the scale of the receding axis used for the depth. A *cavalier oblique* drawing has the receding axis drawn at full scale. A *cabinet oblique* drawing has the receding axis drawn at half scale. A *general oblique* drawing is drawn with the receding axis at a scale other than one-half or full size, typically a three-quarter scale. With all three types of drawings, the receding axis can be drawn at any angle greater than 0° and less than 90° from the projection plane. Common receding axis angles and scale factors for oblique drawings are shown in **Figure 5-3**.

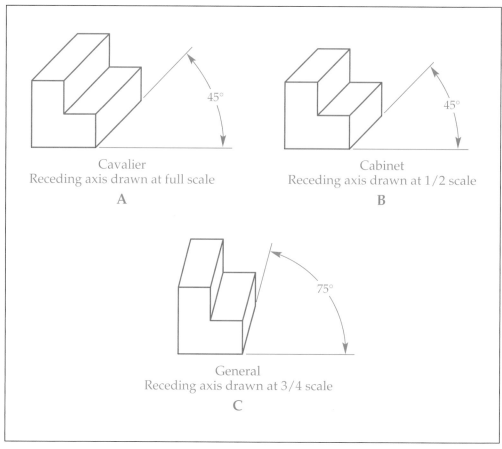

Figure 5-3. Oblique drawings are identified as cavalier, cabinet, or general based on the scale of the receding axis.

Perspective Drawings

Perspective drawings are pictorial drawings that represent what is normally seen by the eye from a given viewing point. In this type of drawing, lines along receding axes converge as they recede. This is different from the other types of pictorial drawing previously discussed, where lines remain parallel. There are three types of perspective drawings. These are *one-point, two-point,* and *three-point perspective* drawings, **Figure 5-4.** The name for each type of drawing refers to the number of points of convergence. A point of convergence is called a *vanishing point.*

Notice in **Figure 5-4** that the depth lines of the L-shaped object recede to a single point. Therefore, this view is a one-point perspective. The width and depth lines of the house recede to two separate vanishing points. This object is a two-point perspective. The two horizontal axes and the vertical axis on the skyscraper each recede to different points. This is a three-point perspective.

Applications of Pictorial Drawings

There are advantages and limitations to each of the different pictorial drawing styles. It is important that you select the most appropriate pictorial method for your illustrations. Proper selection of a pictorial style will improve the viewer understanding and, in many instances, save you considerable drawing time.

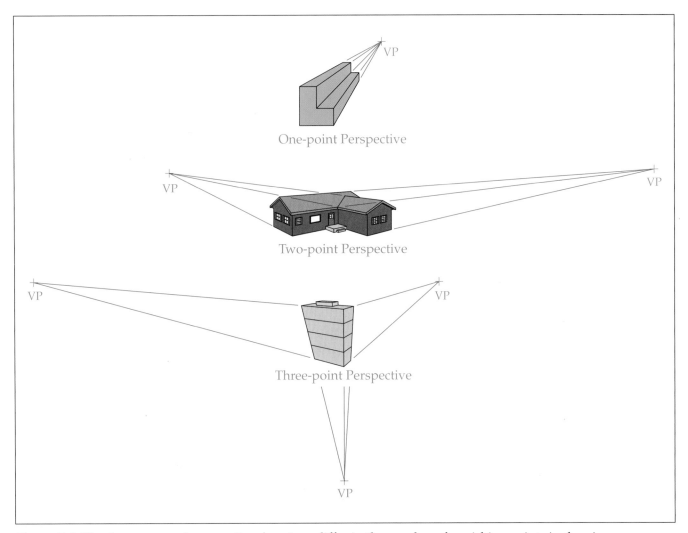

Figure 5-4. The three types of perspective drawings differ in the number of vanishing points in the view.

Axonometric Drawing Applications

The primary application of axonometric drawings is to show objects with rectangular shapes made of flat plane surfaces. Curved and irregular shapes are more difficult to show in an axonometric drawing because all object faces are receding.

Isometric drawings are the most used of all pictorial drawings. The major advantages of isometric drawings include layout speed and ease of scaling. The disadvantages include a slight oversizing and distortion of proportion. Dimetric and trimetric views can produce a more realistic appearance than an isometric view. However, they require more time to develop because of the different measuring scales.

Oblique Drawing Applications

The major use of oblique drawings is to show objects with circular, cylindrical, or irregular shapes. These shapes are easier to draw using a normal orthographic front view, rather than the foreshortened pictorial views of axonometric or perspective drawings. Obliques are also used for objects with one very long axis. See **Figure 5-5**. This illustration shows an object that is best drawn as an oblique drawing because of its shape.

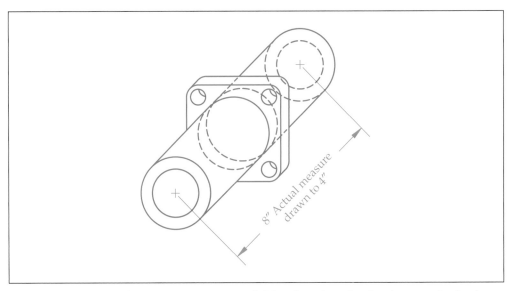

Figure 5-5. Oblique drawings are most useful for showing circular features in the front view and for objects with a relatively large dimension along the receding axis. This object is drawn as a cabinet oblique drawing with the receding axis at half scale.

Perspective Drawing Applications

Perspective drawings are most often used in architectural and interior design applications. These fields often require very realistic views to show a specific viewing location and elevation. Perspective drawings take longer to draw in comparison to other pictorial styles. Perspective drawings are used mostly for flat-surface objects with rectangular shapes.

Advantages and Disadvantages of Pictorials

Drawing an object as a pictorial drawing is the best way to show what the object looks like. The viewer does not need an engineering or drafting background to understand what is being shown. Also, it is often quicker to draw a single view pictorial, rather than a multiview drawing.

Hidden lines are normally omitted on pictorial drawings (unless the lines are necessary to understand the drawing). Labeling part names and placing general notes on a pictorial is much quicker than dimensioning orthographic views.

Pictorials are not good drawings for parts where dimensions are required to manufacture an object. Also, there may be other disadvantages to using a pictorial drawing. Portions of the object may be somewhat distorted due to the viewing angle, hidden details may be difficult to visualize, and complex shapes may make a pictorial view difficult to draw.

Isometric Drawing Review

As previously discussed, the most common type of pictorial drawing is isometric. Isometric drawing techniques were introduced in Chapter 4. An isometric projection results by rotating an object 45° to the frontal plane and tipping it forward 35°16′ from the horizontal plane. See **Figure 5-6.** This

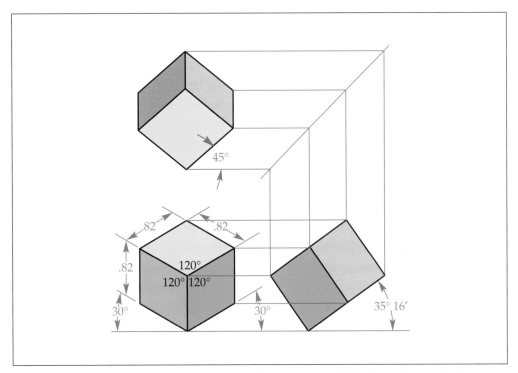

Figure 5-6. In an isometric projection, an object is rotated 45° about its vertical axis and then tipped forward 35°16′. The height, width, and depth measurements are foreshortened to approximately 82% of their true length.

produces a front view with 120° between each of the principal axes. Therefore, lines representing the horizontal planes of an object in an isometric view are at 30° from the X axis. The height, width, and depth measurements are foreshortened to approximately 82% of their true length. Foreshortening all measurements by 82% can make direct drawing of a true-scaled isometric view difficult.

An isometric projection is not the same as an isometric drawing. An isometric drawing is easier to draw because the measurements are drawn at full scale. This means that the 82% foreshortening is ignored. Instead, the object is drawn using isometric lines and the full height, width, and depth measurements, as shown in **Figure 5-7**. In reality, this creates an object that is approximately 1.22 times larger than a true representation in an isometric projection. However, this slight oversizing is acceptable.

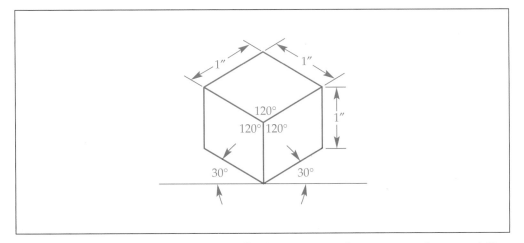

Figure 5-7. An isometric drawing of a 1″ cube. Isometric drawings are drawn to full scale along all three axes.

Types of Isometric Drawings

There are three types of isometric drawings: regular, reversed axis, and long axis. See **Figure 5-8.** Each of these drawing types has a 120° axis layout. However, the orientation of the axes determines the type of isometric drawing.

Most isometric drawings used in industry fall into the regular isometric category. For a *regular isometric drawing*, both the left and right horizontal axes are drawn at a 30° angle above horizontal. A *reversed axis isometric drawing* has its axes drawn in the exact opposite orientation as a regular axis isometric drawing. It has a downward orientation of the horizontal axes and can be used to show features on the bottom of an object. A *long axis isometric drawing* has one major axis aligned horizontally. The other base axis of the object and the vertical axis are inclined at a 60° angle to horizontal. Long axis isometric drawings are used for objects where most of the features are along one axis.

Object Orientation in Isometric Views

Before you begin an isometric drawing, you must analyze the shape and detail characteristics of the object. You need to orient the object to provide the best view of the features you want to emphasize. For example, if major features are on the bottom of the object, use a reversed axis isometric drawing.

As a general rule, an object should be oriented in a position similar to how it normally appears. If it sits on a base, then use the base as the bottom. *Operating position orientation* is the position an object is in when a person is using, controlling, or viewing it in its natural environment. If the "natural" orientation produces many hidden features, you may want to select a different view.

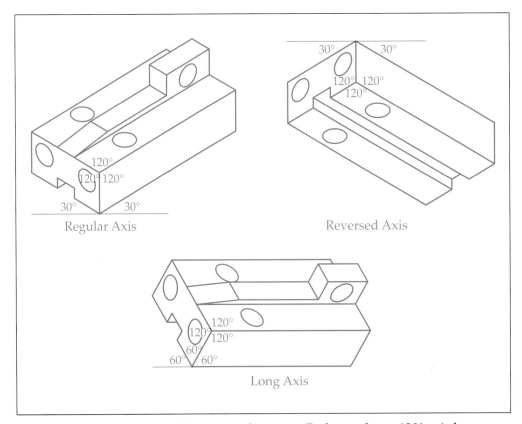

Figure 5-8. The three types of isometric drawings. Each type has a 120° axis layout, but the orientation of the axes varies.

Drawing Isometric and Nonisometric Lines

Horizontal and vertical lines in an isometric drawing are always drawn parallel to the isometric axes. These lines are called *isometric lines.* Any lines not parallel to an isometric axis are *nonisometric lines.* To draw a nonisometric line, you must plot the appropriate location of each endpoint on isometric lines and then connect the endpoints.

The *block-in technique* is a useful method for drawing objects with isometric lines. To use this technique, an isometric block is first drawn with light construction lines and dimensions equal to the width, depth, and height of the object. See **Figure 5-9**. The basic object shown is developed by "blocking in" individual points of the outline and connecting the points with isometric lines until the actual shape is defined. Notice that the step features are first located on the front face of the object. The corresponding points on the back face are then located on the back plane of the block. To complete the object, isometric lines are drawn from the points on the front face to the corresponding points on the rear face. The accuracy of each point should be checked before darkening construction lines.

Blocking in a view greatly improves the overall accuracy of your drawing. The more complex an object's shape, the more useful it is to use this technique.

Nonisometric lines and surfaces can also be drawn using the block-in technique. However, they require some additional steps when measuring features and locating points. The construction of an inclined surface and a skewed surface in an isometric drawing is shown in **Figure 5-10**. The measurements for the nonisometric lines are taken from the top, front, and side orthographic views. First, an isometric block with dimensions equal to the width, depth, and height of the object is drawn. Distances for the surfaces are then measured on each orthographic view and transferred to locations on isometric lines parallel to axis lines on the block. Notice that measurements are always made parallel to one of the three isometric axes. The points are then connected with nonisometric lines to complete the object. Nonisometric lines are not true length because they do not lie on a normal isometric surface. Since they are not on, or parallel to, an isometric plane or axis, they can never be measured directly with a scale.

To help avoid confusion when constructing an isometric object, draw all of the normal lines and surfaces on a layout first. Temporary labels may also help you avoid problems on more complex shapes. Then, select one nonisometric

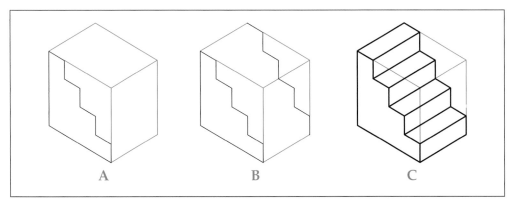

Figure 5-9. Constructing an isometric view using the block-in technique. A—An isometric block with dimensions equal to the width, depth, and height of the object is first drawn. Points are then located on the front face and connected with isometric lines.
B—Point locations are made on the back face of the object and connecting lines are drawn.
C—Points on the two faces are connected with lines to complete the object.

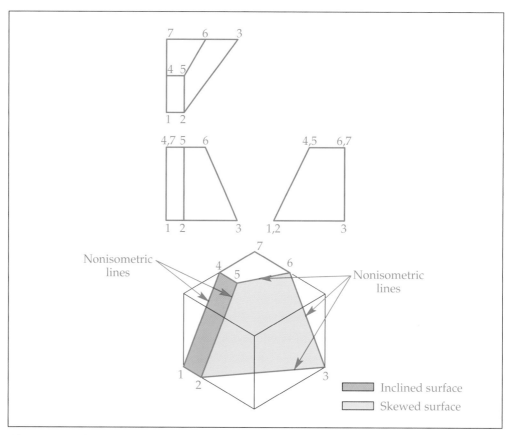

Figure 5-10. Constructing nonisometric lines and surfaces from orthographic views.

surface at a time and locate its corners by measuring along an appropriate isometric axis. Connecting each corner point with lines then defines the nonisometric intersection with the normal or isometric surface.

Drawing Isometric Arcs, Circles, and Irregular Curves

In an isometric drawing, circles and arcs appear as ellipses and partial ellipses. Isometric arcs and circles and irregular curves can be developed on an isometric drawing using the coordinate method. The *coordinate method* locates points on the curve using isometric lines as reference lines. See **Figure 5-11**. This illustration shows how an orthographic view may be necessary to develop coordinate points that can be transferred to the isometric view. The resulting arcs are drawn on the top and bottom faces of the object. Irregular curves can be transferred to an isometric view using the same technique.

To use the coordinate method, the isometric drawing begins with "blocking in" the basic shape of the object. Then, coordinate points are identified on the curved shape in the orthographic projection. Using isometric lines as reference lines, coordinates are transferred from the orthographic view to the corresponding coordinate points in the isometric view. Using an irregular (French) curve or spline, the points are then connected in a smooth curve. In **Figure 5-12**, an irregular curve is developed using the same procedure. Reference lines are used for both the orthographic and isometric views.

It saves unnecessary layout work if you first complete the surface closest to you in the isometric view. Then, locate coordinate points for background features. This can save you from plotting numerous points that are actually hidden.

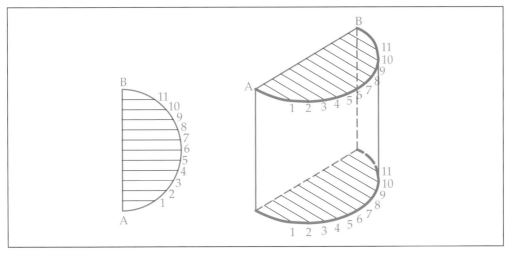

Figure 5-11. The coordinate method can be used to create arcs and curves in isometric drawings. Coordinates are measured in the orthographic view and then transferred to the isometric drawing.

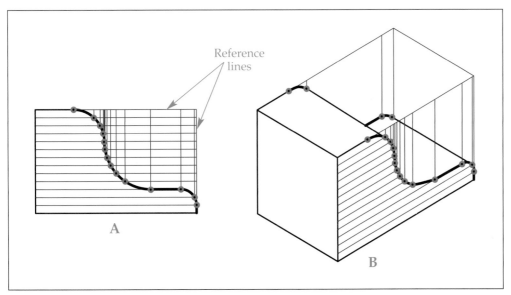

Figure 5-12. Using the coordinate method to create an irregular curve. A—Points are located on the curve in the orthographic view. B—The points are then transferred to the isometric drawing.

The coordinate method can also be used to construct isometric circles, **Figure 5-13**. However, this method is used only when abnormal sizes or surface angles eliminate easier methods of isometric development. Plotting coordinates for a circle is time-consuming because it requires as many points as possible to provide a good definition of the curved shape. If the circle is on an isometric surface, you can use the four-center method or a standard isometric template. The *four-center method* locates center points for arcs defining the isometric circle on an isometric plane. The isometric circles in **Figure 5-14** were developed using this method. The procedure is as follows:

1. Locate the center point of the isometric circle and construct an isometric square around it. Draw the sides of the square with isometric lines. The isometric square will appear as a parallelogram because it is inclined to your line of sight.
2. Locate the midpoint on each side of the isometric square.

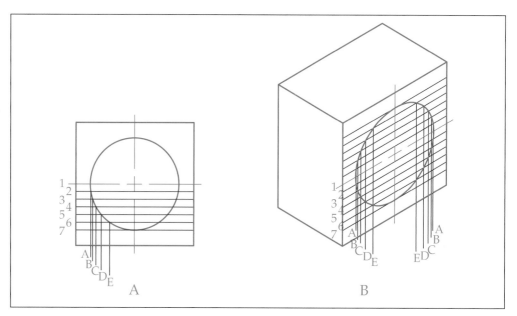

Figure 5-13. Drawing an isometric circle using the coordinate method. A—Points are located along one quadrant of the circle in the orthographic view. B—The coordinates from one quadrant are used to determine point locations for all four quadrants of the isometric circle.

3. Construct lines from each of the two opposing corners to their opposite midpoints, as shown in **Figure 5-14.**
4. Draw the arcs for the sides of the ellipse using the isometric square corners as the center points and the radius value labeled R1. Using light construction lines, draw arcs for the ends of the ellipse using the line intersections as the center points and the radius value labeled R2.
5. Check that the curves come together in a smooth transition at their tangency points before darkening in the small arcs.

A very small amount of error in developing a four-center ellipse can cause problems in arcs meeting at their proper point of tangency. If a large arc does not quite match the tangency point of one of the smaller arcs, reset your compass to the proper radius of the small arc, **Figure 5-15.** Then, swing an arc from each end

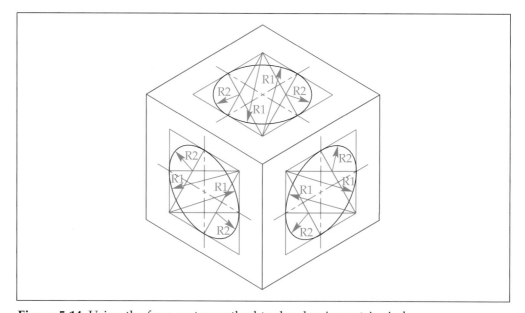

Figure 5-14. Using the four-center method to develop isometric circles.

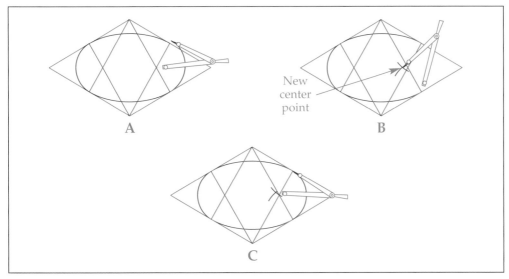

Figure 5-15. To correct minor errors when using the four-center method, establish a new center for a "correction" arc. This arc creates a smooth transition between the existing arcs. A—The compass is set to the radius of the small arc. B—Construction arcs are drawn from the ends of the large arcs. The intersection of the arcs establishes the new center point for the correction arc. C—The correction arc meets the larger arcs at the ends to create a smooth curve.

of the large arcs to establish a new center location for the small arc. This new center for the "correction" arc provides a point of rotation that adjusts the small arc to provide a smooth transition curve to each of the larger arcs.

Centering Isometric Views

Centering an isometric drawing in a given drawing area is easy if you follow the proper procedure. The following steps are used for centering a regular isometric drawing with normal surfaces.

1. Locate the center of the drawing area by drawing diagonal construction lines from the corners of the drawing area.
2. Drop a vertical construction line from the center of the drawing area one-half the maximum height of the object, **Figure 5-16A.**
3. Draw a construction line one-half the maximum width of the object at a 30° angle to horizontal. If you want to orient the isometric view so that the front face of the object is on the left in the frontal plane, draw the line to the right. If you want to orient the isometric view so that the back face is shown, draw the line to the left.
4. Draw a construction line one-half the maximum depth of the object at a 30° angle and in the opposite direction of the line drawn in Step 3. The endpoint of this line is the bottom front or bottom rear corner of the object.

The following procedure is used to center a reversed axis isometric drawing.

1. Find the center of the drawing area with diagonal lines.
2. Draw a vertical construction line upward from the center of the drawing area one-half the maximum height of the object, **Figure 5-16B**.
3. Draw a construction line one-half the maximum width of the object at a 30° angle to horizontal. Draw the line to the right if you want to orient the isometric view so that the front face is on the left in the frontal plane. Draw the line to the left if you want to orient the isometric view so that the back face is shown.

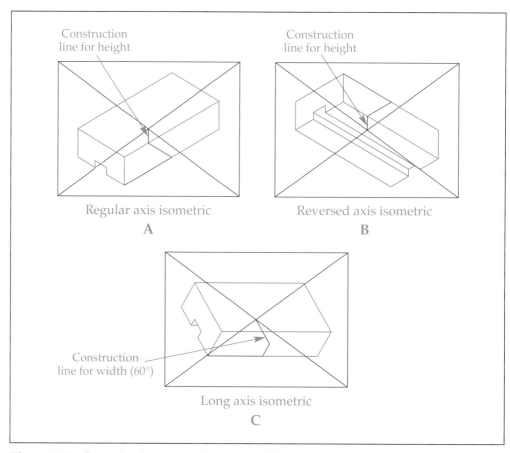

Figure 5-16. Centering isometric drawings. After construction lines are drawn for the height, width, and depth, each object is drawn using isometric lines.

4. Draw a construction line one-half the maximum depth of the object at a 30° angle and in the opposite direction of the line drawn in Step 3. The endpoint of this line is the top front or top rear corner of the object.

The following procedure is used to center a long axis isometric drawing.

1. Locate the center of the drawing area.
2. Draw a construction line downward at a 60° angle to horizontal from the center of the drawing area one-half the maximum width of the object. See **Figure 5-16C**. If you want to orient the isometric view so that the front face of the object is on the left in the frontal plane, draw the line to the right. If you want to orient the isometric view so that the back face is shown, draw the line to the left.
3. Draw a construction line one-half the maximum height of the object at a 60° angle to horizontal in the opposite direction from the line drawn in Step 2.
4. Draw a horizontal construction line one-half the maximum depth of the object in the same direction as the line in Step 3. The endpoint of this line is the bottom front or bottom rear corner of the object.

Isometric Drawing Aids

There are many different types of isometric drawing aids available to simplify and speed up the drawing process. Chapter 2 introduced you to a number of these drawing aids. There are more types of templates for isometric drawing than for any other single type of drawing. These templates are available in both US Customary and metric units. The three major types of templates are the isometric ellipse template, angle ellipse template, and isometric protractor.

Isometric Ellipse and Angle Ellipse Templates

Like other templates, isometric ellipse and angle ellipse templates are made of sheet plastic with precise holes for you to trace the shape. Isometric ellipse templates are used to draw ellipses on normal isometric surfaces. Since the dimensions on isometric drawings are approximately 1.22 times larger than the true dimensions found on an isometric projection, the holes on an isometric ellipse template are 22% oversized. No calculations are necessary to use an isometric ellipse template. Simply select the appropriate shape and trace the ellipse. Isometric ellipse templates are labeled *Isometric* or *35°16'*.

If a circular shape is inclined or skewed to the principal orthographic planes, an angle ellipse template must be used. All sizes on the angle ellipse template are true size. Therefore, you need to compensate for isometric oversizing by using a larger ellipse. For the US Customary system, find the size by multiplying the true size by 1 1/4 (1.25). This factor is used because it is easier to work with than 1.22. If necessary, round up or down to the closest ellipse size you have available. If working with metric measurements, multiply the true size by 1.22 and then round off to the closest size available.

Full angle ellipse template sets normally come with a series of 10 separate templates. They range from a very flat viewing angle of 10° to the almost circular view at 55°. A 10-template set has increments of 5° between each ellipse template. Selection of the appropriate angle ellipse template for an isometric surface is done with an isometric ellipse protractor. This tool is explained later in this chapter.

Using ellipse templates

Isometric and angle ellipse templates are oriented by aligning the template axis lines with construction lines on the drawing. The proper axis alignments for the standard isometric planes are shown in **Figure 5-17**. Notice that each isometric ellipse must have three-axis alignment for proper orientation. The center of the ellipse is first aligned with the center of the hole or cylinder. At the same time, it is turned to align with the vertical axis and with the appropriate horizontal axis. Markings on an isometric ellipse template include four axis lines crossing from one side of an ellipse to the other.

On vertical isometric surfaces, the minor (narrow) axis of the ellipse is parallel to the horizontal axis on the other side of the vertical axis. If the template is correctly aligned with the center point of the circle and the minor axis is aligned with the horizontal axis on the drawing, one of the angled axis lines to the side of the major axis will be in a vertical orientation. An ellipse on a horizontal surface is aligned so that the minor axis is vertical and the major axis is horizontal.

The proper alignment of an angle ellipse template is shown in **Figure 5-18**. First, orient its center on the construction line center point on the drawing. Then, rotate it until the minor axis is parallel to the centerline or thrust line drawn through the middle of the hole or cylindrical shape. In the example shown, a 1" ellipse is used for a hole measuring .8125" in diameter. The true size (.8125") is multiplied by 1.22 and the resulting value (1.01) is rounded off to 1".

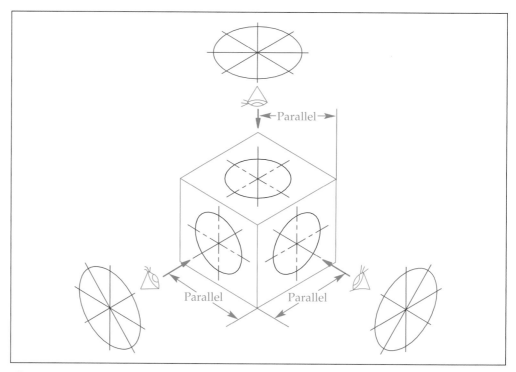

Figure 5-17. Proper axis alignments for isometric ellipses.

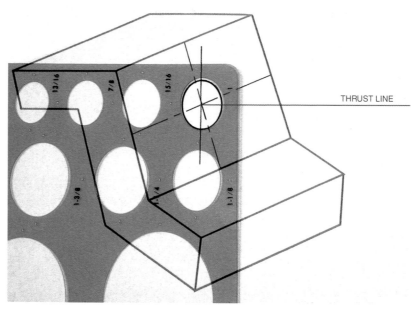

THRUST LINE

Figure 5-18. Aligning an angle ellipse template. After calculating the size of the ellipse, the minor axis is aligned with the thrust line of the hole.

The difficult part in using an angle ellipse template is determining the viewing angle when selecting the appropriate angle ellipse and then drawing the centerline of the hole. This centerline is often referred to as the *thrust line.* When a hole or cylinder is inclined or skewed to the normal isometric planes, developing the thrust line can be complicated. Most applications of the angle ellipse template in isometric drawing require the use of an isometric ellipse protractor to determine what degree angle ellipse is needed. The isometric protractor is also used to draw the thrust line.

Other ellipse template uses

Another useful application of the isometric ellipse template is to measure along nonisometric lines. Direct measurements with a scale can only be made along one of the three isometric axes. However, if an isometric ellipse lies on the same plane as a nonisometric line, a measurement corresponding to the ellipse size can be transferred to the line using the template. See **Figure 5-19**.

Since an isometric ellipse represents an isometric view of a circle, any point on the perimeter of the ellipse represents the same distance from the center. In **Figure 5-19,** the 2″ diameter ellipse is in its correct orientation on the vertical isometric plane, with its center point aligned on the desired axis point for the nonisometric line. When the perimeter of the ellipse crosses the nonisometric line, a distance of 1″ is established. Other examples of using isometric ellipses to measure distances on nonisometric lines and surfaces are shown in **Figure 5-20**. In this example, a 1″ diameter isometric ellipse is used to establish the thickness of the inclined face. A 1/2″ diameter isometric ellipse drawn tangent to and below the top edge provides a thickness measure for the upper nonisometric face.

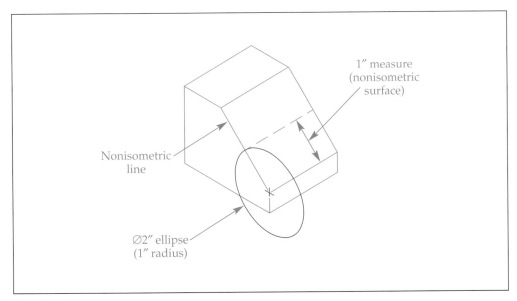

Figure 5-19. Using an isometric ellipse to transfer a measurement to a nonisometric line.

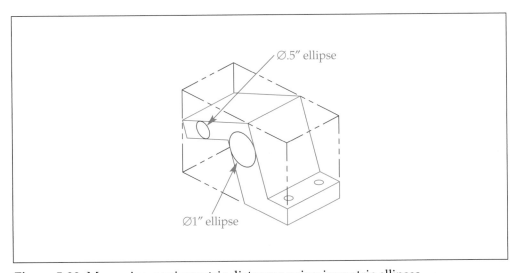

Figure 5-20. Measuring nonisometric distances using isometric ellipses.

The most important thing to remember when using an isometric ellipse template for nonisometric measurements is to maintain the proper orientation of the ellipse and the template. The ellipse must always be oriented so that the minor axis is aligned with one of the horizontal or vertical isometric axes. Notice the holes on the horizontal isometric surface in **Figure 5-20**. The ellipses for the holes are oriented so that the minor axes are aligned with the vertical isometric axis.

Isometric Protractor

An *isometric protractor* is used to measure angles that are inclined or skewed to the principal isometric planes. The protractor provides a direct reading of the angle ellipse template to use, and allows calculation of the thrust line on a nonisometric hole. The isometric protractor is divided into four quadrants. See **Figure 5-21**. The angular scale starts at 0°, progresses to 90°, and then reverses angular scale readings back to 0°. Measurements are made by aligning the minor axis of the template with an appropriate isometric axis. For example, to make measurements on a nonisometric surface in the profile viewing plane, the template should be aligned so that the minor axis is oriented with an isometric axis in the frontal viewing plane.

The inclined surfaces of the object in **Figure 5-18** were developed using an isometric protractor. See **Figure 5-22**. First, the view is blocked in and the normal isometric surfaces are developed. Then, the isometric protractor is centered on the vertex point of the inclined face and the horizontal shelf as shown. The protractor's minor axis is aligned with the right horizontal axis and the desired angle of inclination can be read directly on the angular scale. In this example, the desired angle of inclination is 30° from vertical. A light mark is made at this angle reading to help construct the inclined line on the left isometric plane. Then, a line is drawn from the vertex to the tick mark. This line's intersection with the construction line at the top of the block establishes its endpoint. The horizontal isometric line for the top surface is then drawn from the top vertex point and blocked in. This line is parallel to the horizontal shelf line. The face is completed by drawing a line from the right vertex on the top surface to the shelf line.

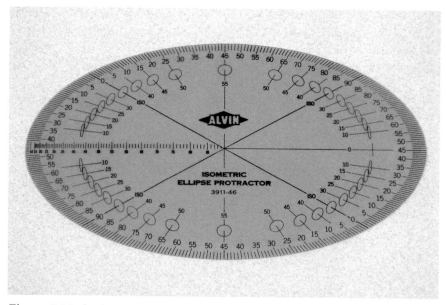

Figure 5-21. An isometric protractor.

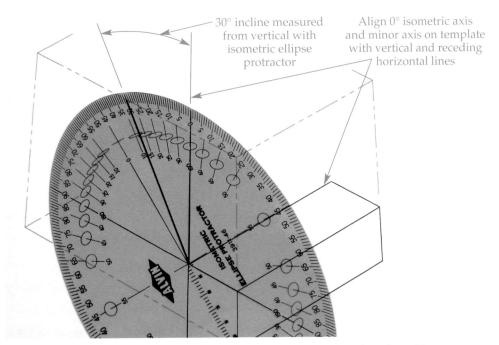

Figure 5-22. Using an isometric protractor to draw an inclined surface. The protractor is aligned with the horizontal and vertical axis lines to mark the desired angle of inclination.

Referring to **Figure 5-18**, the top portion of the object is perpendicular to the inclined surface. The isometric protractor can be located on the top vertex point to draw the perpendicular line. If you make a mistake other than incorrect alignment of the minor axis, it will usually occur in this step. First, align the minor axis of the protractor with the right horizontal axis drawn in the previous step. Then, start at the 30° point used in **Figure 5-22** and count 90° around the protractor. Count the degrees of rotation in tens from the starting angle, then add any single degree increments one at a time. Finally, draw a light reference line and use it to box in a line for the top portion of the object.

Notice that the isometric protractor has a number of different ellipses printed on its face. These are the approximate shapes of the angle ellipses. The angle of each ellipse is also provided. In addition, for each ellipse, a line extends from the outer edge of the protractor through the minor axis of the ellipse. This line is used to align the angle ellipse template with the isometric axis and represents the thrust line. The reading on the outer edge of the protractor indicates the thrust line angle. Some holes and cylinders intersect surfaces at angles that are not perpendicular. Therefore, determining which angle ellipse to use by aligning with the face of the surface, instead of along the thrust line, can lead to errors. After determining the proper angle, an angle ellipse template is used to draw the ellipse.

Notice that some thrust line angles on the outer scale of the isometric protractor do not line up exactly on a specific angle ellipse. In these cases, use the angle ellipse that is the closest to the reading. For example, look at the angle reading 30° up from right horizontal. A thrust line angle at 75° would use an angle ellipse of 25° since it is closer than any other. In a case where the reading is in the middle of two angle ellipses, such as the 60° thrust line reading, use either of the two closest ellipses.

Developing a thrust line and selecting an appropriate angle ellipse for the inclined face of the object in **Figure 5-18** is shown in **Figure 5-23.** Constructing a thrust line on a nonisometric surface requires that the minor axis of the

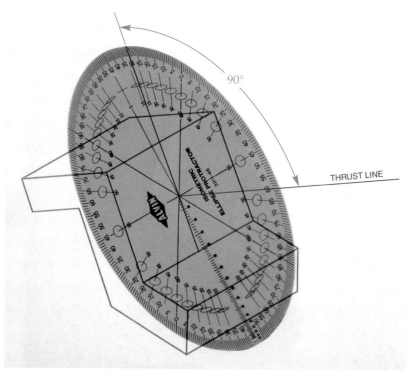

Figure 5-23. Developing a thrust line and selecting the proper angle ellipse with an isometric protractor.

isometric protractor be aligned parallel to a natural (or generated) isometric line on the face of that surface. If the surface is inclined or skewed, it will typically have one or more isometric lines defining where the nonisometric surface intersects an adjacent "normal" isometric surface. Aligning the template with the minor axis parallel to one of these lines establishes the reference point to then calculate the thrust line. In the example in **Figure 5-23**, the adjacent top surface is perpendicular. The protractor is aligned on the center of the inclined surface, and the thrust line is drawn parallel to one of the perpendicular axis lines. Measure the angle of the thrust line from the surface and mark it. If the thrust line is perpendicular to the surface, you can count 90° in either direction. If the intersection of the thrust line is at some other angle, it is very important to calculate the angle in the proper direction. Begin counting degrees from the surface toward the "open air" side of the object. Counting thrust angle degrees from the surface and toward the mass of the object can result in an incorrect inclination of the thrust line. In the example shown, the thrust line angle reading is 60°, and the nearest angle ellipse to this reading is 50°. Therefore, a 50° angle ellipse is used to draw the hole. Refer to **Figure 5-18**.

Isometric Drawing with Computer-Aided Drafting

Many of the isometric drawing techniques previously discussed in this chapter are incorporated into computer-aided drafting programs. Various drawing aids and tools used for creating isometric views are common to most CAD software. For example, drawing grids can be typically set up to simplify drawing isometric lines. An *isometric grid* is a series of dots that align on a 30° pattern to establish the left and right axes, **Figure 5-24**. Alternating dots establish

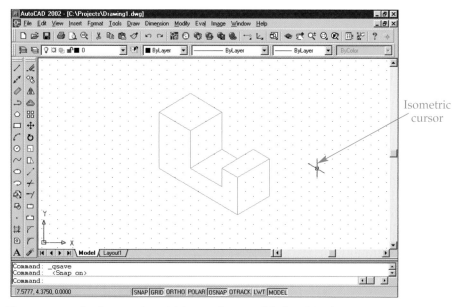

Figure 5-24. An isometric grid is used to establish a 30° dot pattern for the isometric axes in a CAD program. Notice the orientation of the cursor.

the vertical axis. An isometric grid can be used for preliminary layout, blocking in basic object shapes, and developing simple objects.

The distance between grid dots can be set in most CAD programs. Do not set the grid pattern too dense. Set the isometric grid by selecting a distance that can easily divide most width, depth, and height measurements of the major features. If the object has extremely precise measurements that cannot be divided easily, grid dot spacing may not be of great value. In this case, set the grid dot spacing at 1/2″ or 1″ intervals and use it only for edges of the object or centerline layout.

Even if the distances cannot be easily divided, the grid provides a visual reference. Also, most CAD programs use an isometric cursor when the isometric grid is turned on. The *isometric cursor* is a set of crosshairs that are rotated to align with the isometric axis. Refer to **Figure 5-24**. This allows you to see if you begin to stray off the isometric axis when you are drawing a line. Snap mode can also be set on most CAD programs to match the grid. Snap mode is discussed in Chapter 4 of this text.

Isometric Axes

Each of the isometric axes can be drawn with one of two angle entries using a CAD system, **Figure 5-25.** Most CAD systems measure angles from 0° horizontal on the positive X axis. Also, CAD systems measure angles in a counterclockwise direction. For a regular isometric drawing, 30° and 210° can be used for the right axis, 90° and 270° for the vertical axis, and 150° and 330° for the left axis. For a reversed axis isometric drawing, 30° and 210° can be used for the left axis, 150° and 330° for the right axis, and 90° and 270° for the vertical axis. For a long axis isometric drawing, 0° and 180° can be used for the major axis, 60° and 240° for the right axis, and 120° and 300° for the left axis.

Knowing the different isometric axis angles is important because in some instances, grid dots cannot be used. Also, when drawing nonisometric lines, angular values other than those used for the standard isometric axes must be used. This means that you must specify a starting point for a line and then indicate a specific distance and angle.

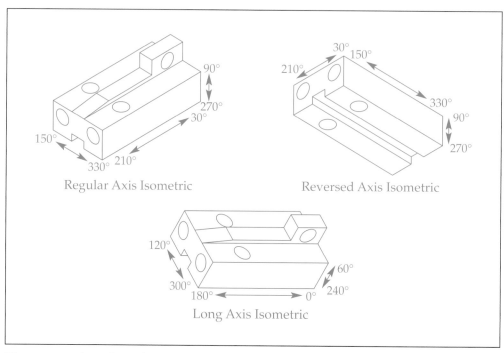

Figure 5-25. Angular values used to draw different isometric axes in a CAD program.

Isometric Line Drawing Functions in CAD

There are three basic ways to specify point coordinates for drawing lines in a CAD program. Coordinates can be specified as polar, relative, and absolute coordinates. Absolute coordinates require that you enter the exact X and Y coordinate for each point. Relative coordinates require an X and Y coordinate based on the last point entered. These entry methods work well for orthographic drawings and some applications of oblique drawing. However, they do not work well for isometric drawing. The polar coordinate entry method is invaluable in pictorial drawing. Polar coordinates require a distance coordinate and an angular value based on the last point entered or the origin.

As discussed in Chapter 4, polar coordinates are typically entered using the format (*distance,angle*). In most cases, polar coordinates are entered as relative coordinates located from a previous point. This method is very useful for drawing isometric lines. For example, the entry @5<30 tells the computer to locate a point 5 units from the previous point at an angle of 30°. The @ symbol is used by most CAD systems to tell the computer to locate a point at a given distance from the last point picked. In this example, the distance is 5 units. The < symbol means "at an angle of." In a typical CAD system, angles are measured counterclockwise from 0° (horizontal). In this example, the angle is 30° from horizontal. Thus, when using the **Line** command, the entry @5<30 specifies to draw a line 5 units from the previous point at a 30° angle from horizontal.

When using a CAD program, drafters and illustrators typically begin an isometric drawing at some point toward the lower-middle portion of the drawing area. Then, they draw all remaining lines using polar coordinates. After the drawing is completed, it is centered in the drawing area.

The regular isometric drawing in **Figure 5-26** was drawn with lines using polar coordinates. The procedure is as follows. First, enter the **Line** command and start at the origin point. Next, draw a three-unit line at 150°. This locates Point 1. For Point 2, draw a two-unit line at 90°. For Point 3, draw a one-unit line at 330°. For Point 4, draw a one-unit line at 270°. For Point 5, draw a

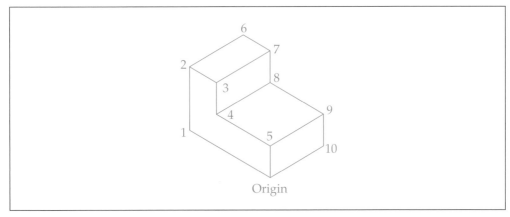

Figure 5-26. The point locations shown are used to draw isometric lines with polar coordinates. The coordinates are entered using a distance and angle from the previous point.

two-unit line at 330°. Finally, use the **Close** option to draw the final segment to the origin or draw a one-unit line at 270°. The left face of the object is now complete.

To draw the top face on the left, start at Point 2. Draw a two-unit line at 30°. This locates Point 6. For Point 7, draw a one-unit line at 330°. For Point 3, draw a two-unit line at 210°. The face is complete. You do not have to draw from Point 3 to Point 2, since this line was created on the left face.

To draw the top face on the right, start at Point 5. Draw a two-unit line at 30°. This locates Point 9. For Point 8, draw a two-unit line at 150°. For Point 4, draw a two-unit line at 210°. This completes the face.

To draw the upper portion of the right face, start at Point 8. Draw a one-unit line at 90°. This locates Point 7. All the other edges have already been drawn, so this one line completes the upper portion. To draw the lower portion, start at the origin. Draw a two-unit line at 30°. This locates Point 10. Then, draw a one-unit line at 90° for Point 9. This completes the face and the drawing.

Isometric Planes

Most CAD systems use the terms *left isoplane*, *right isoplane*, and *top isoplane* to describe the three normal isometric planes. See **Figure 5-27**. This illustration

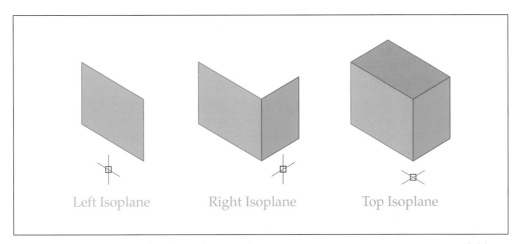

Left Isoplane   Right Isoplane   Top Isoplane

Figure 5-27. Three standard isoplane and isometric cursor orientations are available in most CAD programs.

shows how the three isoplanes appear in a typical CAD program. The cursor crosshairs on the left isoplane include a vertical axis line and an axis line inclined at 150°. The crosshairs on the right isoplane include a vertical axis line and an axis line inclined at 30°. The top isoplane has crosshairs angled at 150° and at 30°. These isoplane cursors can be used to draw a regular or reversed axis isometric drawing.

The standard isoplane crosshairs do not lend themselves to developing a long axis isometric drawing. Some CAD systems allow you to modify the orientation of the crosshairs. For instance, you can set the axis lines to 60° and 120°.

One special function of the isoplane crosshairs is the ability to automatically generate appropriately inclined isometric circles (ellipses). Drawing a circle in any isoplane automatically creates an ellipse properly oriented to that plane. This is discussed in the next section.

Isocircle and Ellipse Functions

A very powerful isometric drawing feature included in CAD programs is the ability to automatically create correct isometric ellipses. Most CAD systems call these ellipses *isocircles* or *isometric ellipses*. To draw an ellipse on an isometric plane, simply select the proper isoplane (left, right, or top), locate the center of the ellipse, and give the radius or diameter. The computer automatically sizes and draws the ellipse in the correct orientation and proportion. CAD systems vary greatly in how nonisometric ellipses are drawn. You should refer to the operating manual or Help system provided by the software manufacturer to determine the correct procedure.

Isometric Editing and Illustrating Techniques

There are a number of editing functions in CAD programs that can be used to construct or modify isometric drawings. Two common editing functions are erasing and trimming lines. *Trimming* or *clipping* is removing a portion of an object using another object as a cutting edge. Trimming can be very helpful in developing isometric drawings that do not show invisible or hidden features. You may not know which features are visible and which are hidden until you finish all constructions. Once all constructions are complete, use the **Erase** and **Trim** functions to remove the hidden features. See **Figure 5-28.**

Remember, technical illustrations are primarily meant to show what an object looks like. There are some shortcut techniques to produce quick drawings that meet the illustration goals. Some of these include rounding off measurements, adding lines, or deleting lines to improve visualization. While these techniques produce a drawing that is visually easier to recognize, the drawing is not technically accurate.

The reality of illustration in industry is that shortcuts are taken to achieve the desired audience visualization within a specific time frame. This does not mean that object design specifications are ignored. Simply, some dimension variations cannot be seen without measuring features on the drawing. Also, adding or deleting lines may make it easier for an untrained viewer to comprehend the message the drawing is meant to convey.

Certain editing functions may work well in orthographic projections but not in isometric views. This may be due to how the CAD system draws objects in isometric mode. A circle may actually be drawn as a series of lines. Problems may also occur if you do not have the correct isoplane selected. Experiment

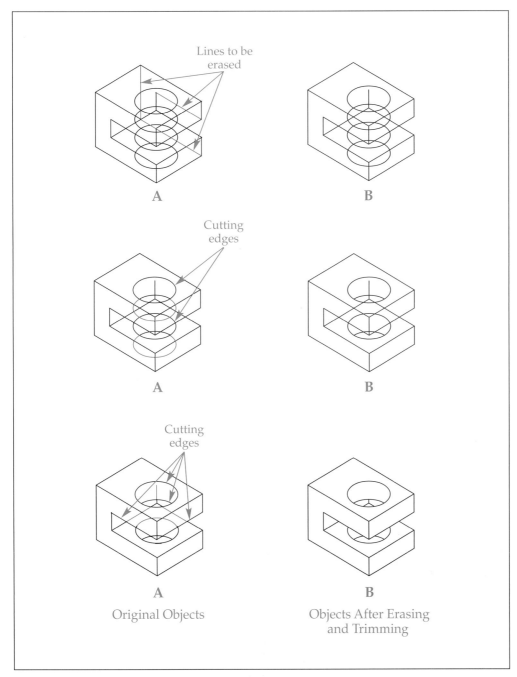

A

B

Cutting
edges

A

B

Cutting
edges

A

B

Original Objects

Objects After Erasing
and Trimming

Figure 5-28. Using erasing and trimming operations to remove hidden lines from an isometric drawing. The objects to be removed are shown in red.

with different combinations of drawing aids and editing tools on your CAD system. Determine which ones work best in isometric mode.

Dimetric Drawing

Developing dimetric drawings is often a good compromise method. An oversized and somewhat disproportionate isometric drawing may not be desirable. Likewise, a more difficult and time-consuming trimetric drawing may be undesirable. In this situation, developing a proportionate and relatively simple dimetric drawing may be your preferred choice.

Dimetric Drawing Principles

Earlier in this chapter, the differences between dimetric drawings and isometric drawings were discussed. The examples in **Figure 5-29** show the major combinations of axis angles and scales normally used to develop dimetric drawings.

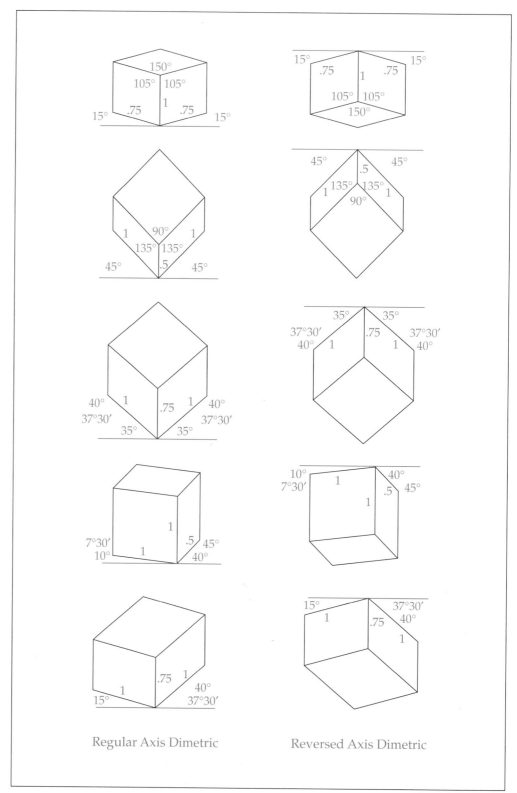

Figure 5-29. Common axis angles and scales normally used for dimetric drawings.

Constructing arcs and circles in a dimetric drawing may require special attention. If the feature lies between two adjacent axes with the same scale factor, you can use an angle ellipse template. However, when the feature lies between two axes with different scale factors, the circular features must be developed using the coordinate layout method.

A simple dimetric shape using the 10° left and 40° right axis layout is shown in **Figure 5-30**. The normal scale factors used for this configuration are 1 for the left and vertical axes and .5 for the right axis. A standard angle ellipse template can be used on the left face of the object because that surface has identical vertical and horizontal axis scale factors. However, ellipses on the right and top faces cannot be drawn with an angle ellipse template because of the distortion created by the scale of .5 along one axis. The coordinate layout procedure described below is common in this situation. Refer to **Figure 5-30** as you go through the example.

1. Draw an orthographic view of the circular shape. Divide it with grid lines to establish coordinate points on the arc perimeter.
2. Draw a grid with light construction lines aligned with the horizontal axis in the dimetric view. Be sure to reduce the distance between grid lines along the foreshortened axis by the appropriate scale factor.
3. Transfer locations of points on the arc perimeter from the orthographic view to the dimetric view.
4. Find the closest ellipse on an angle ellipse template to fit the points marked on the grid. Another way is to draw portions of the ellipse using a series of arc segments from different ellipses, or connect the coordinate points with an irregular curve.

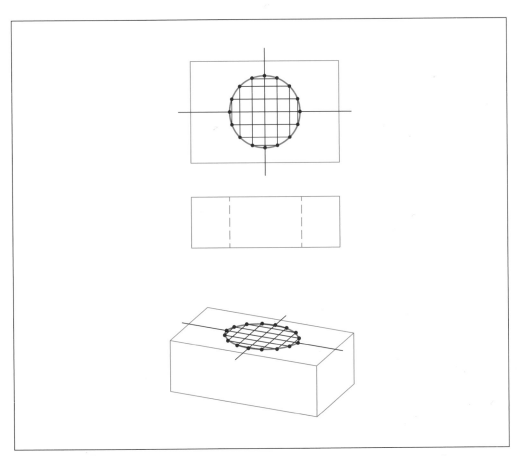

Figure 5-30. The coordinate layout method is commonly used to draw circles in dimetric drawings.

Dimetric Drawing Applications

If developing a fairly realistic appearance and proportion is important and speed is not a large factor, a dimetric drawing may be appropriate for the application. If there are many circular or irregular shapes, a dimetric drawing is probably not the best method. Objects that lend themselves to dimetric drawings are made up of rectangular, planar, and normal surfaces.

Trimetric Drawing

Trimetric drawing is typically the most realistic pictorial method available within the axonometric family of drawing styles. As the prefix *tri* indicates, three different angles are used between the pictorial axis lines, which form a Y-shape at the front of the view layout. Of the three axonometric styles, trimetric drawing is the most complex to develop. Although it is the least used of the axonometric methods, there are many applications in the illustration field to justify study of this technique.

Trimetric Drawing Principles

Trimetric drawings are used for applications that require a pictorial drawing with maximum realism in the proportions of an object. Trimetric drawings have three different angles for the horizontal and vertical axis orientations. In addition, three different scales are used to measure the width, height, and depth axes. The examples in **Figure 5-31** show the two major axis angle combinations preferred in industry. These are the 45°/15° and 35°/25° horizontal axis orientations.

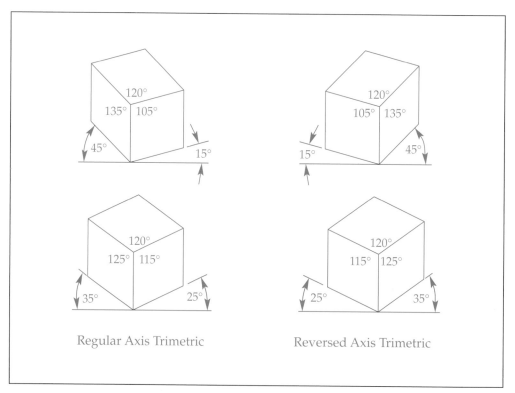

Figure 5-31. Standard axis orientations used for trimetric drawings.

Notice that the examples in **Figure 5-31** do not have specific scale factors identified. Instead, these examples use an alternate method of angle ellipse sizing to establish measures along each axis. The first example has 135° between the vertical and horizontal axes on the side with the 45° receding axis. It has a 105° angle on the side receding at a 15° angle. The angle between the two horizontal axes on top of the object is 120°.

The three angles in the second example are also different. The side receding at a 35° angle has 125° between the vertical and horizontal axes. The other side has an included angle of 115°, while there is 120° between the two horizontal axes on top of the object. Although there are other layout orientations, these two examples are somewhat standard within many areas of industry.

The planar surfaces of a trimetric drawing can be constructed using measurements made with standard ellipse templates. The angle ellipses are used to locate the extents of the "glass box." When developing a trimetric drawing with 45° and 15° receding axes, use a 25° angle ellipse template to measure on the left face, a 35° angle ellipse on the top face, and a 55° angle ellipse on the right face. See **Figure 5-32**. In **Figure 5-32A**, a 25° angle ellipse with a 1″ diameter is used to develop the left side of the glass box. The right and top surfaces of the glass box are developed with 55° and 35° angle ellipses. In a similar fashion, **Figure 5-32B** shows using 30°, 40°, and 35° angle ellipses to create measures on trimetric drawings receding at 35° and 25° angles.

After completing the basic block-in process for the trimetric view, lay out angle ellipse template measurements to locate the desired features on the object. This is done using the same principles applied to locate nonisometric lines with ellipses. See **Figure 5-33**. This illustration shows how other trimetric measurements may be completed using an angle ellipse. In this case, a 3/8″ × 1/4″ chamfer and a 1/2″ step are located and cut out of the block.

An alternate method of trimetric axis development is to use a different scaling factor on each of the three axes. Generally, the vertical axis is given a scale factor of 1. The axis with the smallest receding angle is usually given a scale factor of .75. The axis with the largest receding angle is usually given a scale factor of .5. This is a much quicker way to develop objects without curved surfaces. Holes, fillets and rounds, or cylindrical projections must be developed using the time-consuming coordinate method previously discussed for dimetric

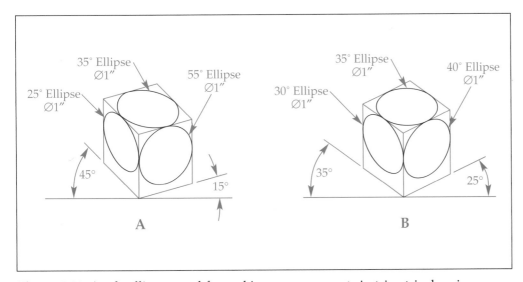

Figure 5-32. Angle ellipses used for making measurements in trimetric drawings.
A—Angle ellipses for a trimetric drawing with 45° and 15° receding axes.
B—Angle ellipses for a trimetric drawing with 35° and 25° receding axes.

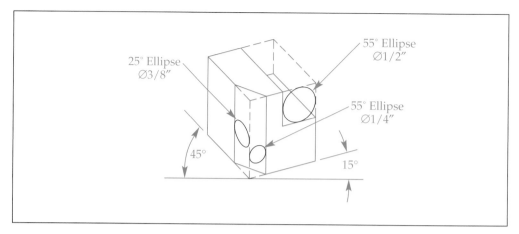

Figure 5-33. Making trimetric measurements with angle ellipses.

drawing. However, using approximate scaling factors does not always produce a true trimetric drawing. This method is recommended for trimetric drawings that use receding axis angles of 45° and 15°. For trimetric drawings that use different receding axis angles, this method can cause distortion in the object. See **Figure 5-34**.

Orienting the object properly is important in developing a realistic trimetric drawing. For example, if an object has one side that is significantly larger, rotate the object so the long side aligns with the smaller of the receding horizontal angles. This creates a more proportionate appearance. See **Figure 5-35.**

Trimetric Drawing Applications

Trimetric drawings are used for situations requiring proportionate pictorial drawings. Trimetrics are often extremely complex to draw and time-consuming. Therefore, trimetric drawing is the least used of any axonometric style.

Trimetrics are most appropriate for objects with primarily rectangular or normal surfaces. When using angle ellipse measuring techniques, trimetric drawings may be more adaptable to objects with measurements not easily subdivided by the standard dimetric scaling factors.

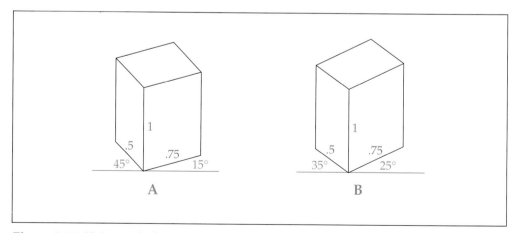

Figure 5-34. Using scale factors to lay out a trimetric drawing. A—Accurate scaling is produced when using the scale factors shown with axis angles of 45° and 15°. B—Using the same scale factors with axis angles of 35° and 25° creates distortion in the height of the object.

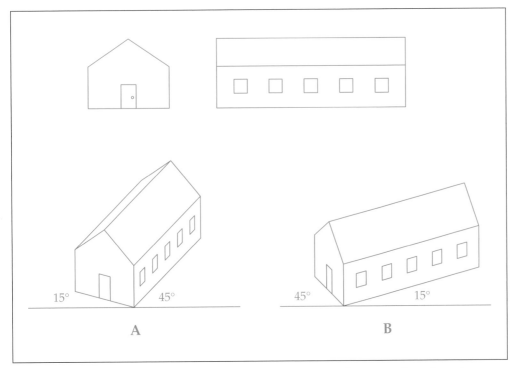

Figure 5-35. Using an orientation that best represents the features of an object is important when developing dimetric drawings. A—This orientation presents a disproportionate view of the side of the building. B—Aligning the long axis with the smaller receding horizontal angle creates the best representation.

Parallelogram Construction of Arcs and Circles

When creating dimetric and trimetric drawings, there are times when you may not be able to calculate the true surface angle or use an angle ellipse protractor to identify the appropriate angle ellipse. In these situations, the parallelogram method of ellipse construction is used to develop an ellipse with an approximately correct shape. The parallelogram method is useful in drawing arcs and circles with different scales on their axes.

To create a circle or arc with this method, first construct a parallelogram with sides parallel to the axis lines, **Figure 5-36A.** For this example, 10° and 40° angles are used for the receding horizontal axes. The scale factors are 1 for the left

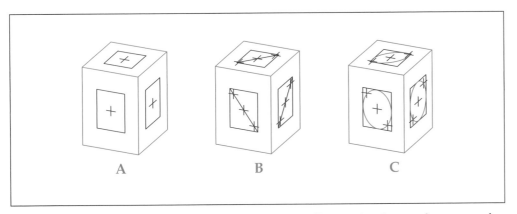

Figure 5-36. The parallelogram method of drawing ellipses. A—Locate the center of the ellipse. B—Mark the major axis. C—Select the appropriate angle ellipse.

and vertical axes, and .5 for the right axis. Next, locate the ellipse center. If the hole is centered on the surface, use diagonal lines between opposite parallelogram corners. The intersection is the center of the ellipse. Then, establish the endpoints of the major axis by using appropriately scaled measurements, **Figure 5-36B.** The major axis is always the longer of the two diagonal lines connecting opposite corners of the parallelogram. Now, simply select an appropriate angle ellipse. The ellipse should have both true scale diameter along the major axis and tangency to each side of the parallelogram, **Figure 5-36C.** Since the actual rotation angle is not important, use trial and error to find the closest fit.

Computer-Aided Dimetric and Trimetric Drawing

Using a CAD program to develop dimetric and trimetric drawings provides numerous shortcuts to the illustrator. Developing ellipses is much simpler and more accurate with a CAD system. Major CAD-based aids useful in dimetric and trimetric drawing include axis rotations of grid and snap configurations, modified scaling functions that allow you to enter measures that are automatically scaled to correct sizes, and numerous editing options. These options are discussed in the following sections.

Establishing Dimetric and Trimetric Axes

Most CAD systems allow you to change snap and grid distances and angles. By rotating the snap and grid to a specific angle, you can easily draw on the horizontal axes of the object.

Develop as much of the object as possible along a single axis angle. Then, change to a vertical alignment or rotate to the other receding horizontal axis angle. Draw the features with that alignment, rotate to the remaining axis, and complete the object.

At first, using these drawing aids may be difficult. However, after practice, they will be easy to use and your drawing efficiency will increase.

Scale Modification Functions

One of the distinguishing features of dimetric and trimetric drawings is their use of different axis scales. One method of scaling on a CAD drawing is to calculate the scaled measures mathematically and then enter the exact distance a line is to be drawn. However, rounding and mathematical errors tend to create problems. Another method is setting the snap to the desired unit of measure for the scale factor, such as .5, and then setting the grid to twice that distance. This produces a scale factor of 1 on the grid pattern and a scale factor of one-half for the invisible snap point between each grid dot.

Some CAD systems allow you to input values as fractions. This allows you to enter the actual measurement for the feature and the scale factor. For example, the polar coordinate entry @3.875/2<40 draws a line from an initial point and at an angle of 40° from the horizon one-half the distance of 3.875. This applies a scale factor of .5 on the 40° receding axis. This method works well when many measurements that are needed do not fall on convenient snap or grid dot intervals. However, for most CAD systems, the fractional dividend and divisor must be entered as whole numbers.

Another way to enter automatically scaled values is to use the CAD system calculator. For example, when drawing features, you can use mathematical expressions to determine distance values at any scale from existing endpoints or center points to other points on the drawing. The calculator function typically allows you to enter decimal values.

Object Snap Functions in Dimetric and Trimetric Drawing

The object snap functions intersection, endpoint, midpoint, and node (point) are very useful for dimetric and trimetric drawings. The tangent object snap function can also be useful in certain situations.

Examples of using object snaps on a dimetric drawing are shown in **Figure 5-37**. The features corresponding to the snap options are labeled to indicate how they were drawn. After drawing the basic shape layout, the inclined surface was generated by drawing its inclined edge on the left face of the object. The origin of the line was established using the midpoint object snap option on the front vertical line. Snapping to the midpoint on the top right edge located the end of the inclined line. The inclined line on the far side of the sloping face was developed by using the copy function. After selecting the first inclined line to copy, the intersection snap was used to select the bottom end of the line and drag a copy across the face to the midpoint of the left vertical line. To construct the horizontal line across the left face of the object, the midpoints of the left and front vertical lines were selected with the midpoint object snap to establish the endpoints.

The cutout feature was also developed using object snaps. To draw the horizontal construction lines, the lower horizontal line was copied twice. The first line was located using the midpoints of the inclined lines. The second line was located using the midpoints of the lines on the upper surface. To draw the inclined construction lines, one of the inclined lines was copied twice. To establish the endpoints, the lower horizontal line was divided into thirds with construction points. The inclined lines were then located as copies by selecting the construction points as endpoints with the node object snap. The lines for the top of the cutout feature were drawn by dividing the upper horizontal lines into thirds with construction points and drawing lines between the points. The remaining lines for the cutout feature were drawn using construction lines, points, and the intersection and node object snaps. Finally, portions of the construction lines were erased or trimmed as necessary to complete the object.

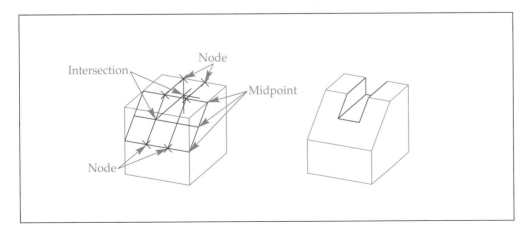

Figure 5-37. Using object snap functions in constructing a dimetric drawing.

Dimetric and Trimetric Drawing Editing Functions

There are a number of CAD editing commands that work well with dimetric and trimetric drawings. These include **Scale**, **Copy**, **Mirror**, and **Erase**. Some CAD systems may have other commands that work well in addition to these. However, all CAD systems should have these four basic editing functions. Also, you should make it a work habit to always save your drawing before, and after, completing any editing functions.

The **Scale** command is useful when all measurements are first drawn full scale. It can then be used to change lines on the receding axes to the proper scale, such as half scale or three-quarters scale. Using this method reduces the chance for human error due to mathematical calculations.

The **Copy** command is also useful for editing functions with dimetric and trimetric drawings. For example, once a line or arc is accurately constructed, it is easier to copy that item than to draw it again. Use snap functions when copying features to improve accuracy.

The **Mirror** command is useful if the receding angles and surface measurements on the left and right axes of a dimetric drawing are equal. For example, after drawing one side of the basic layout to a given scale factor, simply mirror that side along the vertical axis. This saves you from redrawing the first side. However, look closely at the orientation and inclined direction of specific features before duplicating them with the **Mirror** command. The mirrored copy will be the exact opposite of the original items selected in relationship to the axis line identified.

When using the **Mirror** command, the most common error is selecting an incorrect mirror line. This is the line about which the objects are copied. If an appropriate line has already been drawn, use its endpoints rather than other snap points or grid dots when defining the axis. A slight error in picking the correct snap or grid dot may go undetected until you try to match item alignments with other features. If needed, draw a temporary construction line to use as the mirror line.

The **Erase** command can be used to remove construction lines and unwanted features. You will likely need to draw many construction lines in developing a CAD drawing. One common practice when using a CAD system with layers is to draw all construction lines on one layer. When the drawing is completed, the construction layer can be deleted. Layer functions also allow you to "turn on" and "turn off" the construction layer whenever needed.

The **Trim** and **Extend** commands in CAD programs are also useful in many different dimetric and trimetric drawing situations. As previously discussed, trimming is a function used to remove unwanted portions of an object while leaving the segments you want to keep. By identifying the line, arc, or other feature you want to establish as a cutting edge, or boundary, you can trim off elements inside or outside of the boundary.

Whether the trimming command is called **Trim**, **Clip**, **Remove Section**, or **Cut**, this editing function is helpful in modifying construction lines and arcs into the precise measures and contours of object lines. A similar editing function is accomplished with the **Extend** command. This command allows you to add to an object so that it extends to a boundary edge. The extend and trim functions allow you to draw numerous temporary construction lines and arcs and then quickly modify them into finished object lines.

Summary

There are three major types of pictorial drawings used in technical illustration. These are axonometric, oblique, and perspective drawings. The axonometric drawing types include isometric, dimetric, and trimetric. The oblique drawing types include cavalier oblique, cabinet oblique, and general oblique. The perspective drawing types include one-point, two-point, and three-point.

Isometric drawings are the most used of all pictorials. They are easy to lay out and can be drawn quickly. Oblique drawings are generally used for objects with circular or irregular shapes. Perspective drawings are most often used in architectural and interior design applications.

There are three types of isometric drawings. These are regular, reversed axis, and long axis. Most isometrics are drawn as regular isometrics. A reversed axis isometric drawing is similar to a mirror image of the regular isometric. The long axis isometric is used for objects that have one side that is much longer than the others.

Circles appear as ellipses in isometric drawings. Circular features are drawn using the coordinate method or isometric ellipse templates. The angle ellipse template and isometric protractor are also very useful for isometric drawings.

When creating pictorial drawings on a CAD system, there are several commands and functions that aid the drawing process. The standard isometric grid provides a series of dots on a 30° pattern and is used to help draw on the left and right isometric axes. The isometric cursor also helps you keep track of your orientation when drawing. Most CAD systems call the three isometric drawing planes the left isoplane, right isoplane, and top isoplane. These different drawing planes are very useful when drawing isometric circles because the computer automatically calculates the correct ellipse. Dimetric and trimetric drawings can also be developed on a CAD system. While there are features that help in developing these types of drawings, most CAD systems do not have specific dimetric and trimetric functions.

Review Questions

1. Identify the three pictorial drawing classifications and the three drawing styles in each classification. Sketch your answer as a "family tree."
2. Describe the principal applications of each pictorial drawing classification.
3. What are the advantages and disadvantages of pictorial drawings?
4. Explain the major differences between isometric, dimetric, and trimetric drawings.
5. What are the major advantages of isometric drawing over other forms of pictorial drawing?
6. Explain the difference between an isometric projection and an isometric drawing.
7. Describe what a viewer sees when an object is drawn as a regular isometric, a reversed axis isometric, and a long axis isometric.
8. What is a nonisometric line?
9. List the steps used to center an isometric drawing.
10. What are the primary differences between an isometric ellipse template and an angle ellipse template?
11. Explain the proper ellipse axis alignments for each of the three standard isometric planes when using an ellipse template to orient circular features.
12. In relation to isometric drawing, what is a *thrust line*?
13. Explain how to use an isometric protractor to determine the proper angle ellipse size for an inclined surface.
14. What are the three normal isometric planes typically called in a computer-aided drafting program?
15. When using a CAD program, at what angles are the normal isometric axes drawn to develop a regular isometric drawing?
16. Which coordinate entry method works best for locating points on an isometric CAD drawing?
17. Briefly explain how to create an ellipse on a normal isometric plane in a CAD drawing. Why is this method more efficient than manual techniques?
18. What types of objects are best developed as dimetric or trimetric drawings?
19. Identify and describe three different CAD editing commands that are useful when developing dimetric and trimetric drawings.

For Problems 1–8, create regular isometric drawings using the orthographic views shown. Draw the objects at full scale using the dimensions given. Use isometric grids and other drawing aids necessary to assist you in the drawing process.

1. Step Block

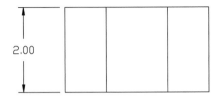

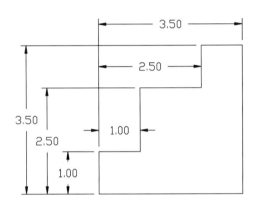

2. V-Block

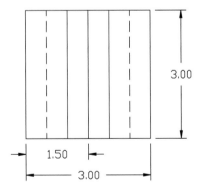

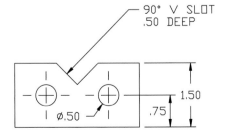

3. Connecting Bracket

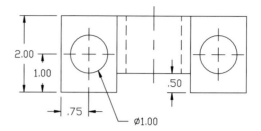

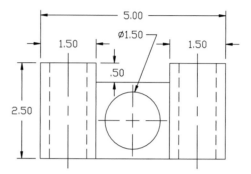

4. Angled Stop

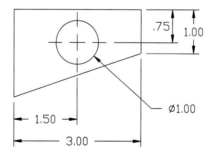

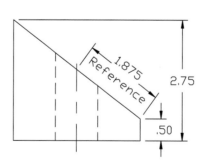

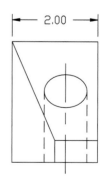

5. Ratchet Pawl

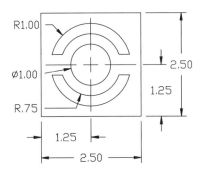

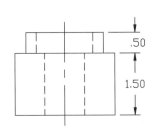

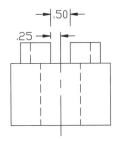

6. Pin Mount

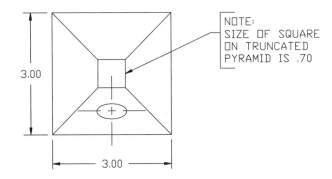

NOTE:
SIZE OF SQUARE
ON TRUNCATED
PYRAMID IS .70

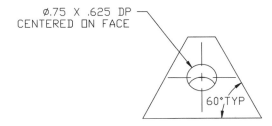

Ø.75 X .625 DP
CENTERED ON FACE

60°TYP

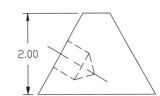

2.00

7. Beveled Connector

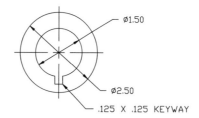

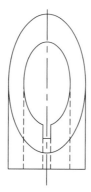
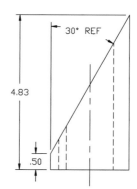

8. Slot Box

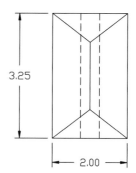

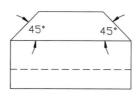
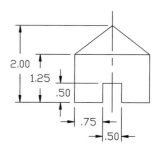

9. Create a dimetric drawing of the fork clip using the orthographic views shown. Use the dimensions provided and the appropriate scale factors for the dimetric axes. Orient the view to create the most realistic representation.

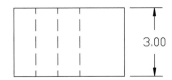

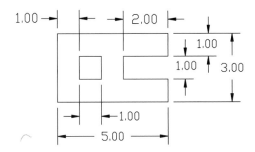

10. Create a dimetric drawing of the beveled guide using the orthographic views shown. Use the dimensions provided and the appropriate scale factors for the dimetric axes. Orient the view to create the most realistic representation.

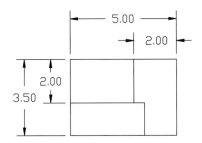

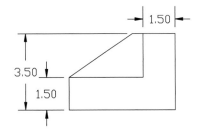

11. Create a trimetric drawing of the shoe stop using the orthographic views shown. Use the dimensions provided and the appropriate scale factors for the trimetric axes. Orient the view to create the most realistic representation.

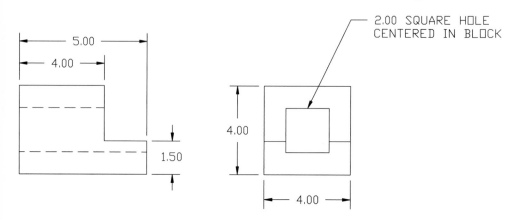

12. Create a trimetric drawing of the wedge block using the orthographic views shown. Use the dimensions provided and the appropriate scale factors for the trimetric axes. Orient the view to create the most realistic representation.

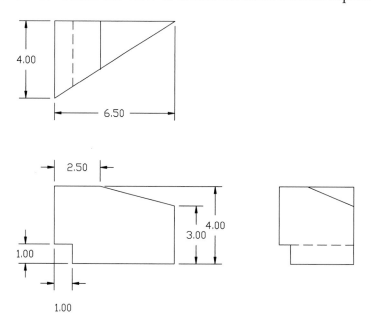

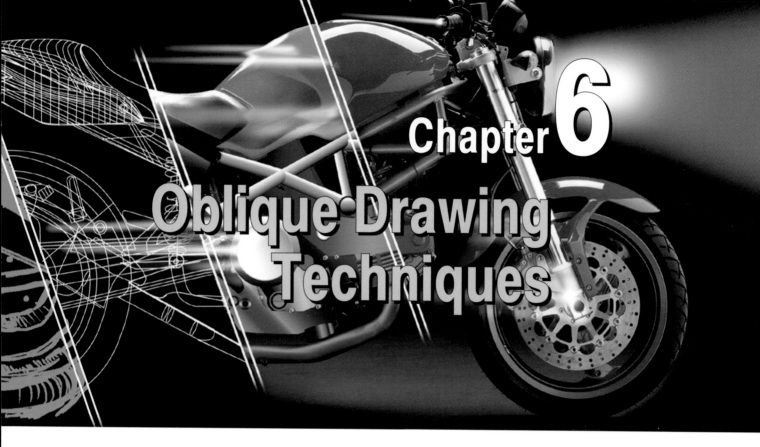

Chapter 6

Oblique Drawing Techniques

Learning Objectives

At the conclusion of this chapter, you will be able to:

☐ Explain the similarities and differences in relation to the cavalier, cabinet, and general oblique drawing styles.

☐ Determine the best orientation for an object drawn as an oblique.

☐ Plan, lay out, and draw oblique views of an object using the cavalier, cabinet, and general oblique drawing techniques.

☐ Explain the basic rules and procedures of oblique drawing.

☐ Identify the major computer applications used in oblique drawing development.

Introduction

Oblique drawing is a form of pictorial drawing in which a normal front view parallel to the projection plane is used with top and side views to describe an object. As discussed in Chapter 5, there are three different types of oblique drawings. These are cavalier oblique, cabinet oblique, and general oblique. The primary difference between each of the oblique drawing types is the scale of the receding axis. See **Figure 6-1**. In this type of drawing, the receding axis may be drawn at any angle from horizontal. However, this axis typically recedes along a 30°, 45°, or 60° angle. Refer to **Figure 6-1**.

Each type of oblique drawing has an orthographic front view. This front view is the same no matter which type of oblique is being drawn. The depth axis recedes from the front view to create a three-dimensional pictorial view.

A major application of oblique drawing is to show objects with circular, cylindrical, or irregular shapes. Using a normal orthographic front view to draw these shapes is often easier than developing the foreshortened pictorial views in an axonometric or perspective drawing. Obliques are also appropriate for objects with one very long side.

The examples in **Figure 6-1** are called *regular axis obliques*. Their receding axis lines are drawn upward from the front of the object. In *reversed axis obliques*, the receding axis is drawn downward from the front of the object. Reversed axis obliques are used to emphasize the bottom of an object and are similar to reversed axis isometric drawings. Additionally, you can draw the receding axis lines toward the right or left to emphasize the side features.

Oblique drawings are very simple to create. An orthographic front view modified into an oblique pictorial view can save both time and complex drawing procedures. Without the oblique style of drawing, it would be very difficult to develop pictorial views of many circular, cylindrical, and irregularly shaped objects.

Cavalier Oblique Drawings

Cavalier oblique drawing is generally accepted as the quickest method of oblique development. Cavalier obliques do not require calculation of reduced scale measurements. No matter which angle is used for the receding axis, the scale used along the receding axis is always the same as the scale used for the front face. This simplifies the drawing process and is an advantage of the cavalier style over the other oblique styles.

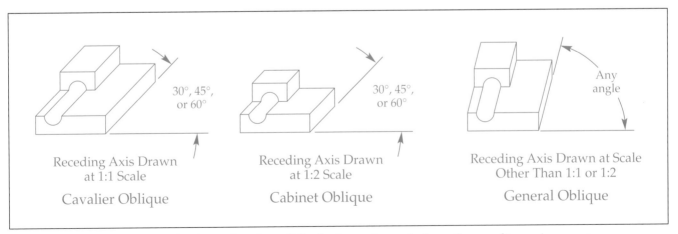

Receding Axis Drawn at 1:1 Scale

Cavalier Oblique

Receding Axis Drawn at 1:2 Scale

Cabinet Oblique

Receding Axis Drawn at Scale Other Than 1:1 or 1:2

General Oblique

30°, 45°, or 60°

Any angle

Figure 6-1. The three types of oblique drawings differ in the scale used along the receding axis.

Creating a Cavalier Oblique

To draw a cavalier oblique, first draw a normal orthographic front view. Receding lines are then drawn from each of the visible intersections on the front view. If you want to emphasize features on the side of the object, use a 30° angle. If you want to equally emphasize the top (or bottom) and a side, draw the receding lines at a 45° angle. Using a 60° receding angle places more visible emphasis on the top or bottom surface.

There are some special advantages to drawing a cavalier oblique with a 30° receding axis. The 30° receding axis is the same angle used on isometric drawings and allows you to use a standard isometric ellipse template to draw arcs and circles on the top, bottom, and sides of the object. There is some visual distortion when isometric circles are drawn close to normal orthographic circles on the front surface. However, a 30° cavalier oblique is much easier to draw than a 45° or 60° cavalier oblique.

Regardless of whether an object is drawn to a larger or smaller scale, if the height, width, and depth measurements use the same scale factor, the drawing is a cavalier oblique. For example, an object measuring 8.75″ tall, 9.25″ wide, and 4.25″ deep is normally drawn on a standard A-size sheet using the 1/2 scale on a mechanical engineer's scale or architect's scale. All height, width, and depth measurements are equally scaled to half of their actual size. Since the scale relationship of the receding axis measurements is the same scale used for the height and width, the drawing still meets the definition of a cavalier oblique.

Objects Appropriate for Cavalier Oblique Drawing

There are certain types of objects that especially lend themselves to the cavalier oblique style. Objects with curved shapes on the receding faces are best drawn as cavalier obliques. This aids in coordinate layout of features without having to calculate reduction scales. For example, the object in **Figure 6-2** has an irregularly curved shape along one side. By dividing the orthographic view of the curve with a coordinate grid and transferring these measurements along

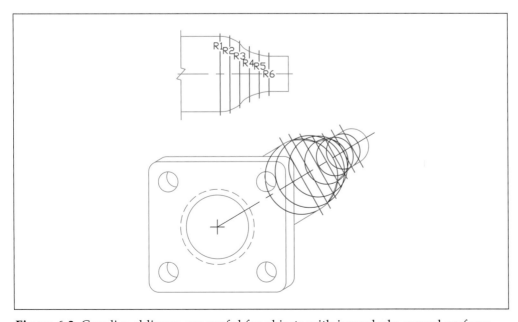

Figure 6-2. Cavalier obliques are useful for objects with irregularly curved surfaces.

the same grid spacing on the receding axis, a series of points can be developed. By connecting these points with an irregular (French) curve, the visible shape can be easily developed.

Cavalier obliques are very useful in drawing objects with considerable shape and detail. Not having to calculate scale reduction along each of the depth measurements simplifies the process and decreases the chances for error. An object with a relatively small depth measurement is best drawn using the cavalier oblique technique. Objects that have considerably greater width and height than thickness create rather realistic cavalier views. When the depth of an object is equal to or greater than its height or width, use a cabinet oblique instead of a cavalier oblique. Cabinet obliques are discussed next.

Cabinet Oblique Drawings

Cabinet oblique drawings are developed for objects with considerable depth. Such objects appear unnatural or distorted if they are drawn as cavalier obliques. The same guidelines used for the receding axis angle when drawing cavalier obliques apply to cabinet obliques. The only difference is the scale of the receding axis. A cabinet oblique is developed using a receding axis scale that is one-half the scale used for the front face.

Cabinet obliques have a more realistic appearance than cavalier obliques. The foreshortened depth measures are more noticeable in relation to the width and height measures since the eye perceives the width and height at a more perpendicular viewing angle than features along the receding axis. A comparison of the cavalier and cabinet oblique styles with the same object is shown in **Figure 6-3**. Notice that the cabinet oblique drawing appears less distorted than the cavalier oblique drawing.

Creating a Cabinet Oblique

Drawing a cabinet oblique is almost identical to drawing a cavalier oblique. A front orthographic view is drawn and receding axis lines are developed. The angle used for the receding axis lines depends on whether you want to emphasize the top, side, or top and side equally. In many applications, receding axis lines are drawn at 60°, 30°, or 45° angles. The measurements on the receding axis lines are drawn at one-half the scale used on the front face.

The following technique may help you remember whether the cavalier or cabinet oblique has a half-size scale on the receding axis. Visualize the overhead kitchen cabinets in your home. The depth of the "cabinets" is about half their height. This relationship is similar to the half-size scale relationship used for the receding depth axis in "cabinet" oblique drawing.

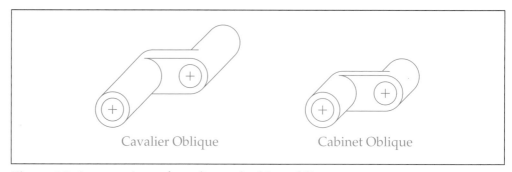

Cavalier Oblique   Cabinet Oblique

Figure 6-3. A comparison of cavalier and cabinet obliques.

Objects Appropriate for Cabinet Oblique Drawing

Many types of objects are well suited for the cabinet oblique drawing style. Generally, any "deep" object appears more realistic when drawn as a cabinet oblique. The less detail the object has along its depth axis, the easier it is to draw as a cabinet oblique. On the other hand, when an object has much detail on its depth axis, it is better drawn as a cavalier oblique. Also, avoid drawing objects that have warped, concave, convex, or irregular curves as cabinet obliques. A simple rectangular object is the ideal candidate for a cabinet oblique.

General Oblique Drawings

Any oblique drawing that uses a scale other than a full or half-size scale on the receding axis is called a general oblique. The receding axis is normally drawn at an angle other than 30°, 45°, or 60°. However, any angle can be used for the receding axis. Generally, the greater the receding axis angle, the smaller the scale factor used for measures along the axis. Using this guideline, a greater foreshortening of measures occurs as the rotation angle increases.

For convenience of calculation, most oblique drawings use a full or half-size scale for the receding axis. For example, the most common oblique drawings use either a full scale factor on a 30° receding axis or a half-size scale factor on a 60° receding axis. If you select a significantly higher angle of rotation, using a half-size or one-quarter scale will improve the realism of the drawing.

Objects with complex shapes and features require many calculations to develop as general obliques. Therefore, these objects should not be drawn as general obliques. The best objects for general oblique drawings are simple shapes without much detail. Objects with irregular curves, arcs, and circles should be avoided because every curved line that is not on the front surface must be drawn with the coordinate method. This is generally not worth the extra time and effort. For this type of object, use a cavalier oblique.

Oblique Drawing Techniques

Specific techniques are required to develop oblique drawings. First, you need to be very careful in deciding how to orient the object. Improper selection of which face to use as the front view can completely negate the advantage of ease in drawing arcs, circles, and irregular curves. Second, it is also important to follow basic layout procedures to make sure that the object is drawn correctly. The following sections provide an overview of selecting object views, applying layout procedures, and developing angles and curves on oblique drawings.

View Selection

There are three fundamental rules to apply when deciding which face of an object should be used as the front view. These rules are listed below:
1. Orient objects so that surfaces with the most arcs, circles, and irregular curves are located on, or parallel to, the front view. This is shown in **Figure 6-4.**
2. Orient objects so that the longest measurement is on, or parallel to, the front view. This is shown in **Figure 6-5.**

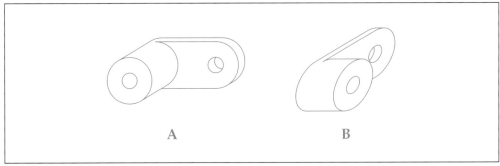

Figure 6-4. Orienting the object so that the most circles, arcs, and curves are parallel to the front view. A—This object is oriented correctly. B—Avoid this type of representation in an oblique drawing.

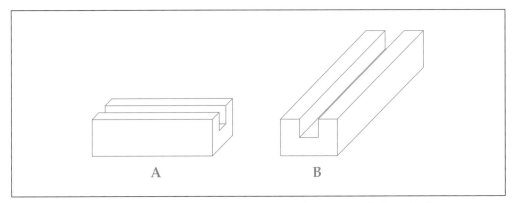

Figure 6-5. Orienting the object so that the longest face is parallel to the front view. A—Correct orientation. B—Incorrect orientation.

3. When an object creates a conflict between the first two rules, orient the object so that circular shapes appear on the front view, as shown in **Figure 6-6.**

Oblique Layout Procedures

The first step in drawing an oblique drawing is the same as that used with other types of drawings. Lay out the border and title block in construction lines and then find the center of the drawing area. The following steps explain how to center a cavalier oblique pictorial in the drawing area for a circular object. Refer to **Figure 6-7** as you go through the procedure.

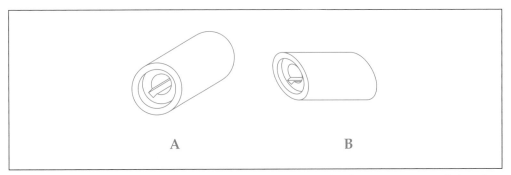

Figure 6-6. Orienting the object so that most of the circular features are parallel to the front view. This guideline should be used when both the circular features and the longest face cannot be drawn in the front view. A—Correct orientation. B—Incorrect orientation.

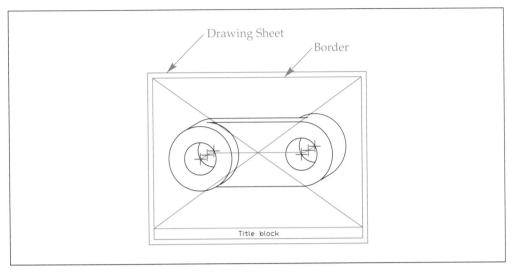

Figure 6-7. Centering a cavalier oblique drawing for a circular object.

1. Beginning at the center of the drawing area, draw a horizontal construction line one-half the width of the object. Draw the line to the left if the axis recedes to the right. Draw it to the right if the axis recedes to the left. In most applications, the right side of the object is displayed. The first layout line is drawn to the left in **Figure 6-7.** This allows you to show the right side of the object and construct the features in the foreground first.
2. Draw a construction line from the end of the horizontal line at a distance of one-half the total depth of the object and at a given angle. This line establishes the angle of the receding axis. For the example shown, a 30° angle from horizontal is used. The endpoint of the line in the example is the center point for the circle nearest to you.
3. Locate and mark the center points for each of the arcs and circles on the object along the receding axis. Start at the circle nearest to you and work back to each consecutive center.
4. Draw construction arcs and circles corresponding to each center point. Connect the appropriate straight lines at their tangency points. Finally, determine which lines, arcs, and circles are visible.
5. Repeat the process for the other side of the object. Draw a horizontal construction line in the opposite direction, draw a construction line for the receding axis line, and locate center points along the receding axis. Draw the arcs, circles, and straight lines needed to complete the object.
6. Darken the visible object lines.

The procedure used to center a rectangular oblique object is a little different than that for a circular object. The following steps are used to center a cavalier oblique drawing for a primarily rectangular shape. Refer to **Figure 6-8** as you go through the steps.

1. Beginning at the center of the drawing area, draw a construction line downward one-half the maximum height of the object. Draw the line upward if you are developing a reversed axis oblique.
2. If you want to develop an oblique that shows the right side of the object, draw a horizontal construction line one-half the width of the object to the right. To show the left side of the object, draw the line to the left.
3. Draw a construction line from the end of the horizontal line for the receding axis. Draw the line equal to one-half the maximum depth of the object. In the example shown, the axis line is drawn at a 30° angle from horizontal.

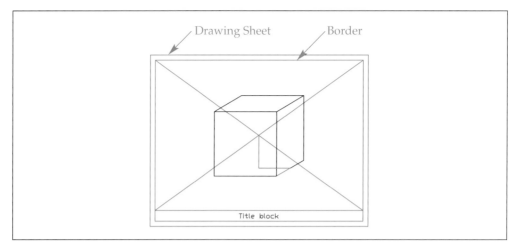

Figure 6-8. Centering a cavalier oblique drawing for a rectangular object.

4. The endpoint of the receding axis construction line is the starting point for the cavalier oblique view. This point is the bottom corner between the front and receding side of the rectangular shape.

5. After drawing normal horizontal and vertical lines from the starting corner to define the orthographic front view, draw the depth lines from the corners along the receding axis. These lines define the top, or bottom, of the object and one side. Finally, draw the remaining lines needed to complete the object.

Angle Construction on Oblique Drawings

Angles on oblique drawings can only be measured if they are parallel to the front plane. Angled lines on the top, bottom, or side surfaces are similar to nonisometric lines. Therefore, constructing inclined or skewed lines on an oblique drawing is similar to constructing nonisometric lines on an isometric drawing. First, the object is blocked in. Using the appropriate orthographic view, the coordinate point for each endpoint is located. Then, these points are connected to form the line. See **Figure 6-9**. This illustration shows how an inclined surface can be defined on an oblique drawing by transferring the coordinates from the orthographic view.

If the drawing has a receding axis angle of 30° and the same scale factor for the front and receding faces, you can use an isometric protractor to construct the angle lines. The drawing in **Figure 6-9** is a cavalier oblique with a receding axis angle of 30°. As discussed in Chapter 5, the minor axis and one other primary axis of an isometric protractor must align with two isometric axes. For the cavalier oblique in **Figure 6-9**, the receding and vertical axes can be used to align the protractor. You can measure angles on the top or bottom surface by aligning the minor axis vertically. However, there is not an axis line receding at an opposing 30° isometric angle on the left side. You can still use the protractor to measure angles on the receding surface. As **Figure 6-10** shows, simply align the isometric protractor with the center on the vertex of the angle and the 0° line of the template along the 30° baseline on the object. This alignment orients the minor axis with a 30° isometric angle. To draw the 60° angle, count 30° clockwise from the vertical 90° angle reading on the template and mark the reading. Remove the template and draw the angled line from the vertex to the angle mark. You may find that this method of angle development is much quicker than transferring coordinates from an orthographic view.

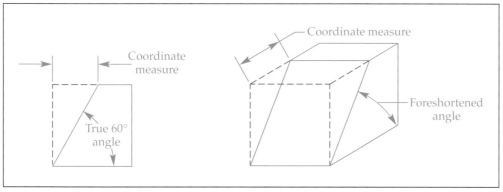

Figure 6-9. Defining an inclined surface in an oblique drawing by transferring coordinates from an orthographic view.

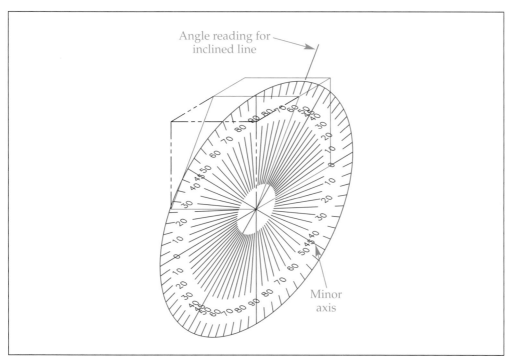

Figure 6-10. An isometric protractor can be used to locate angled lines on a cavalier oblique drawing when the receding axis angle is 30°.

Constructing Oblique Curves

As previously discussed, arcs and other circular features in the front view of an oblique drawing appear true size. Circular features on the top, bottom, and side faces appear as ellipses. When drawing a cavalier oblique with a 30° receding axis, you can use an isometric ellipse template to draw curved shapes. Ellipse angles can be determined using an isometric protractor. The same alignment procedure for measuring inclined angles is used to measure and draw ellipses. For oblique drawings with other receding angles and scale factors, curved shapes must be developed using coordinate layout methods.

When laying out coordinates to draw oblique curves, a coordinate grid is used to transfer points from the orthographic view to the oblique surface. See **Figure 6-11**. As with dimetric, trimetric, and perspective pictorial drawings, it is necessary to use an irregular (French) curve, spline, or trial and error when fitting the closest angle ellipse to draw arcs connecting the coordinate points.

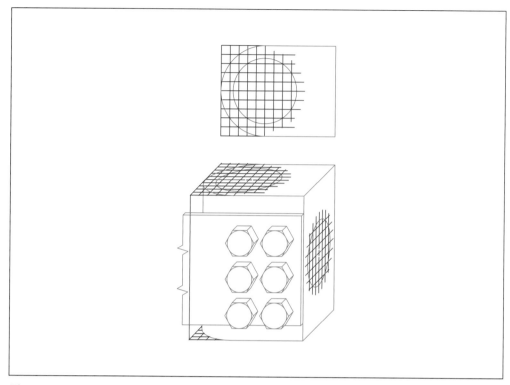

Figure 6-11. Using the coordinate method to create curved shapes on the receding faces of a cabinet oblique drawing. Points are transferred from the orthographic view to the oblique surfaces.

The drawing in **Figure 6-11** is a cabinet oblique with a 45° receding axis. Notice that the features in the front view are drawn true size. The features on the top and side surfaces are foreshortened. When drawing the coordinate grids, a half-size scale is used for the receding axis grid lines on the oblique surfaces.

Computer-Aided Oblique Drawing Applications

Most of the computer-aided drafting functions useful in developing isometric, dimetric, and trimetric drawings can be applied to oblique drawings. These functions are discussed in Chapter 5. They allow you to easily develop cavalier, cabinet, or general oblique drawings. The following sections discuss specific CAD applications for oblique drawing development.

Establishing a Receding Axis

As discussed in Chapter 5, most CAD programs allow you to change the angle of the grid and snap configurations. However, before changing the angle settings, develop the border and title block, locate the center of the drawing area, lay out the starting point of the oblique drawing, and develop the orthographic front view. Then, rotate the grid and snap to the appropriate angle for the receding axis. After changing the angle settings, features along the receding axis can be conveniently drawn.

Always make sure you rotate the snap and grid functions using the same base point. This makes it easy to rotate everything back to the original orientation if needed. This also ensures that the grid dots and snap points stay aligned whether drawing on a frontal plane, horizontal plane, or profile plane.

Scale Modification Functions

There are two primary methods of modifying scales along the receding axis in a CAD system. These include changing the grid and snap spacing and using scale factor modifiers when entering coordinates for drawing lines. These methods are discussed in the next sections.

Grid and snap spacing

One way to scale measures along the receding axis is to set the snap and grid spacing at specific increments. For example, reducing the grid spacing from 1/2″ to 1/4″ allows you to develop a cabinet oblique with a .5 scale factor on the receding axis. Four grid dots in the original grid spacing is equal to 2″. Four grid dots after reducing the spacing is equal to 1″.

Scale factors

Another way to scale measures along a receding axis is to directly input coordinates at different scale increments. As discussed in Chapter 5, polar coordinates are invaluable for drawing axis lines. Most CAD systems allow you to enter the real measure as a fraction numerator and a value for the scale factor as the denominator. For example, after picking a starting point at a corner of the front view, you may enter the polar coordinate @6/2<40. This draws a 6-unit line at a one-half scale on a 40° receding axis.

This method typically requires you to enter measures as whole numbers or decimals. Thus, if the distance measurement is 3 1/2″, you must enter 3.5 so that the computer will calculate the measure. You may find a decimal equivalency chart helpful when using fractional values. In some cases, manual calculations may be necessary. However, do not let this drawback prevent you from using this method. If *most* of the measurements are whole units, this still may be the quickest method.

Snap Functions in Oblique Drawing

Rotating the snap grid and changing the snap spacing are two ways of using snap for an oblique drawing. Other snap functions that are useful in oblique drawing are the object snaps center, quadrant, intersection, endpoint, midpoint, node (or point), and tangent.

As **Figure 6-12** shows, the center object snap function is especially useful for developing centerline layouts. The circular surface for one end of a cylinder is developed in **Figure 6-12A.** The view in **Figure 6-12B** is then developed as follows:

1. Copy the inner circle labeled Circle 1. For the base point, snap to the center of the circle. Enter the new location as @1/2<30. This places the copy 1/2″ from the last point selected at an angle of 30°. The last point selected is the center of the inner circle.
2. Draw Circle 3, as shown in **Figure 6-12B.** Since this circle has the same depth as Circle 2, snap to the center of Circle 2 to locate the center of Circle 3. Then, specify the radius of Circle 3 to draw the circle.

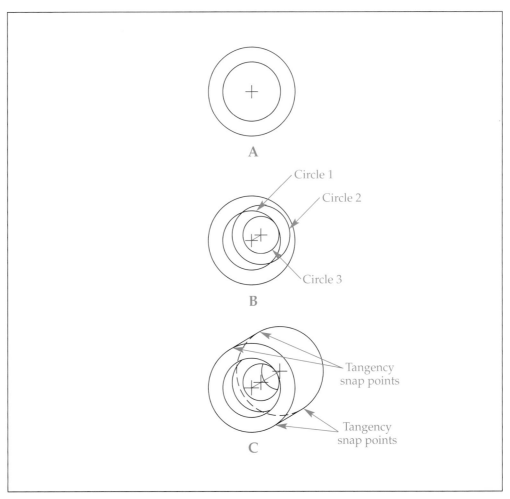

Figure 6-12. Using object snap functions with circles to develop a cylinder. A—Two circles are drawn for the first face of the object. B—The center object snap is used with polar coordinates to create two additional circles for the depth of the object. C—The drawing is completed by creating a final circle and using the tangent object snap to connect lines to the rear face.

The tangent snap function is used to complete the cylinder in **Figure 6-12C**. First, the circle forming the front end of the cylinder is drawn using the center point shown. Then, lines are drawn connecting this circle to the circle on the rear end of the cylinder. This is done by entering the **Line** command and snapping to the tangent of the front circle. For the second point, snap to the tangent of the rear circle by picking a location close to where the line would attach to the circle. Repeat the procedure for the line on the other side of the cylinder. To complete the object, arcs are added and trimmed as necessary.

Oblique Drawing Editing Functions

Some CAD editing commands are used more frequently than others in oblique drawing. Among the most useful editing commands are **Copy, Erase, Extend, Trim**, and **Move**. It is easier to copy identical features than to draw new ones. The **Erase** command allows you to remove features, such as construction lines, when they are not needed. The extending and trimming functions allow you to edit line and arc segment lengths. The **Move** command is used to change the location of an object in the drawing. All of these editing functions provide great flexibility when developing any of the oblique drawing types.

Summary

The oblique pictorial drawing types include the cavalier oblique, cabinet oblique, and general oblique types. All obliques have a normal orthographic front view with a receding axis drawn at a given angle. The three oblique styles differ in the scale factor used on the receding axis. Typical angles used for the receding axis include 30°, 45°, and 60°.

The most commonly used oblique is the cavalier oblique. Its equal scaling factors on each axis eliminates the need for calculating scale reduction on the receding axis. An isometric ellipse template and isometric protractor can also be used for this style when the receding axis is set at a 30° angle.

The cabinet oblique uses a receding axis scale that is one-half the scale used on the front face of the object. The general oblique style uses a receding axis scale other than full or half size.

There are several applications where oblique pictorials are best to use. Objects with a number of arcs and circles concentrated on one plane, objects with irregular curves, and objects with relatively large measurements along one axis are good candidates for the oblique style.

When drawing an oblique, it is very important to select the most appropriate surface for the front view. Three rules govern view selection. The first rule is to orient objects so surfaces with the most arcs, circles, and irregular curves are located on or parallel to the front view. The second rule is to orient objects so their longest measurement is on or parallel to the front view. The third rule states that when object shapes create a conflict between the first rule and the second rule, follow the first rule.

When creating an oblique drawing, first locate center points or starting corners on the front face of the object. Then, develop each visible surface from front to back. In this way, visibility is established as you develop the drawing. Also, time and effort lost to developing unnecessary hidden lines and surfaces is eliminated.

Most angles cannot be directly measured on the receding surfaces of an oblique drawing. However, due to the full scaling on each axis, a cavalier oblique using a 30° receding axis can have angles on the top, bottom, or side measured with an isometric protractor. In all other situations, angles on the receding surfaces of oblique drawings must be developed using coordinate layout methods. Likewise, arcs and circles on a cavalier oblique with a 30° receding axis can be drawn with an isometric ellipse template. All arcs, circles, and curves drawn on oblique surfaces with any other receding axis angle or variable scale must be developed using coordinates transferred from an orthographic view.

Oblique drawings can be developed quickly and efficiently using the tools of a CAD system. The receding axis lines can be scaled by varying the snap and grid spacing and rotation or by having the system automatically scale measurements. Snap functions are also useful when constructing oblique drawings. The center, intersection, endpoint, midpoint, and tangent object snaps are commonly used on oblique views. The copy, move, trim, extend, and erase editing functions also help in creating oblique drawings.

Review Questions

1. Describe the basic characteristics of a cavalier oblique drawing.
2. Describe the basic characteristics of a cabinet oblique drawing.
3. Describe the basic characteristics of a general oblique drawing.
4. List two major applications of oblique drawing in relation to object shapes.
5. Which view is first drawn when creating an oblique drawing?
6. Name two advantages of drawing a cavalier oblique with a 30° receding axis as compared to the other oblique drawing types.
7. What types of objects are best drawn with a cabinet oblique?
8. Explain the three primary rules that apply to object orientation in oblique drawing.
9. Outline the procedure used to center a rectangular oblique drawing in the drawing area.
10. How are angles and circles drawn on receding surfaces of cabinet and general oblique drawings?
11. Identify two ways to modify scale measures along receding axis lines in an oblique drawing using a CAD system.

Drawing Problems

For Problems 1–11, create cavalier oblique drawings using the orthographic views shown. Use the dimensions provided and the proper scaling. Orient the front view to create the most realistic representation. Choose a receding axis angle that portrays the side of the object accurately.

1.

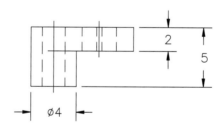

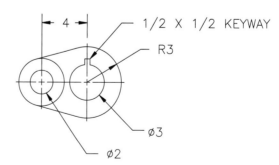

2.

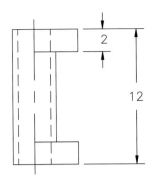

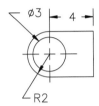

3. To create the shading pattern for the knurling on the collar, draw diagonal lines intersecting at 90° angles or use a CAD-generated crosshatch pattern. Orient the pattern so that the lines do not align with one of the major oblique axes.

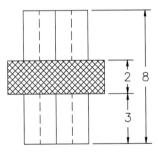

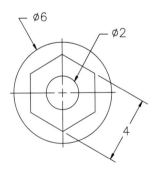

4.

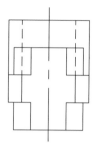

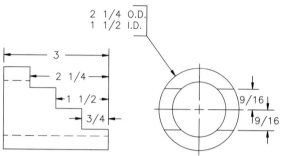

5.

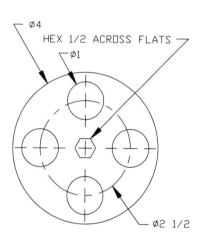

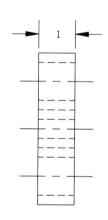

6.

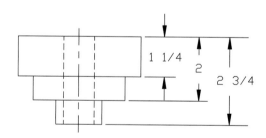

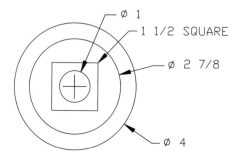

7.

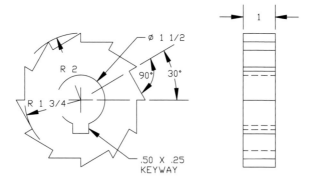

8.

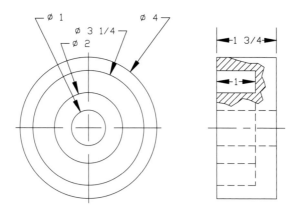

9.

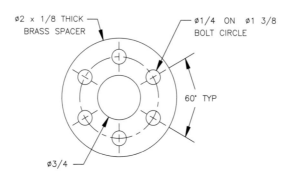

10.

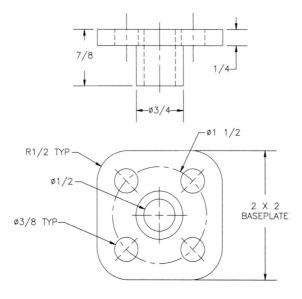

11.

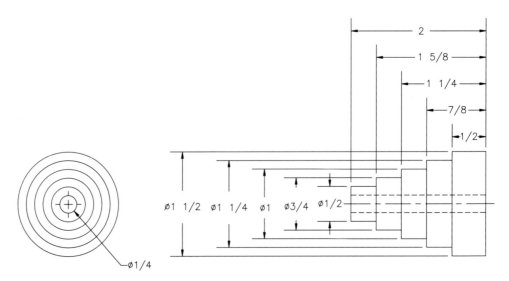

12. Create a cabinet oblique drawing of the object shown using the orthographic views provided. Use a 60° receding axis angle. Use the dimensions provided and apply the proper scaling. Orient the front view to create the most realistic representation.

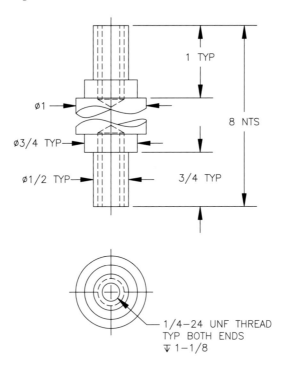

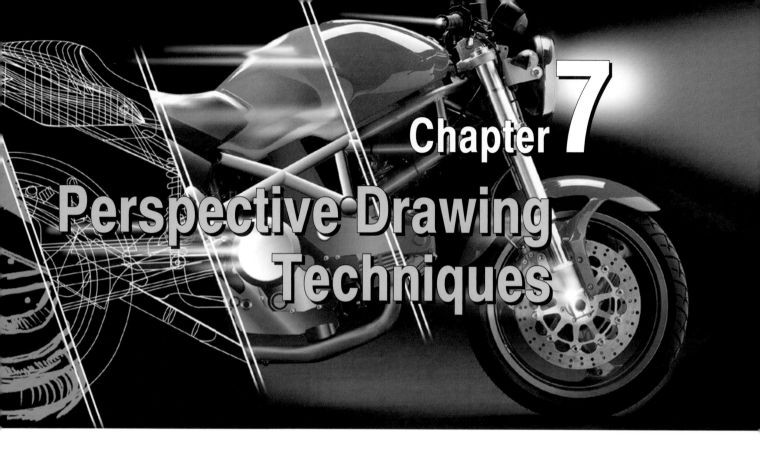

Chapter 7

Perspective Drawing Techniques

Learning Objectives

At the conclusion of this chapter, you will be able to:

- ☐ Explain the common terminology used in perspective drawing.
- ☐ Describe the three types of perspective drawing.
- ☐ List the primary applications of each type of perspective drawing.
- ☐ Identify and explain the drawing aids used in manual perspective drawing.
- ☐ Describe the common computer-aided drafting functions used to develop perspective drawings.
- ☐ Plan, lay out, and develop the three types of perspective drawings.

Introduction

Perspective drawing is one of the most realistic methods of manual drawing. As with other types of pictorial drawing, all three-dimensional relationships are presented in a single view. However, the lines on the receding axes converge at vanishing points. This creates a much more realistic appearance of an object. Perspective drawings are most often found in interior design, commercial art, architecture, and technical illustration. This chapter covers the different types of perspective drawing and their applications.

Perspective Drawing Overview

There are three types of perspective drawings. They can be classified as one-point, two-point, or three-point. The name refers to the number of vanishing points in the drawing. For example, a one-point perspective has one vanishing point, while a three-point perspective has three vanishing points. Vanishing points are discussed in the following section on perspective drawing terminology.

A one-point perspective is often called a *parallel perspective.* A two-point perspective is often called an *angular perspective.* A three-point perspective is often called an *oblique perspective.* See **Figure 7-1**.

The principle that makes a perspective drawing appear realistic is that objects further from the viewer appear smaller, or foreshortened. This can be seen by looking at railroad tracks. Although you know that the distance between the tracks stays the same, it appears to decrease as the tracks recede into the distance. See **Figure 7-2.** This proportional foreshortening is one characteristic of perspective drawings.

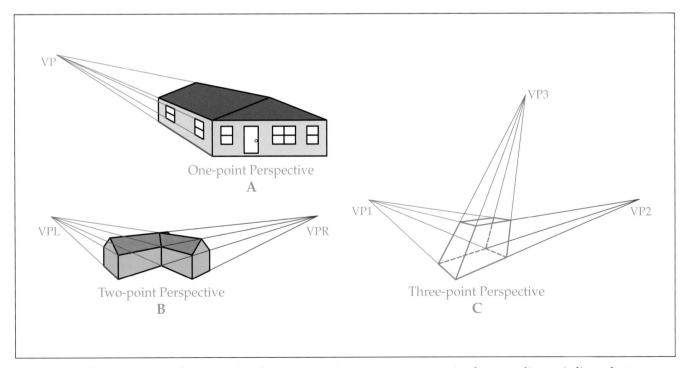

Figure 7-1. The three types of perspective drawing. A—A one-point perspective has receding axis lines that converge to a single vanishing point and is often called a parallel perspective. B—A two-point perspective has two vanishing points and is often called an angular perspective. C—A three-point perspective has three vanishing points and is often called an oblique perspective.

Figure 7-2. Railroad tracks appear to converge as the viewing distance increases. This principle is used to develop a perspective drawing. (Jack Klasey)

Perspective Drawing Terminology

There are several important terms that apply to perspective drawing. These terms are listed and explained below. Each term has a common abbreviation. See **Figure 7-3**. As shown in the illustration, a perspective drawing typically includes a plan view, an elevation view, and the perspective view. Lines are projected between points to create the drawing.

The *station point* is the point where the viewer is "standing" in relation to the object. The abbreviation SP is normally used to identify the station point.

The *picture plane* is a theoretical plane onto which an image is projected. Projection lines are developed from the various points on the object to the station point. As the projection lines pass through the picture plane, they determine the foreshortening on the receding axes. The plan view of the object is oriented so that a major vertical intersection is on the picture plane. This aids in proper vertical scaling of the drawing. You should be aware that portions of the object that project from a position ahead of the picture plane in the plan view are distorted and drawn oversized. Therefore, make sure that the major plan view features of the object are located either on or behind the picture plane. The picture plane is identified with the abbreviation PP. As shown in **Figure 7-3**, the front-left corners of the house are on the picture plane.

Vanishing points are the points to which the receding axis lines project. The vanishing point in a one-point perspective is designated VP. The vanishing points in a two-point perspective are identified as *vanishing point right* (VPR) and *vanishing point left* (VPL). Vanishing points on one-point and two-point perspectives are located on the horizon line. Refer to **Figure 7-3.**

The *horizon line* indicates the location of the horizon with respect to the elevation of the ground line. The scaled or projected height of the horizon line in relation to the ground line determines the elevation from which the viewer is looking at the object. This elevation is where receding lines converge and can be thought of as "eye level." The horizon line is abbreviated HL. Three typical orientations of the horizon line with respect to the same object are shown in **Figure 7-4.**

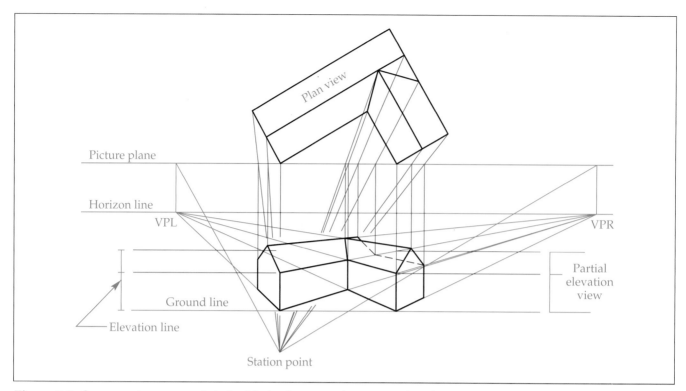

Figure 7-3. Common terms associated with the features of a perspective drawing.

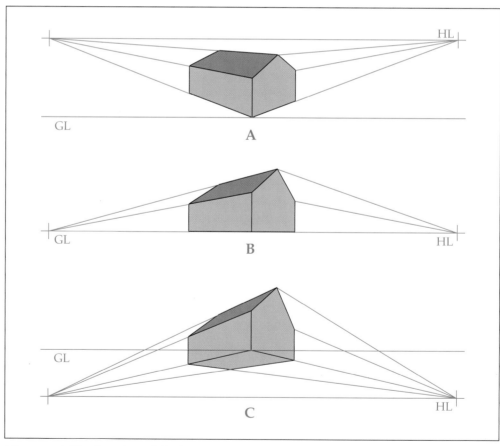

Figure 7-4. The location of the horizon line relative to the ground line determines the elevation from which the object is viewed. A—Locating the horizon line above the ground line creates a "bird's-eye" view. B—Locating the horizon line on the same level as the ground line establishes a ground level view. C—Locating the horizon line below the ground line creates a "worm's-eye" view.

The *ground line* represents the elevation where the object sits in relation to wherever it touches the picture plane. Placing the ground line below the horizon line places the viewing level above the object. Moving the ground line above the horizon line places the viewing level underground, or beneath the object. The ground line is abbreviated GL.

Planning a Perspective Drawing

The first step in preparing a perspective drawing is to visualize what particular features you want to emphasize to the audience. This determines how you will need to orient the object. Locate special features along the axis nearest the picture plane in the plan view. It is necessary to develop the plan view to locate these features on the face of the object closest to the picture plane. You can think of the picture plane as an edge view of the frontal plane as you look down on the object. Draw a few trial sketches to find the best view. If the object is small enough and available, you can rotate it in your hands to visualize the best view.

After identifying what to emphasize, you need to determine the best type of perspective to use. To do so, you need to look at the advantages and disadvantages of each type. The level of realism increases from a one-point perspective to a three-point perspective. However, as the realism increases, so does the drawing time required. A two-point perspective is a good compromise between realism and reasonable production time.

Perspective Drawing Layout

The first step in perspective layout is to draw a plan (top) view of the object. A one-point perspective typically has the plan view parallel to the picture plane. Two-point and three-point perspectives normally rotate the plan view to provide a three-dimensional effect in the perspective view. Many illustrators draw the plan view for a two-point or three-point perspective on a different sheet of paper than the perspective view. This allows them to rotate the view to a variety of different angles and see how they affect the perspective view. Any angle of rotation can be used. However, at least one vertical feature should be located on the picture plane so the scaled vertical height can be developed from the ground line in the perspective view. Traditional rotation angles used for the plan view are 30°, 45°, and 60°.

The effect of rotating the plan view in a perspective drawing is shown in **Figure 7-5**. In **Figure 7-5A**, a plan view and a very simple perspective view of a house are shown. As you can see, there is very little detail to show the features of the drawing. By rotating the plan view, you can show a view of the roof intersection in the perspective view, **Figure 7-5B.** The plan view of the house is rotated at a 30° angle to the picture plane. The plan view is also oriented differently with the picture plane. The resulting view shows more features of the house and adds an element of interest to the drawing.

When orienting the plan view, the picture plane is also drawn. The picture plane should be parallel to the bottom edge of the drawing sheet. Most illustrators prefer to locate the picture plane on a front surface or corner of the object in plan view. This reduces excessive foreshortening of features behind the picture plane and distortion of features in front of the plane. Any part of an object located in front of the picture plane appears larger than other object features. The farther behind the picture plane a feature is located, the smaller it appears in the perspective view. In **Figure 7-5B**, only the corner of the roof line extends in front of the picture plane.

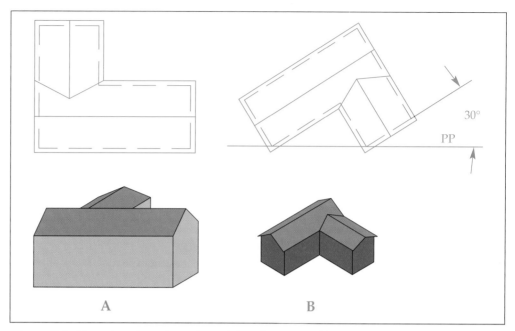

Figure 7-5. Changing the orientation of the plan view allows you to show more features in a perspective drawing. A—The plan view is oriented so that it is parallel to the picture plane. B—The plan view is placed in a different orientation and rotated 30°.

The ground line and the horizon line are added next to establish the layout for the perspective view. Place the ground line above or below the horizon line. Placing it above the horizon line creates a view from below the object. Placing it below the horizon line creates a view from above. The vertical distance between the ground line and the horizon line should always use the same scale used by the plan view. This will maintain proper proportion in the drawing. In **Figure 7-6**, the location of the ground line establishes a perspective view seen at approximately eye level.

The station point is added next. It is normally located by projecting a vertical line down from a major feature located on the picture plane. This is shown as Line A in **Figure 7-6**. A vertical intersection is commonly used since this establishes the intersection as a true scale elevation line.

One-point perspectives often have the station point shifted to the right or left of the center of the object. Two-point and three-point perspectives usually have the station point located directly below the vertical intersection lying on the picture plane. For two-point and three-point perspectives, shifting the station point left or right creates a skewed view that may visually distort very "tall" objects.

With the vertical projection line drawn, the station point can be located in a variety of places. However, there is a general rule for placing the station point in a preliminary location. The cone of vision lines from the station point to the extreme edges of the object should not exceed 30°. These lines are labeled B in **Figure 7-6**. In most cases, a maximum angle of 30° keeps the vanishing points at a reasonable working distance and results in a realistic perspective view. In some cases, a 30° angle may locate the station point so far away from the picture plane that the vanishing points are off the drafting table. In this instance, find the maximum horizontal distance of the plan view and locate the station point 2 1/2 to 3 times that distance below the picture plane.

After preliminary location of the station point, the vanishing points need to be located. In a one-point perspective, a location is selected on the horizon line in vertical alignment with the station point. Two-point and three-point perspectives

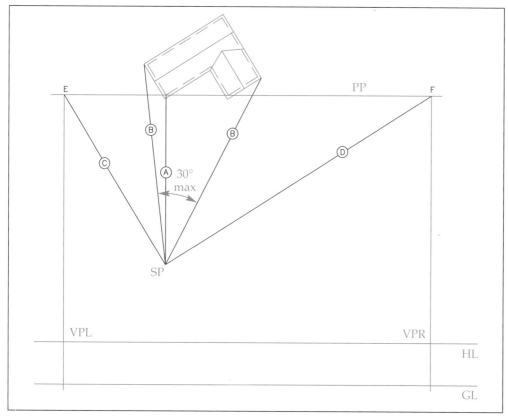

Figure 7-6. Locating the station point and vanishing points for a two-point perspective.

have a vanishing point right (VPR) and a vanishing point left (VPL). These points are located by first drawing construction lines from the station point parallel to the sides of the object in the plan view to intersect the picture plane. In **Figure 7-6,** these lines are labeled C and D and the intersections with the picture plane are labeled E and F. The vanishing points are located by projecting vertical lines to intersect the horizon line. These intersections are VPR and VPL.

After locating the vanishing points, you may need to change the location of the station point to move the vanishing points closer to the drawing area. It is common for the vanishing points to fall completely off the drawing. This is not a problem as long as the points are on the drawing table. If the vanishing points are off the drawing sheet, write down their horizontal distances from the station point along the horizon line so that they can be relocated for multiple drawing sessions. A piece of drafting tape can also be placed on the table and the location of the vanishing points marked on the tape. The tape can then be left on the drafting table until the drawing is finished.

After locating the vanishing points, receding lines for the horizontal axes can be drawn, and you can begin developing the perspective view. The next section discusses the drawing procedures used in developing one-point perspective drawings.

One-Point Perspective Drawings

A *one-point perspective* has only one vanishing point. The location of the vanishing point on the horizon line determines the direction of the receding horizontal axis. Different receding views in relation to the vanishing point are shown in **Figure 7-7**. The disadvantage of a one-point perspective is the distortion

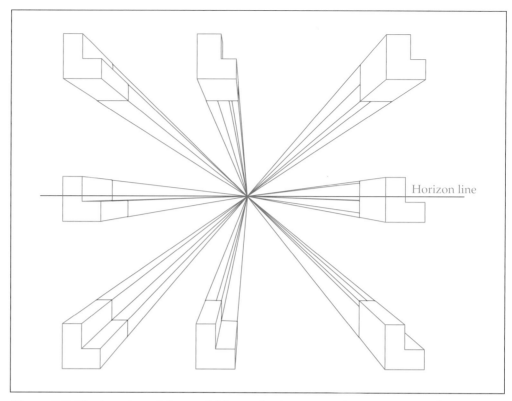

Figure 7-7. The location of the vanishing point in a one-point perspective drawing determines which receding features are visible.

created in the front view. In reality, this view recedes to a vanishing point located somewhere outside the outline of the object, as in two-point or three-point perspective drawing.

Notice that a one-point perspective is similar in appearance to an oblique drawing. One-point perspectives have many similarities to oblique drawings. A one-point perspective is useful when illustrating an object with a significant number of cylindrical shapes, holes, or radii. The rules of view selection for oblique drawings also apply to one-point perspectives. The object should be oriented so that the front view contains the most circular shapes or the longest measurements of the object. When an object is both circular and long, the circular orientation should be developed as the front view.

Primary Applications

A one-point perspective is simple and quick to draw. Therefore, one-point perspectives are used by interior designers, architects, and mechanical designers to provide a three-dimensional representation with little time investment. A one-point perspective appears more realistic than an isometric or oblique drawing.

Interior designers often use a one-point perspective to help visualize what a design will look like as you walk into the room. See **Figure 7-8**. This illustration shows a very basic example of an interior with the vanishing point located within the drawing area.

One-Point Perspective Drawing Procedures

Begin a one-point perspective by drawing the plan view. Then, draw a normal orthographic front view directly below the plan view. Leave ample

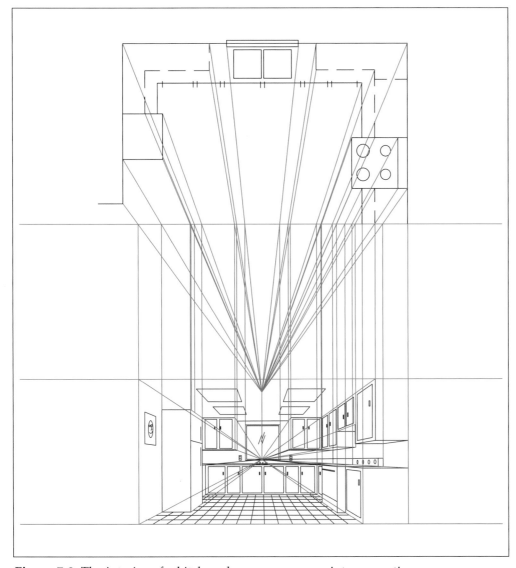

Figure 7-8. The interior of a kitchen drawn as a one-point perspective.

space between the views to avoid crowding when the receding features are developed. Locate the station point and draw projection lines from features behind the front face of the object to the station point. Then, locate the vanishing point and draw projection lines to it from the outermost edges of the front view. The vanishing point and the station point should be vertically aligned. See **Figure 7-9.**

The drawing in **Figure 7-9A** represents a blocked-in perspective view of the object. Notice how the depth of the object in the perspective view is determined with projection lines drawn from the rear corners and the center line in the plan view to the station point. The intersections of the projection lines with the picture plane determine the depth of the object in the perspective view. Simply drop vertical projection lines from the intersection points until they meet the appropriate line receding to the vanishing point. Hidden lines are shown in this figure to assist you in visualizing how the one-point perspective is developed. Normally, hidden lines are omitted in perspective drawings.

In **Figure 7-9B**, the object is redrawn using arcs for the front and back surfaces. In this drawing, the station point is located to the left of the center of the object to provide a view of the left side. As is the case in **Figure 7-9A**, vertical projection lines are used to develop the depth of the object.

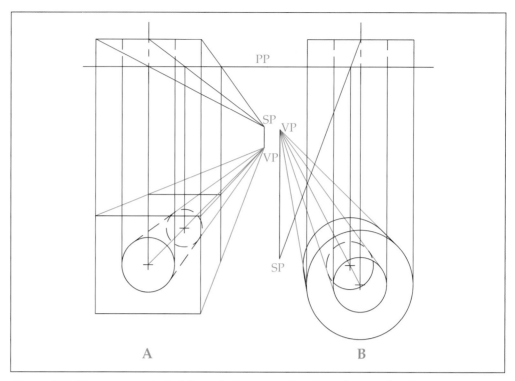

Figure 7-9. Lines are projected from features on the plan view to develop a one-point perspective. A—A blocked-in perspective view. B—A perspective view showing the left side of the object.

Two-Point Perspective Drawings

As the name implies, a *two-point perspective* has two vanishing points. In this type of drawing, the perspective view has decreased size along each axis as features on the object recede toward each of the vanishing points. The visual effect of this scale reduction causes the two-point perspective to appear more realistic than the one-point perspective. Although it is a bit more complex to develop than a one-point perspective, the two-point perspective is still a relatively simple view to develop.

Primary Applications

A two-point perspective is fairly easy to create and produces a realistic view. Therefore, two-point perspectives are used by architects, interior designers, mechanical designers, and illustrators to produce realistic pictorial drawings with only a moderate time investment.

Two-Point Perspective Drawing Procedures

To develop a two-point perspective, first draw and orient the plan view on a picture plane. Draw the ground line and the horizon line. Then, locate the station point and the vanishing points. The vanishing points are located by drawing lines from the station point parallel to the sides of the object in the plan view until they intersect the picture plane. Vertical lines are then drawn from these intersection points to the horizon line. The intersections of the vertical lines locate vanishing point left (VPL) and vanishing point right (VPR).

Next, draw a vertical construction line from any edge or intersection lying on the picture plane until it intersects the ground line. See **Figure 7-10.** In this example, two corner features on the inside walls lie on the picture plane. After projecting the vertical lines to the ground line, measure the height of these features vertically from the ground line by projecting a horizontal line from the corresponding features in the elevation view. The resulting two lines are parallel to the picture plane and appear in the drawing as true length. Next, draw projection lines from the top and bottom of each feature to the vanishing points. These lines establish the perspective planes.

This same procedure can be used to establish the remaining vertical lines for the walls. See **Figure 7-11.** To project these features to the perspective view, establish a line of vision from the station point to each of the hidden wall features in the plan view. At the point where the vision line passes through the picture plane, drop a vertical projection line to the perspective view. The intersections of each line with the perspective planes can be used to establish the limits of the walls on the receding axes. Lines connecting the corners of the walls along the receding axes establish the wall surfaces. A similar procedure can be used to develop the roof lines to complete the perspective view. This is discussed in the next section.

Inclined surfaces

Developing inclined surfaces in a perspective drawing can be complex. They are usually developed by dropping a projection line from the plan view until it intersects a horizontal projection line from the elevation view or an elevation line along a receding axis. By identifying the location of each corner of the inclined surface, you can draw lines connecting the corners to define the surface. The roof surfaces in **Figure 7-12** were developed in this manner. First, draw projection lines from the station point to each of the visible corners on the

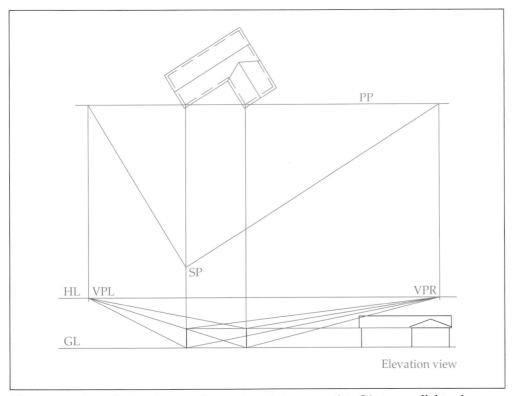

Figure 7-10. Developing features in a two-point perspective. Lines parallel to the picture plane are drawn by projecting vertical lines from the features intersecting the picture plane in the plan view to the ground line. The true height is determined by projecting a line from the corresponding features in the elevation view.

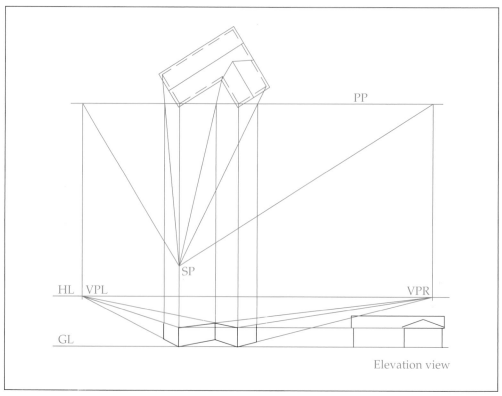

Figure 7-11. The vertical wall lines are defined by projecting lines from intersections on the picture plane to the receding axes. The corners of the walls are connected with lines to complete the wall surfaces.

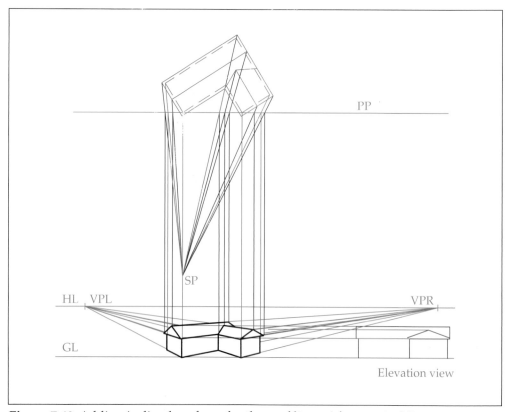

Figure 7-12. Adding inclined surfaces for the roof lines. After vertical lines are projected from the features in the plan view, horizontal lines are projected from the corresponding features in the elevation view. Lines are then drawn along the receding axes to complete the roof features.

roof. Where each line intersects the picture plane, drop a vertical projection line to the perspective view. Then, draw horizontal projection lines from the elevation view at the right. When a horizontal projection line intersects a vertical projection line from a feature on the picture plane, the true perspective height of that feature is identified. From that point, draw a line to the appropriate vanishing point to develop the elevation of points behind the picture plane.

In some cases, it is necessary to develop an imaginary elevation point in front of the perspective view and then develop a receding elevation line back to the appropriate vanishing point. Then, any point with that elevation can be developed by dropping a vertical projector from the plan view until it intersects the receding elevation line. The intersection of the receding line and the vertical projector locates the point in space. The inclined surface is then developed by locating each point on the inclined surfaces and connecting them with edge defining lines. Make sure to project front elevation lines to the correct vanishing point. The corners of the house located on the picture plane in **Figure 7-12** have an L-shaped wall located between them. Notice that all features in the plan view that recede to the right of the intersecting point on the picture plane are projected to the right vanishing point in the perspective view. All features receding behind the picture plane and to the left in the plan view are projected to the left vanishing point in the perspective view. Most errors in perspective layout involve projecting to the wrong vanishing point.

As you develop perspectives with inclined surfaces, it helps to label points to keep track of intersections and lines. Some illustrators use different colors of lead when developing perspective drawings with inclined or skewed surfaces to keep track of lines. The basic drawing shapes are then traced on a clean sheet of media for the final shading, rendering, and small details.

Developing perspective drawings with circular features by hand is not common practice because of the level of complexity. For this reason, isometric views are recommended for objects with circular features. In some perspective drawing applications, circular features can be "blocked in" with construction lines and drawn by experimenting with different angle ellipses to achieve an approximate shape.

Three-Point Perspectives

A *three-point perspective* drawing is the most complex type of perspective drawing to develop. It is generally avoided when the object to be drawn is highly detailed, or it uses curved shapes on an inclined plane or on more than one principal plane. You will find that the three-point perspective generates the most realistic view of the perspective drawing methods. You will need to carefully evaluate whether your need for realism is great enough to justify the increased time it will require to develop a three-point perspective.

Primary Applications

The primary applications for three-point perspectives are presentation drawings of tall objects. The viewer may be looking up or down at a building or object. Illustrators, architects, and engineers use this style of perspective to present tall buildings, towers, and similar structures. An application of three-point perspective drawing in the mechanical drafting field is illustrating tall machinery with a rectangular shape. This style of perspective can be the most realistic of all pictorial drawing methods.

A simple three-point perspective is shown in **Figure 7-13**. The surfaces recede along the left and right axes, as with a two-point perspective. However,

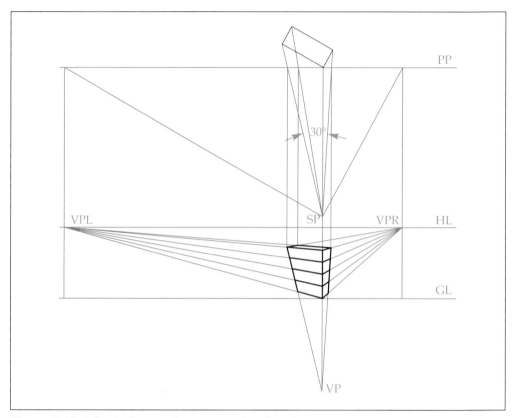

Figure 7-13. A basic three-point perspective drawing.

the third vanishing point causes the width to recede as well. This adds realism to the drawing. A three-point perspective is the most realistic type of perspective drawing. However, development of three-point perspectives can be difficult if the object has curved features, holes, cylindrical shapes, or a significant amount of fine detail.

Three-Point Perspective Drawing Procedures

Developing the plan view, picture plane, station point, horizon line, ground line, and left and right vanishing points for a three-point perspective involves the same procedures used in two-point perspective drawing. The station point should be aligned with the center of the object to avoid excessive distortion. The third vanishing point, labeled VP, is located along a vertical projection line from the station point at a distance of approximately one and a half times the height of the object. If the third vanishing point is offset to the left or right of the object, the perspective view is tilted in the direction of the vanishing point.

After establishing the vanishing points, locate the features in the three-point perspective view using the same drawing procedures in two-point perspective drawing. Vertical lines recede to the third vanishing point.

Manual Perspective Drawing Aids

There are a variety of manual drawing aids that may assist you in developing perspective drawings. The principal manual drawing aids include extra wide or extended wing drawing boards, perspective grid sheets, and several types of commercial perspective drawing boards.

Extended Manual Drawing Boards

When drawing a perspective view, vanishing points may be located off the actual drawing sheet. Therefore, it may be necessary to create an oversize drawing surface using a 4′ × 8′ sheet of plywood, particle board, or tempered fiber board. You may find it helpful to insert a map pin or very thin finishing nail into the extended drawing board at each vanishing point. Then, you can use a long straightedge to draw lines that project to the vanishing point.

Perspective Grids

Perspective drawing grids can save considerable time by eliminating all the preliminary layout tasks needed to locate the vanishing points and other visual aids. Perspective grids have a vertical scale line, a horizon line, and a series of lines that recede to vanishing points. Grids are available in a wide variety of scales and viewing angles. They allow large drawings to be created without having to actually locate vanishing points that may be several feet from the drawing area.

To use a grid, vertical lines are aligned on the drawing board and the grid sheet is taped down. As shown in **Figure 7-14**, the grid lines show through the drawing media. A triangle, straightedge, T-square, or drafting machine scale can then be used to trace grid lines receding to a vanishing point.

Commercial Perspective Drawing Boards

Commercial perspective drawing boards allow quick construction of one-point and two-point perspectives. These drawing boards have a variety of printed scales and vanishing points.

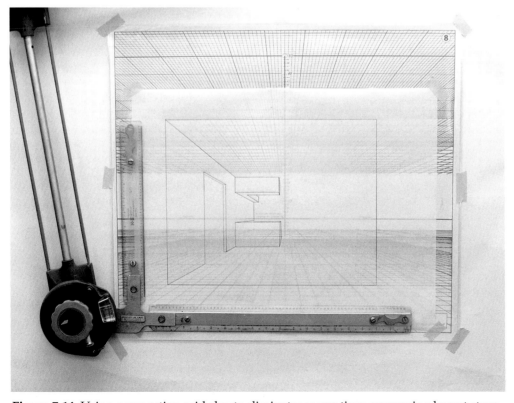

Figure 7-14. Using perspective grid sheets eliminates many time-consuming layout steps.

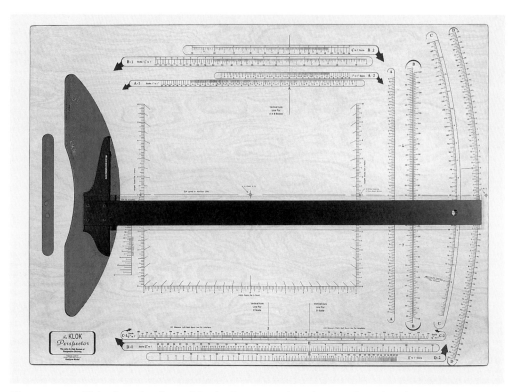

Figure 7-15. This drawing board includes several scales used to develop perspective drawings. (The Utley Company, Inc.)

The drawing board shown in **Figure 7-15** provides a selection of eight different scales on the right of the board. The left vanishing point is located with the special T-square that slides in a curved groove. The horizon line is in the center of the board. Five different holes are located along its length. These are especially useful as one-point perspective vanishing points. The drawing paper is taped to the surface of the board and a pin is inserted in the appropriate vanishing point(s).

Computer-Aided Perspective Drawing Applications

Many computer-aided drafting software programs allow you to automatically create a perspective view from a three-dimensional model. This feature can save a great amount of drawing time. Two common functions that permit automatic perspective generation are dynamic viewing and 3D perspective viewing. These features are discussed in more detail in Chapter 8. However, a CAD program without automatic perspective viewing capability can still be used to draw a perspective in much the same way drawings are created manually. The drawing tools in a CAD program can dramatically reduce the time spent on developing perspective views.

Dynamic and 3D Perspective Viewing of Objects

Dynamic viewing in a CAD program allows you to display a three-dimensional model at any viewing angle by rotating the model dynamically. This tool simplifies the task of creating a perspective view and is useful for displaying hidden features of the object. Most CAD programs also have preset views that can be used to orient the model in a standard pictorial view. Another common viewing tool allows you to identify an angle, distance, and elevation from which to view a 3D model. The program then calculates the coordinate locations of points, lines, and surface features to generate a true perspective view. Hidden lines and surfaces are normally removed automatically by the program.

Layer and Color Applications in Perspective Drawing

The layer and color assignment capabilities of a typical CAD system can be very useful in developing perspectives. Layers are called "levels" in some CAD systems. Layers in a CAD system are similar to a stack of transparent sheets of plastic drawing film used in manual drafting. The layer function is useful because a perspective drawing may contain many different types of drawing elements.

Using layers and colors, a plan view can be drawn in a specific color and on a specific layer. It can be easily rotated and moved to a location at the top of the drawing area. Then, the picture plane, station point, ground line, horizon line, and vanishing points can be drawn on different layers in different colors. Construction lines can also be drawn on other layers using different colors. Finally, the perspective view can be drawn on yet another layer in a specific color. When the drawing is completed, you can instruct the program to turn off, or hide, all but the object layer. None of the lines on the hidden layers will show on the monitor or on the printed drawing. This eliminates the time-consuming process of erasing construction lines. In addition, using separate colors for the different layers allows you to instantly see which layer an object is on. When using a CAD system with automatic perspective drawing and viewing capabilities, layers and colors are not as important since fewer construction lines are required.

Establishing Vanishing Points with a CAD System

Vanishing points only need to be located when using a CAD system that does not have automatic perspective drawing and viewing capabilities. They are located by projecting lines from the station point to the picture plane and then downward until the projection line intersects the horizon line. This is the same process used in manual layout.

Once the vanishing points are located, record their coordinates. Then, when drawing a line to a vanishing point, simply enter the XY coordinates. This is quicker than moving the cursor to the vanishing points since they normally fall beyond the visible drawing area. Using coordinates allows you to keep the screen zoomed-in on the perspective view. If the vanishing points fall in the visible drawing area, an object snap function such as point, endpoint, or intersection can be used.

Constructing and Using an Elevation Line

An *elevation line* is a line drawn to the side of a perspective view used as a substitute for a side view of the object. Points representing the vertical measurements of all features are marked and labeled on this line. This saves the time of drawing the whole side view.

To draw an elevation line, begin by drawing a line vertically from the ground line using the same scale applied in the plan view. Place the line to the side of the perspective view. Then, draw points on the line at the specific elevation of each of the object's features. Label these points with the text or dimensioning function to keep track of what features the points represent. The elevation line is now ready to be used. See **Figure 7-16**.

As shown in **Figure 7-16,** projecting a horizontal line from each point on the elevation line to a true length vertical line in the perspective view transfers true scale measurements to the perspective view. As previously discussed, a vertical line parallel to the picture plane in the plan view is a true scale line in the perspective view. Foreshortened height is established by drawing projectors from the picture plane to their corresponding intersections on height lines receding to the vanishing points.

Using Orthographic Projection in Perspective Drawing

Most CAD systems have orthographic drawing and viewing modes. When working in orthographic view, only vertical or horizontal lines can be drawn. This function is helpful in projecting points from the picture plane, elevation view, or elevation line to the perspective view. When working in orthographic mode, the points are projected at precisely 90° from the picture plane and elevation to the perspective view. This improves the appearance of the drawing. More importantly, it improves the accuracy of your vertical and horizontal projections.

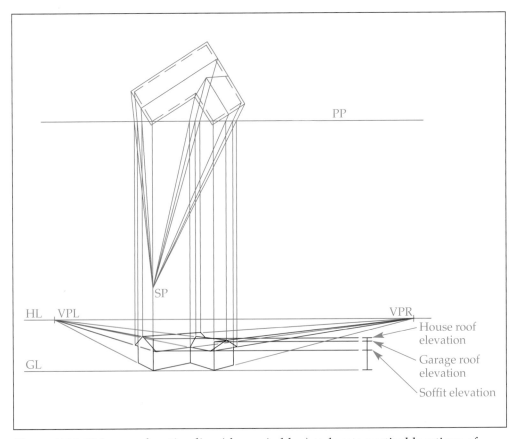

Figure 7-16. Using an elevation line (shown in blue) to locate vertical locations of features in the perspective view.

Moving Points and Views

Most perspective illustrations require that points, lines, features, and even complete views be moved at some point in the drawing process. A CAD system simplifies this process. Most programs have two primary ways to relocate features. The first method is to make a copy of the original and place it at another location. This function is usually called the copy or translate function. The original is then erased. The second method involves the move function. This function allows you to relocate the object to any new coordinate location.

You should always save your work before moving any objects or views. Then, if a problem occurs that you cannot undo, simply open the saved drawing and correct it. Another way to reduce problems in moving objects from one location to another is to select known points and assign them as the base point and displacement point. These known points may be intersections, midpoints, endpoints, or center points. This also makes it easy to move the object back to its original location.

Summary

Perspective drawings are based on the principle that the farther an object is from a viewer, the smaller it appears. Objects that are parallel in a perspective view often appear to converge to a vanishing point as they recede into the distance.

There are three types of perspective drawings. The three types differ in the number of vanishing points used. A one-point perspective has a front view parallel to the picture plane and an axis receding to a single vanishing point. It is the easiest type of perspective to draw. A two-point perspective has vanishing points to the right and left of the object. This produces a more realistic appearance than a one-point perspective. A three-point perspective has three different vanishing points and produces an extremely realistic view of an object. However, manually drawn three-point perspectives are often complex and time-consuming to produce. Therefore, three-point perspective is the least used of the perspective drawing techniques.

There are a wide variety of tools available to help illustrators create perspective drawings. These tools include perspective grid sheets and drawing boards. Computer-aided drafting systems can dramatically reduce the amount of time required to draw a perspective view. Many CAD systems have automatic perspective drawing and viewing functions for 3D models. Other standard CAD drawing tools, such as object snap functions and editing commands, simplify the perspective drawing process as well.

Review Questions

1. Identify the three types of perspective drawings and describe how they differ.
2. What are the primary applications for each of the perspective styles of drawing?
3. Define the following terms: *station point, horizon line, ground line, picture plane*, and *vanishing point*.
4. Explain the relationship between the horizon line and the ground line on a perspective drawing.
5. Describe the basic procedure used to lay out a one-point perspective.
6. What are the general guidelines for locating the station point on a perspective drawing?
7. In a two-point perspective, how are the vanishing points located?
8. What is the advantage to drawing the plan view on a separate sheet when developing a two-point or three-point perspective drawing?
9. Which features in the plan view project to true length when drawn in the perspective view?
10. Where is the vertical vanishing point located on a three-point perspective drawing?
11. What is the purpose of using a plywood board to create an oversize drawing surface when developing a perspective view?
12. Identify two manual drawing aids that are typically used in perspective drawing.
13. What is an *elevation line*?
14. Describe two primary computer-aided drafting functions that aid in the development of perspective drawings.

Drawing Problems

Using the orthographic views shown, develop the following problems as perspective drawings. For each problem, use your own dimensions and orient the object to create the most realistic representation. Drawing grids are provided to help you determine dimensions. Using the methods discussed in this chapter, develop a plan view and elevation view for each problem and lay out the necessary drawing guidelines corresponding to the type of perspective drawn. Locate features by projecting points from intersection points on the picture plane. The type of perspective to develop is identified in Problems 1–3. For Problems 4–8, determine the best type of perspective to draw depending on the object and the time available.

1. One-point perspective

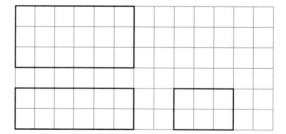

2. Two-point perspective

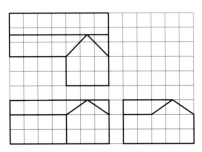

3. Three-point perspective

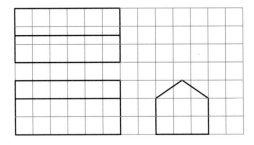

For Problems 4–8, determine the best type of perspective to draw or complete the problem as directed by your instructor.

4.

5.

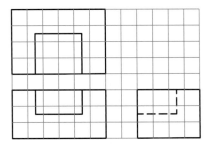

6.

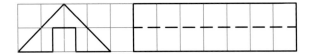

7.

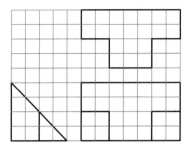

8.

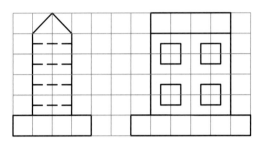

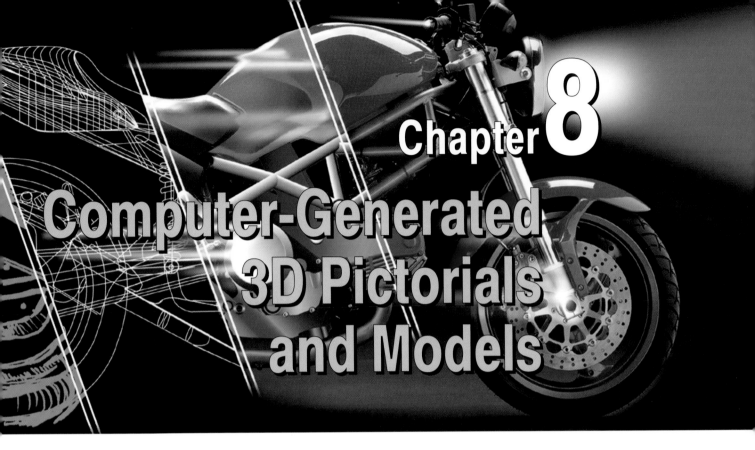

Chapter 8

Computer-Generated 3D Pictorials and Models

Learning Objectives

At the conclusion of this chapter, you will be able to:

- [ ] Explain how two-dimensional (2D) drawings differ from three-dimensional (3D) drawings.

- [ ] List the procedures used to apply thickness to 2D objects to create 3D representations.

- [ ] Describe the world coordinate system and explain the purpose of a relative coordinate system.

- [ ] Explain how 3D views are generated using preset views, viewpoint functions, and dynamic viewing tools.

- [ ] Define surface modeling and identify its applications in illustration.

- [ ] Describe how to construct edge-defined surfaces, revolved surfaces, ruled surfaces, and tabulated surfaces.

- [ ] Identify the predefined 3D objects commonly available in a CAD system and explain how they are used to develop surface models and solid models.

- [ ] Explain the applications for solid modeling and describe the various CAD drawing techniques used to create solids.

Introduction

Pictorial drawings create an illusion of depth. True three-dimensional representations are only possible either by making a physical replica of the object or by defining its three-dimensional characteristics with a computer-aided drafting (CAD) system. This chapter presents the fundamental concepts of three-dimensional computer-generated pictorial drawing. It focuses on the various 3D drawing functions, viewing tools, and modeling techniques used in a CAD system.

Two-Dimensional Vs. Three-Dimensional Drawing

Two-dimensional pictorial drawings and three-dimensional models can be created and displayed in a number of different ways using a CAD system. A computer-generated isometric drawing of a space shuttle is shown in **Figure 8-1A**. In this view, the shuttle is seen from a perpendicular viewpoint. It is oriented as if you are looking straight down at a sheet of drawing paper. The same drawing is shown in **Figure 8-1B** with the viewpoint shifted lower and to the right. This drawing is oriented as if you are looking from an angle at a sheet of drawing paper laying on a desk. As the viewing angle is shifted lower and to the right, more distortion of the view results. Both drawings in **Figure 8-1** are 2D representations. All coordinates for the objects are located on the XY drawing plane. The viewing angle in **Figure 8-1A** creates a 3D effect.

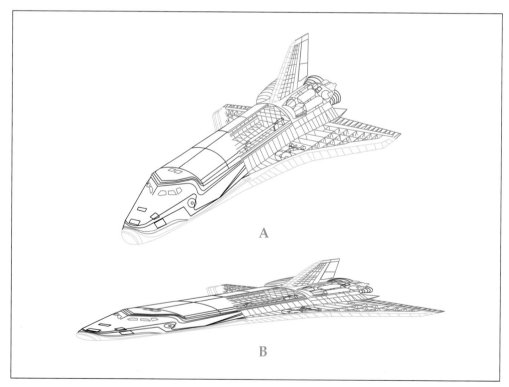

Figure 8-1. A two-dimensional drawing of a NASA space shuttle displayed at different viewing angles. A—An isometric "cutaway" view. B—When the viewpoint is changed, it is clear that the isometric drawing is a 2D representation, with no thickness applied.

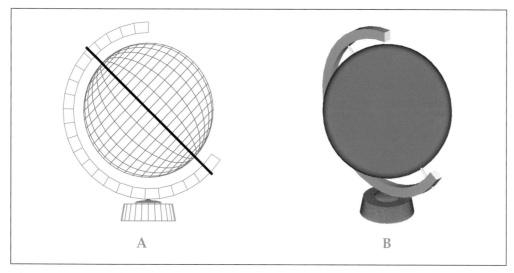

Figure 8-2. A true three-dimensional model of a globe. A—The front view of the object. B—When the viewpoint is changed, it is apparent that the object has three dimensions.

The drawing in **Figure 8-2A** represents the front view of a true 3D model of a globe. Look at the elevated and rotated view of the object in **Figure 8-2B.** Notice how this true 3D model differs from the 2D pictorial in **Figure 8-1B.** The model of the globe is drawn using a third coordinate, the Z coordinate, for the third dimension. The added third, or depth, dimension of a 3D drawing causes the object to appear three-dimensional in any view other than one of the principal views.

Simulated 3D Drawings

A *simulated 3D drawing* is a computer-generated drawing that gives the illusion of a 3D surface or solid model. Simulated 3D drawings are similar to manually created pictorials. In reality, simulated 3D pictorials are computer-generated views that fall somewhere between 2D and true 3D representations. Generally, they are created in one of two ways. They are either developed as projected views or as 2D shapes with "thickness" applied. The next section discusses how 2D shapes can be given thickness to create 3D views.

Applying Thickness to 2D Objects

A simulated 3D drawing can be created by drawing a normal orthographic view or profile of an object and giving it thickness (or height). For example, the drawing in **Figure 8-3A** shows the plan (top) view of four city blocks. The objects are drawn using XY coordinates only. The lines making up the buildings can be given thickness by assigning a base elevation for the plan view and then specifying height values for the corresponding features. The resulting objects have XYZ coordinates and a three-dimensional appearance when the viewpoint is changed. See **Figure 8-3B**. This drawing shows a pictorial view of the objects with "thickness" applied and hidden lines removed. Notice that for two of the buildings, different thicknesses have been assigned to the lines making up the objects in the plan view to create the resulting features. In addition, any lines making up hidden features have been removed. This is an easy way to develop 3D surfaces on objects.

When modifying surfaces on simulated 3D objects, you will encounter certain limitations. The surfaces can be erased, copied, moved, rotated, or modified with most CAD systems. However, these types of shapes often do not work well with snap functions when using the trim, extend, stretch, break, divide, or fillet operations. These operations are often needed to revise basic shapes into finished contours.

Applying thickness to 2D objects is one of the easiest ways to create simulated 3D drawings. However, this method also has some drawbacks in relation to the finished view. For example, a simple cube, rectangle, or cylinder may be generated from a 2D shape by giving it thickness. However, a wireframe view of the resulting object shows surface edges and intersections that would normally be hidden from view. In addition, there are other features that do not display correctly.

Referring to **Figure 8-3B**, lines that would normally not be visible have been removed from the view. In many CAD systems, you can remove hidden lines from wireframe views by using the **Hide** command. This command regenerates the view and removes any wireframe lines that are behind the nearest viewing surface. The computer calculates the surface limits of the shape and redraws the object with only visible object lines showing. Some CAD systems provide added options beyond the simple removal of hidden lines. For example, certain systems recognize the separate surfaces defined by the lines that are drawn when thickness is applied to a shape. They will, upon command, hide individual lines you would normally not see, hide portions of textured surfaces, and automatically shade the surfaces to further aid in visualizing what is visible and what is hidden.

Using the **Hide** command to remove hidden lines from wireframe views is shown in **Figure 8-4**. The cube, rectangle, and cylinder shown in **Figure 8-4A** were created as 2D shapes with thickness applied. The same objects after using the **Hide** command are shown in **Figure 8-4B**. Notice that in both the wireframe and hidden views in **Figures 8-4A** and **8-4B**, the cube and the rectangle appear without their top surfaces. This is because the objects were created from lines and surfaces with height, rather than solid shapes. The cylinder in **Figure 8-4B**

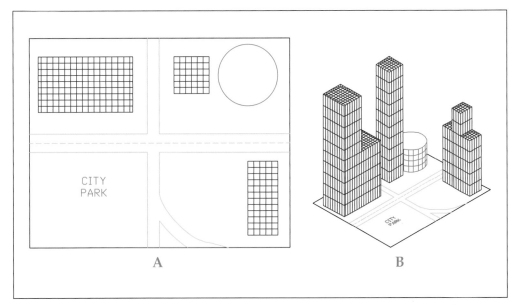

Figure 8-3. A—A plan view of four office buildings. The lines making up each object are given thickness to create a simulated 3D drawing. B—The resulting objects can be seen when the viewpoint is changed. Hidden lines are removed in this view to give the appearance that the buildings have three-dimensional surfaces.

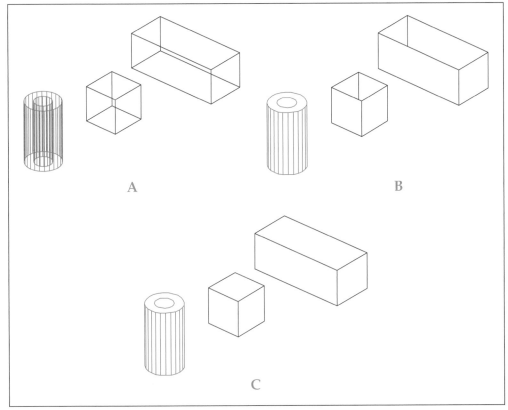

Figure 8-4. Removing hidden lines from objects with thickness applied in a simulated 3D drawing. A—A wireframe view with hidden features shown. B—The objects after using the **Hide** command. C—Three-dimensional surfaces are added to the top of the cube and rectangle.

is made up of two circles with thickness applied. It appears with a top surface because 2D circular shapes are treated differently from lines when thickness is applied and hidden lines are removed. The cube and rectangle can be modified by drawing a 3D surface on the top of each object. See **Figure 8-4C**. The resulting objects then appear to have three-dimensional surfaces. Constructing objects with 3D surfaces is discussed in greater detail later in this chapter.

If the disadvantages of creating a 3D drawing from 2D shapes make this type of construction too inconvenient, it may be better to draw the object as a surface model or a true solid. This is especially true if the resulting object requires more than three or four additional surfaces drawn on upper portions. If the object is relatively simple, however, applying thickness to a 2D construction is a highly useful tool. Surface modeling and solid modeling tools are discussed later in this chapter.

The object in **Figure 8-5** was developed from 2D shapes with thickness applied. The following steps were used to develop this object. The specific steps will vary depending on what software you are using. In the final step, the viewpoint is changed to obtain the 3D view. Viewing tools for 3D objects are discussed later in this chapter.

1. Set the elevation, or altitude. This is how far from the base XY plane the object will be drawn. For the first part of the object, set the elevation to 0.
2. Set the thickness, or height. This is how "tall" the object will be. For the first part of the object, set the thickness to 7/8". In **Figure 8-5A**, this value matches the height dimension for the sides of the base.
3. Draw the rectangle and the 1/2" diameter hole.
4. Next, draw the cylinder on top of the base. Since the base is 7/8" thick, the cylinder needs to start 7/8" above the XY plane. Set the elevation to

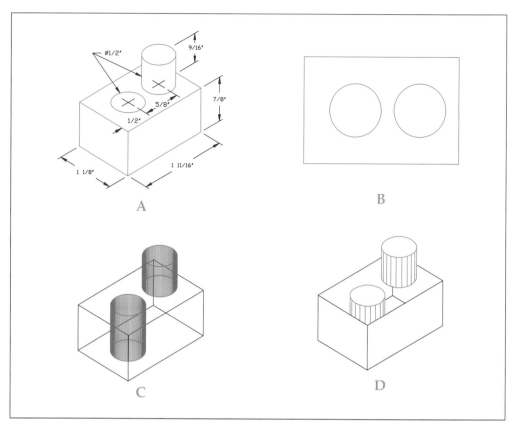

Figure 8-5. Applying thickness to 2D shapes to create a 3D drawing. A—The dimensioned isometric view used to create the drawing. B—The plan view of the object. C—The resulting object with thickness after changing the viewpoint. D—The view after removing hidden lines.

7/8". Also, the cylinder is 9/16" tall. Therefore, set the thickness to 9/16". In the plan view shown in **Figure 8-5B**, both the hole and the cylinder appear identical because they have the same diameter.

5. Change the viewpoint to create the pictorial view. Refer to **Figure 8-5C**. The object shown is displayed without hidden lines removed. In **Figure 8-5D**, the object is shown after hiding lines. Notice that in both figures, the top surface of the base is missing. This is because the object was created with lines and surfaces for the sides. A better way to create a true representation of the object is to draw it as a solid model. Solid modeling techniques are discussed later in this chapter.

Keep in mind the final application of your illustration when choosing between a 2D and 3D CAD drawing. If the drawing will require rendering or shading and different viewpoints, and it is not highly complex in shape, you should consider developing the object as a simulated or true 3D drawing. Developing a 3D view of a simple to moderately complex shape usually takes no more time than developing the object using 2D methods. Therefore, plan ahead as you decide on how to generate the pictorial. Use the method that will save the most work and produce the best results in the long run.

Using 3D Coordinate Systems

When creating a 2D drawing with a CAD system, objects are drawn using coordinates along the X and Y axes. In a 3D drawing, a third coordinate axis, the Z axis, is used. Three-dimensional coordinates are entered as (X,Y,Z). The Z axis projects vertically from the XY plane and specifies points along the depth of the object. Positive Z coordinates project above the XY plane and "out" of the drawing on screen. Negative Z coordinates project behind the XY plane and "into" the drawing on screen.

In a typical CAD system, the planes defined by the default X, Y, and Z axes in the Cartesian coordinate system make up the *world coordinate system*. This is an absolute coordinate system based on the world origin (0,0,0). The world coordinate system is typically sufficient for 2D drawing. When developing a 3D drawing, however, it often helps to change the coordinate system from the world coordinate system so that it is oriented with specific planes or features on an object. After locating features on the default XY plane, for example, you may want to change the viewpoint and move the coordinate axes to a different plane in order to draw on a different surface of the object. In doing so, points can be located along the X and Y axes in relation to other features on the same plane. A coordinate system created in this manner is called a *relative coordinate system*. You can establish a relative coordinate system in any orientation in 3D space.

Axis icons are typically used to indicate which coordinate system is currently in use. See **Figure 8-6**. These icons identify the direction of the axes with respect to the current drawing plane and help establish the viewing orientation for the drawing. Notice the difference between the world coordinate system icon and the relative coordinate system icon. The relative coordinate system icon does not have a square at the origin. When the world coordinate system is active, points are located in relation to the world origin (0,0,0). When a relative coordinate system is active, points are located in relation to the origin of that coordinate system. For example, you may choose to orient the coordinate axes so that the intersection of the X and Y axes is located on the corner of an object in a 3D view.

When working in a 3D view, an icon representing the three axes is displayed. Refer to **Figure 8-6C**. This icon is extremely useful when viewing an object that has been rotated from the plan view so that your line of sight is not perpendicular to any of the principal planes. The direction of the axes indicates the orientation of the current drawing plane. The direction of the Z axis establishes the current depth, or height, direction with respect to the current XY plane. Understanding the relationship of the axes will help you visualize how a three-dimensional object is oriented in space.

The different appearances of coordinate system icons in relation to 2D and 3D views are shown in **Figure 8-7**. The world coordinate system is used to draw the plan view of the object in **Figure 8-7A**. The square shape is drawn with all lines parallel to the X and Y axes. In this example, the lines are given

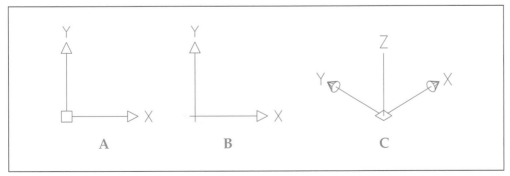

Figure 8-6. Common axis icons used to represent the current coordinate system. A—A world coordinate system icon. B—A relative coordinate system icon. C—A three-dimensional icon indicating the orientation of the axes in 3D space.

thickness to develop the sides of the cube. You will not see lines drawn parallel to the Z axis in the plan view. However, inclined or skewed lines appear as foreshortened lines in the plan view. In **Figure 8-7B**, the viewpoint is rotated to show the object in three dimensions. In addition, hidden lines are removed. Notice the different orientation of the world coordinate system icon. The icon identifies the XY drawing plane and the direction of the Z axis. In **Figure 8-7C**, a relative coordinate system is created to establish a new XY drawing plane. The coordinate system is moved to the top of the cube so that the top surface can be drawn. Notice that the coordinate system icon has changed. The resulting object after completing the top surface is shown in **Figure 8-7D**. In this example, the top surface is drawn as a 3D surface. This drawing function is discussed in the following section *Creating 3D Surfaces*.

The example in **Figure 8-7** shows how features in 3D space can be easily constructed with a relative coordinate system. After drawing a simple profile in 2D space and giving it thickness, the viewpoint can be changed to show the object in three dimensions. A new coordinate system can then be established to locate features in relation to a surface in 3D space. This is often useful for drawing features on angled faces. For example, radial features projecting from an inclined surface can be drawn in this manner. See **Figure 8-8**. In this example, a relative coordinate system is established on the inclined surface of the object in a 3D view so that the cylindrical features can be drawn.

The sides for the angled wireframe object in **Figure 8-8** are drawn using 3D coordinates rather than lines with thickness. This object has a four-unit square base and a height of four units at one end. In **Figure 8-8A**, the square base is drawn in plan view using the world coordinate system. Because the base lies on the XY drawing plane, the coordinates for each corner have the same zero value for the Z coordinate. The viewpoint is then changed in **Figure 8-8B** so that the three-dimensional object can be developed. Notice the orientation of the coordinate axes and the direction of the Z axis. To develop one side of the object, lines are drawn using the XYZ coordinates shown. The inclined face is completed by drawing a diagonal line connecting the two corresponding endpoints with object snaps. This side is then copied four units along the base using object snaps to establish the other side, **Figure 8-8C**. Object snaps can also be used to connect the two sides with lines, **Figure 8-8D**.

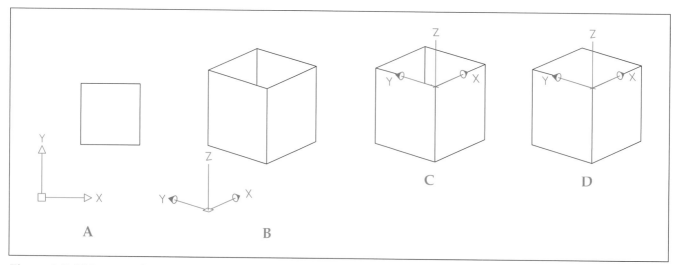

Figure 8-7. Using coordinate systems to draw the features of a cube. A—The base of the cube is drawn in plan view using the world coordinate system. B—Changing the viewpoint displays the cube in three dimensions. Notice the appearance of the coordinate system icon. C—A relative coordinate system is established on top of the cube. D—The top surface is drawn and hidden lines are removed to complete the object.

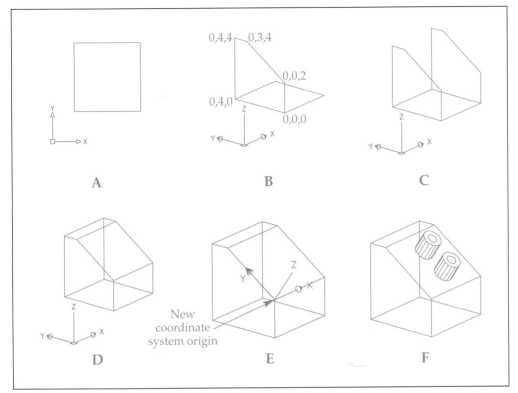

Figure 8-8. Using a relative coordinate system to construct three-dimensional features on an inclined face. A—The base surface is developed in the plan view. B—One side is drawn with 3D coordinates after changing the viewpoint. C—The side is copied to the other end of the object. D—The sides are connected with horizontal lines. E—A new coordinate system is created on the angled surface to draw the radial features. F—The final wireframe model with hidden lines removed after adding the cylinders.

Next, a relative coordinate system is created to establish a new XY drawing plane on the inclined surface of the object. See **Figure 8-8E**. This coordinate system establishes a drawing plane that can be used to construct the radial features. As shown in **Figure 8-8E**, the origin of the new coordinate system is located at the bottom-left corner of the inclined surface. This simplifies the task of locating the circular features. The first (lower) cylinder is drawn on the surface as a pair of circles with thickness applied. The center point for the circles is located 2 units from the origin along the X axis and 1 unit along the Y axis. One circle is drawn with a diameter of 1 unit, and the other is drawn with a diameter of 0.5 units. Each circle is given a thickness of 1 unit to establish the height in relation to the inclined surface. Finally, the circles are copied 1.5 units along the Y axis to create the other cylinder.

In **Figure 8-8F**, hidden lines are removed. This is why the cylinders appear solid. However, the rest of the object appears as a wireframe because it is constructed from lines rather than 3D surfaces or other modeling methods. Although this is a wireframe model, it shows how coordinate systems can be used to establish drawing planes for locating features in 3D space.

Creating 3D Surfaces

Once you understand how to use coordinate systems and basic 3D drawing and visualization techniques, you can begin to apply some of the more powerful CAD functions in developing 3D drawings. One of the easiest functions to master in 3D drawing is the construction of 3D surfaces, or 3D faces. This function is enabled with the **3Dface**, **Surface**, or **Face** command, depending on which software you are using. These commands and other similar commands allow you to draw a flat mesh surface on a 3D object. The surface may be normal, inclined, or skewed as long as you can identify its dimensions in relation to other features. Even complex-shaped surfaces can often be developed with these commands if they are subdivided and then drawn as adjacent rectangular or triangular shapes. The resulting surface can be shaded for presentation purposes. However, the surface has no thickness and cannot be used with other 3D surfaces to define a solid. For example, a cube made up of six separate 3D surfaces is not a solid object, but the sides can be shaded to make it appear solid. Objects constructed from 3D surfaces can be used in surface modeling. Surface models are discussed later in this chapter.

A drawing constructed from 3D surfaces with 3D coordinates is shown in **Figure 8-9**. First, the triangular surface for the base is drawn on the XY plane at zero elevation in the plan view. Therefore, the coordinates for each corner of the base triangle (labeled as Point 1, 2, and 3) have the same zero value for the Z coordinate. See **Figure 8-9A**. Another 3D surface is drawn for the top of the object. This surface can also be drawn in the plan view. Each corner of the top surface has the same Z coordinate to place it 1 1/8″ above the base. The drawing is completed by changing the viewpoint and drawing five separate 3D surfaces for the sides. See **Figure 8-9B**. These surfaces connect the base triangle to the top of the object. Notice that the view in **Figure 8-9B** is shown with hidden lines visible. In **Figure 8-9C**, the object is shown with hidden lines removed. One advantage of constructing objects from 3D surfaces is that the sides can be shaded and displayed from any angle to give the object a true 3D appearance. Three-dimensional viewing functions are discussed next.

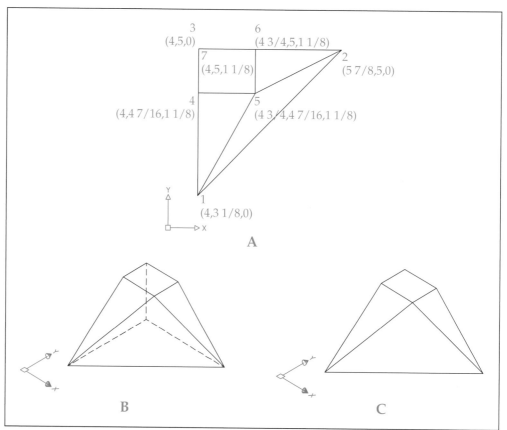

Figure 8-9. Using 3D surfaces and 3D coordinates to draw an object with inclined faces. A—The base surface and top surface are developed in the plan view. B—The five sides are drawn as 3D surfaces after changing the viewpoint. C—The object after hiding lines.

Viewing 3D Objects

One of the most valuable CAD drawing tools for the illustrator is the ability to draw the plan view of a 3D object and then view it from any orientation. In a typical CAD system, there are a number of commands and functions available for establishing viewing directions in a 3D drawing. These are discussed in the following sections.

Orthographic and Isometric Viewing Functions

Most CAD systems have basic view selection options that allow you to display the standard orthographic views of a drawing. These are the front, back, top, bottom, right, and left views. These are preset views that can be selected at any time to orient your view with a principal plane of projection. This allows you to quickly establish one of the orthographic views when it is necessary to draw features parallel to that view. Preset views can usually be accessed with the **View** command.

Preset isometric views are also typically available for displaying 3D drawings. See **Figure 8-10.** These views are based on the position of the viewer and allow you to view an object in an isometric orientation with one of the primary faces on the right or left in the frontal viewing plane. The resulting view has axis lines drawn at 30° to horizontal. Isometric views are useful for visualizing the object during the drawing process or establishing a different coordinate system. When working on a 3D drawing, for example, it is often useful to alternate between isometric and orthographic views to locate features correctly.

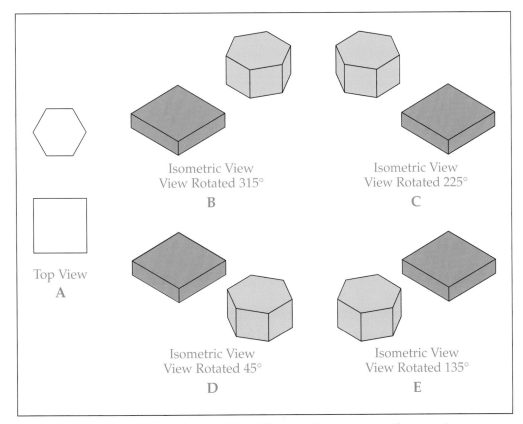

Figure 8-10. Using preset orthographic and isometric views. A—The top view. B—An isometric view seen from the right (southeast). C—An isometric view seen from the left (southwest). D—An isometric view seen from the northeast. E—An isometric view seen from the northwest.

Most CAD systems also allow you to display an isometric view and one or more orthographic views of the object at the same time. This is typically accomplished using viewports. A *viewport* is a viewing area containing a preset or user-defined viewpoint. A 3D drawing on screen can be divided into several viewports to display different views of the same object. In a four-viewport orientation, for example, the isometric view can be placed on the right or left of the display, with the top, front, and side orthographic views on the other side of the display. Different viewing configurations can be established depending on the application.

Establishing a 3D Viewpoint

There are often situations when it is useful to observe an object from a specific viewpoint for construction or visualization purposes. Viewing orientations can usually be set in a CAD system with the **Viewpoint** or **Vpoint** command. This is similar to using a predefined isometric view, but the number of viewing orientations that can be set is virtually unlimited. When using this function, it is important to remember that you are changing the location of the *viewer* rather than the location of the *object*. There are a variety of methods that can be used when establishing the position of the viewer with the **Viewpoint** command. These include the *rotation method*, the *coordinate method*, and the *axis compass method*.

To use the rotation method, first enter the angle of viewing rotation *in* the XY plane. Then, enter the angle of rotation *from* the XY plane. For example, if you want to position your view from the right (southeast) with the object in an isometric orientation, specify a 315° angle to rotate your viewing position 315° in the XY plane. Refer to **Figure 8-10B.** The next step is to enter the angle of your view from the XY plane. In a manually drawn isometric view, the object is normally tipped forward 35°16′ from the horizontal plane. When defining your viewing position in a CAD drawing, however, the object is not rotated from the horizontal plane. Instead, your viewpoint is rotated 35°16′ from the XY plane. To specify this rotation angle, you can enter the decimal equivalent of 35.2667. As an alternative, you can use a rounded angular value of 35°. This will not significantly change the view generated. If you want to view the object from the left (southwest) in an isometric orientation, specify a 225° angle for the rotation angle in the XY plane and a value of 35.2667 or 35° for the angle from the XY plane. Refer to **Figure 8-10C.** Northeast and northwest isometric orientations are shown in **Figure 8-10D** and **Figure 8-10E.**

The coordinate method of designating a 3D viewpoint requires you to enter the coordinate location of the viewer. In a plan view, for example, this coordinate location is defined as 0,0,1 since the viewer is looking "down" at the object on the XY plane from the positive Z axis. You can establish a different viewpoint by entering coordinates for the viewing location. Entering the coordinate 1,–1,1 shifts the viewing location 1 unit to the right along the X axis, 1 unit downward along the Y axis, and 1 unit away from the object along the Z axis. This corresponds to an isometric viewpoint oriented with the viewer positioned to the right.

The table in **Figure 8-11** lists the rotation angles and coordinate entries used to generate isometric and orthographic views with the **Viewpoint** command. These values can be used as a reference when establishing a different 3D viewpoint.

The axis compass method of establishing a viewpoint uses an axis compass icon with another icon representing the coordinate axes. As shown in **Figure 8-12,** the axis compass icon is somewhat like a bull's-eye target. By moving the cursor crosshairs to various locations within the compass, you can change the orientation of the XYZ axes. As the cursor is moved from one point on the compass

Desired View	Rotation Angle in XY Plane	Rotation Angle from XY Plane	Coordinate Location
Southeast Isometric	315°	35°16′	1,-1,1
Southwest Isometric	225°	35°16′	-1,-1,1
Northeast Isometric	45°	35°16′	1,1,1
Northwest Isometric	135°	35°16′	-1,1,1
Top View	0°	90°	0,0,1
Front View	270°	0°	0,-1,0
Right View	0°	0°	1,0,0
Left View	180°	0°	-1,0,0
Rear View	90°	0°	0,1,0
Bottom View	0°	270°	0,0,-1

Figure 8-11. Rotation angles and coordinate locations used to generate 2D and 3D viewpoints.

to another, the coordinate axis icon rotates to indicate the orientation of the X, Y, and Z axes. Although this method does not produce a view with a precisely specified XYZ coordinate, it does provide a quick way to place your viewpoint at an approximate location. In many applications, it is not necessary to have an

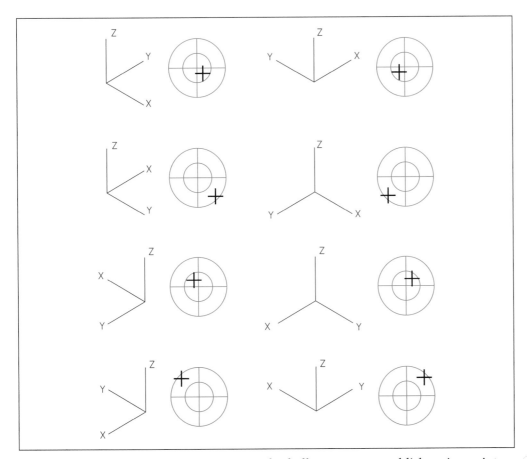

Figure 8-12. The axis compass viewing method allows you to establish a viewpoint by selecting a location within the compass icon. The location selected determines the orientation of the coordinate axis icon.

exact viewing orientation for an object. You may simply want to view it from a location that will allow you to see it in 3D.

The **Viewpoint** command offers a quick and useful way to establish 3D views. In other situations, however, more advanced viewing tools are used to observe objects in 3D dynamically. Orbit viewing tools, discussed in the next section, provide this capability.

Dynamic Viewing Functions

One of the most useful viewing methods for objects in 3D involves the use of *dynamic viewing*. Dynamic viewing methods provide a way to specify exactly how you will display a 3D model. In more advanced CAD systems, a special type of dynamic viewing known as *orbit viewing* is available. This is discussed in the following section.

Orbit viewing

Orbit viewing provides real-time control over the display of an object. It allows you to rotate a view dynamically in 3D space. Orbit viewing is typically accessed with the **Orbit** command. Several options are available, including 3D pan and zoom options. These are very similar to the normal pan and zoom options used with 2D views. However, the most powerful function of this command allows you to move a cursor on screen to orbit the view. You can even use orbit viewing to view fully rendered models.

When you are in orbit view, a large circle with a grip at each quadrant appears. See **Figure 8-13**. In this view, the orbit circle overlays the model you are viewing. This is shown in **Figure 8-13A**. You can orbit the viewing position by placing the cursor on one of the grips, inside the circle, or outside the circle. Placing the cursor on a grip at the upper or lower end of the orbit circle and holding down the left mouse button allows you to rotate the object around a horizontal axis. This will rotate the object forward or away from you. Placing the cursor on a grip on the left or right side of the orbit circle and holding down the left mouse button allows you to rotate the object around a vertical axis. This will spin the object to the left or right. Moving the cursor inside the orbit circle and dragging allows you to orbit the view in any direction. Moving the cursor outside the orbit circle and dragging allows you to orbit the view about an axis projecting perpendicular from the view on screen. With a small amount of practice, you will find it easy to drag and orbit objects to any viewing position you desire.

A number of other display modes are available in orbit viewing. For example, you can orient a 3D view relative to the perspective of a camera. When using the camera function, you can adjust the distance from the camera to the object with the cursor to make the view appear larger or smaller. This is similar to zooming the view. Dragging the camera upward on screen makes the image larger. Dragging it downward reduces the size of the view. You can also "swing" the camera by dragging the cursor to rotate one side of the view away from the camera. This is similar to walking a video camera around the object. For instance, you may want to raise or lower the camera to capture a greater bird's-eye or worm's-eye view.

You can use orbit viewing to view a portion of a model by applying clipping planes. *Clipping planes* are front and rear boundaries that slice through a model to remove certain portions from the view. Portions outside the boundaries are removed. This is similar to creating a section. Using clipping planes takes some practice, but this tool can be very useful once you master the technique. When you apply clipping planes, a viewing window displays a top view of the object with an edge view of the planes. The front and rear planes are displayed as horizontal lines. This is shown in **Figure 8-13B**. Moving the lines up or down with the cursor allows you to adjust the planes and remove portions from the view. Features in front of the front clipping plane are sliced away, and any features behind the rear clipping plane are also removed. In **Figure 8-13B**, the front clipping plane is adjusted to remove features near the front of the block. Since the rear clipping plane is behind the object, the background features remain unchanged. The resulting view is shown in **Figure 8-13C**. In this view, the block has had a portion in front "sectioned" away to show details of the counterbore in the large hole. If desired, you can adjust the clipping planes again to create a different section, or you can restore the original view.

If you look carefully at the side and upper edges in the view in **Figure 8-13A**, you will notice that a small amount of perspective taper toward a single vanishing point can be seen in the orbit view. Although this view appears to have a slight foreshortening of lines, it is set to parallel projection. When you are in orbit view, you can change the type of projection to perspective so that the sides of the object converge toward a vanishing point. See **Figure 8-13D**. This is a quick way to generate a perspective view of a model. While still in orbit view, you can zoom, pan, or orbit the perspective view.

One of the more powerful orbit viewing tools is the ability to "spin" a model into continuous rotation. This display mode is known as *continuous orbit*. To generate a continuous orbit, the cursor is moved in the direction of desired rotation in orbit view. When the mouse button is released, the object continues

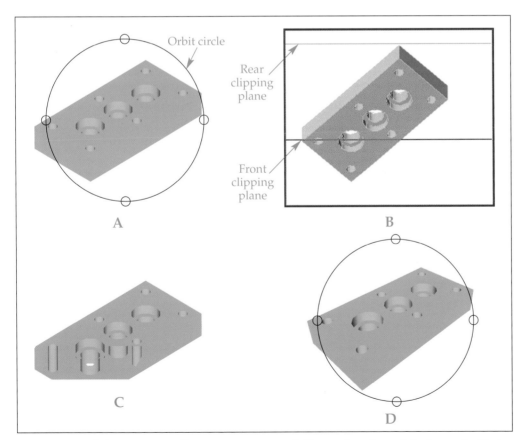

Figure 8-13. Orbit viewing allows you to rotate fully rendered models dynamically in 3D space. A—The orbit circle appears in orbit view and provides different viewing functions based on the use of the cursor. B—A portion of a model can be viewed by "slicing" through the object with clipping planes. Here, the front clipping plane is adjusted to create a "sectioned" view. C—Internal features can be viewed after clipping the model. D—The view is set to perspective projection.

to rotate around the same path. This allows you to view features that may not be apparent in a single view as the model orbits in 3D space. You can also change the angle or speed of rotation at any time during the viewing session.

Surface Modeling Applications

Surface modeling is one of the primary ways to produce a realistic 3D drawing. A *surface model* is a 3D representation that uses *mesh surfaces* to define the exterior surfaces of an object. Surface models are useful for presentation purposes because they can be shaded or rendered to create a very realistic appearance. Computer shading and rendering techniques are discussed in greater detail in Chapter 10. Mesh surfaces making up the features of a surface model can be thought of as "skin" applied over the entire object. While the object appears solid when rendered, it is not a true solid. Solids are created using solid modeling tools and are discussed later in this chapter.

Many software programs allow you to develop surface models using a variety of mesh surfacing tools. Each software program operates a bit differently. However, for the most part, mesh surfaces are created in a similar fashion. The most common types of meshes are edge defined surfaces, revolved surfaces, ruled surfaces, and tabulated surfaces. These are discussed in the following sections.

Edge Defined Surfaces

An *edge defined surface* is a mesh surface that fills an area between four connected edges or boundaries. These boundaries must physically define the edges of the surface, but they do not have to be in the same plane. Also, they may be a combination of adjacent arcs, curves, and lines. The number of line segments creating the mesh is usually controlled by a system variable. This variable will differ between software programs, but most operate similarly. Higher variable values produce a smoother surface. However, be cautious when increasing the number of line segments. Setting the value too high can slow down system performance.

A simple surface-modeled object created from two edge defined surfaces is shown in **Figure 8-14**. The base curves and lines for the surfaces are shown in

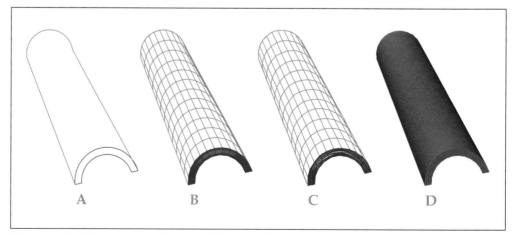

Figure 8-14. Creating a surface model from edge defined surfaces. A—The base curves to be surfaced. The curved top surface is created by selecting the lines and arcs making up the entire length of the object. The curved front surface is created by selecting the semicircular arcs and horizontal lines. B—Mesh surfaces are applied to give the object "skin." C—After removing hidden lines. D—Rendering completes the surface model.

Figure 8-14A. Each surface has four edges made up of two arcs and two lines. Notice that the two surfaces share a common boundary. Also, the boundaries defining the edges are closed. The edges must be connected in this manner to apply the edge defined surface. For each surface, the objects making up the edges are selected using the appropriate surfacing command. After selecting edges, the mesh surfaces are generated automatically. See **Figure 8-14B**. When hidden lines are removed, the object appears as shown in **Figure 8-14C**. The object appears as a true surface model after rendering, **Figure 8-14D**.

Revolved Surfaces

A *revolved surface* is a mesh surface created by rotating a profile about an axis of revolution. This type of surface is very useful when you need to develop a symmetrical 3D object with a centerline axis. The profile, or path curve, may be open or closed. It can be drawn using lines, arcs, or closed objects such as circles or ellipses. You can control the number of line segments making up the surface. The shape can be revolved an entire 360°, or you can specify a different angle for a partial rotation.

Two examples of creating objects from revolved surfaces are shown in **Figure 8-15**. The upper profile shape is a single arc. The lower profile shape uses two arcs and a line to define the curves of the surface. In both cases, the axis of rotation is oriented with the profile shape and the shape is rotated 360°. The resulting surfaced objects are shown in **Figure 8-15B**. Notice that the viewpoint is rotated to establish the best view. In **Figure 8-15C**, the objects are shown with hidden lines removed. This allows you to see the interior and exterior surfaces. The final rendered surface models are shown in **Figure 8-15D**.

To create a revolved surface, the appropriate surfacing command is entered and the curve to be revolved is selected. Then, the axis line of rotation is selected. Finally, the starting and ending angles for the rotation are specified. A full circular rotation from 0° to 360° is the most common application. However, you may have a model where you want to show the interior features in better

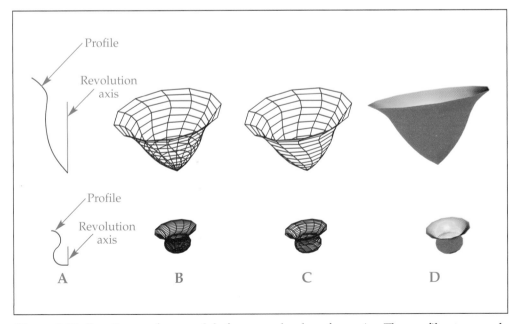

Figure 8-15. Creating surface models from revolved surfaces. A—The profile curve and axis of rotation are drawn. B—Each profile is revolved 360° to generate a mesh surface for the entire object. C—Removing hidden lines provides a better definition of the interior and exterior surfaces. D—The rendered surface models.

detail. In this case, indicate the degrees of rotation needed to draw the surface so that the side closest to the viewer is "open."

Ruled Surfaces

A *ruled surface* is a mesh surface that fills an area between two objects. The objects can be lines, points, arcs, circles, or other combinations of lines that define two ends of the surface. Examples of creating ruled surfaces are shown in **Figure 8-16**.

The procedure used to develop a ruled surface is similar to that used for an edge defined surface. After entering the appropriate surfacing command, select the two lines or shapes. The mesh surface is automatically generated between the two objects. The number of line segments making up the surface is usually controlled by a system variable.

If you join surfaces or areas that share one single defining boundary between two other 3D mesh surfaces, you will often have to zoom in to the view to select the boundary. If you do not zoom in, it is difficult to select the correct boundary line instead of the adjoining 3D mesh surface.

Tabulated Surfaces

A *tabulated surface* is developed by generating a surface from an object along a direction path established by a second line. This type of surface is similar to a ruled surface, but one object forms the outline of the surface and the direction and length is determined by the direction path. See **Figure 8-17**. Objects such as lines, circles, and arcs can be used for the surface outline. In

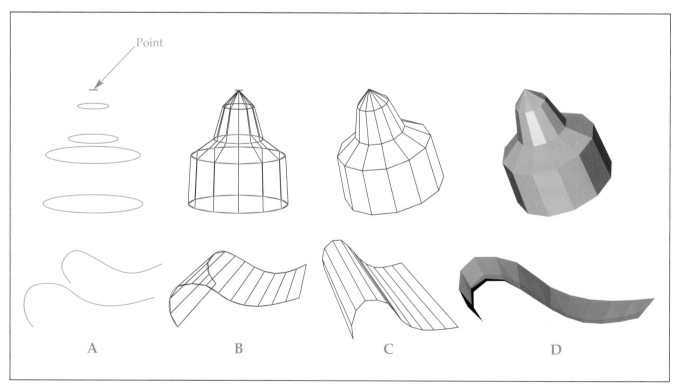

Figure 8-16. Creating surface models from ruled surfaces. A—In the top example, the construction point and four circles define the object to be surfaced. The lower object consists of two curves. B—The mesh surfaces are generated by selecting the objects in order. C—The objects after removing hidden lines and changing the viewpoint. D—The rendered surface models with the viewpoints slightly changed.

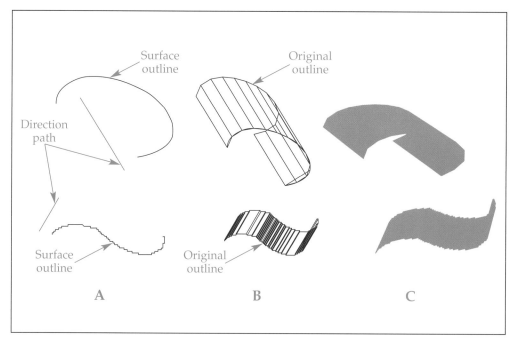

Figure 8-17. Creating tabulated surfaces. A—The surface outline and direction path are drawn. B—Each mesh surface is generated in the direction of the path. The length of the direction path determines the length of the surface. C—The surfaces after rendering.

Figure 8-17, the surface outline in the top example is an elliptical arc. The surface outline in the lower example was developed using the sketch drawing function. Notice the resulting steps in the surface.

To create a tabulated surface, the appropriate surfacing command is entered and the outline curve is selected. This curve defines the shape of the surface. Then, the direction path is selected. The length of this line defines the length of the surface. The direction in which the surface is generated is determined by the location where the direction path is picked. The endpoint that is closest to the pick location is established. Then, the surface is generated in the opposite direction from the endpoint. The number of line segments making up the surface is usually controlled by a system variable.

Constructing 3D Objects

Many CAD software programs provide a number of three-dimensional shapes that can be used as building blocks during the drawing process. These are predefined objects that are designed to simplify the construction of 3D models. You are required to specify certain dimensions, and the computer automatically draws the object. Regardless of the type of software used, you are asked to input the specific location and dimensions appropriate to that shape. In some CAD systems, many of the shapes can be created as surface-modeled objects or as true solids. Predefined solid objects called *solid primitives* can be very useful when creating solid models. Solid modeling is discussed later in this chapter.

Ten common objects available in most 3D modeling programs are shown in **Figure 8-18**. Normally, these objects are drawn in plan view. You are prompted to provide a location for the object and values for the necessary dimensions. An alternate way of drawing predefined 3D objects is to establish two or more viewports so that a plan view and front view or isometric view are displayed

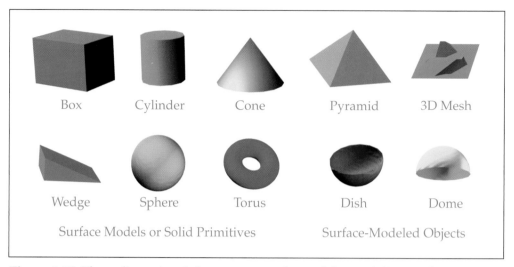

Figure 8-18. Three-dimensional shapes commonly used for modeling applications. Depending on the program you are using, they are drawn as surface or solid models. The objects are drawn by specifying a location and the appropriate dimensions. These objects are shown rendered.

at the same time. This allows you to see the elevation relationships while drawing the objects.

A plan view of the 10 objects is shown in **Figure 8-19.** Each object was drawn by locating either a corner or center point and specifying dimensions. If you plan to draw complex shapes by combining various 3D objects, it is very important to properly locate each object in relation to others. Use different views, relative coordinate systems, and the appropriate drawing tools when necessary to help draw features. A sketch of the finished object with location points for individual 3D shapes will help you visualize the drawing in relation to how it can be developed.

The following sections discuss the basic drawing procedures and dimension requirements for the objects shown in **Figures 8-18** and **8-19.** Refer to these figures as you go through each section. Although procedures vary among different software programs, these general guidelines will provide you with an understanding of the basic information needed to define the shapes.

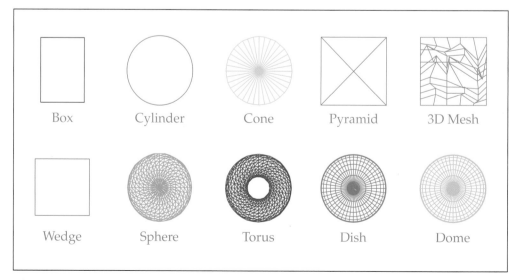

Figure 8-19. A plan view of the predefined three-dimensional shapes shown in Figure 8-18.

3D Box Construction

A box can be drawn as a 3D cube or rectangle, depending on what dimensions you provide to the computer. Typically, a box can be drawn as a surface-modeled object or as a solid primitive. The process begins by accessing the appropriate 3D drawing command and the box option. First, you will be asked to specify the starting location. The starting point is typically the front, lower-left corner of the box. You can specify the location by entering absolute XYZ coordinates, picking a point with the cursor, or indicating a position relative to an established feature.

Once the starting point is identified, the length of the box is defined. You can enter a value or move the cursor to a point establishing the length. Next, you are asked to enter the width of the box. If you want to draw a cube, enter the same value as the length.

Finally, enter the height of the box. In some programs, you are asked for the rotation about the Z axis. Unless you instruct the computer to draw the object at a different elevation, it will be drawn with the base on the current drawing plane. Unlike a box constructed from lines with thickness, surface-modeled and solid boxes do not have an open top. They appear with a surface skin on all sides.

3D Cylinder Construction

A cylinder is a circular object with height. It can be drawn as a surface model or true solid. After entering the appropriate 3D drawing command and option, you must specify the center point of the base. You can enter coordinates or use the cursor. Then, you must enter the diameter or radius of the base. Finally, the height is specified. As with a box, the cylinder is drawn with the base sitting on the current drawing plane.

3D Cone Construction

A cone is similar to a cylinder. However, the top of the object is pointed or truncated. A truncated cone has a circular top with a smaller radius than that of the base. A pointed cone has a top with a radius value of zero. The cone function can be used to draw a cylinder by specifying the same radius value for the base and the top.

A pointed cone can be drawn as a surface model or true solid. A cone primitive with a truncated top is typically drawn as a surface model. After entering the appropriate 3D drawing command and option, you must specify the center point of the base. Then, you must enter the diameter or radius of the base. Next, you are asked for the diameter or radius of the top of the cone. A value of zero creates a cone with a pointed top. To create a truncated surface-modeled cone, enter a value smaller than that of the base diameter or radius but greater than zero. Finally, the height of the cone is specified. You may also be asked to enter the number of line segments used to represent the sides of the cone. Refer to **Figure 8-19**. The cone is drawn with the base sitting on the current drawing plane.

3D Pyramid Construction

A pyramid has a three-sided or four-sided base and a pointed, ridged, or truncated top. There are different types of pyramids you can construct, depending on the program you are using. See **Figure 8-20**. Pyramids are usually drawn as surface-modeled objects.

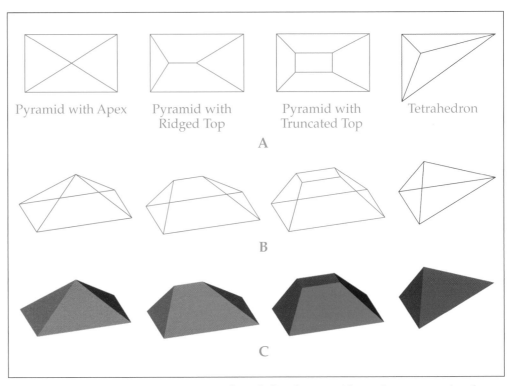

Figure 8-20. A—Four common types of predefined pyramids as they appear in plan view. A tetrahedron has a three-sided base. B—A 3D view of the objects. Hidden lines are shown for clarity. C—The objects after rendering.

A number of entries are required to draw a pyramid. After entering the appropriate 3D drawing command and option, you must locate a first, second, third, and, when appropriate, fourth base corner. A pyramid with a three-sided base is called a *tetrahedron*. The base is drawn on the current drawing plane. Once the base is drawn, you complete the object by defining the top. The values specified depend on the type of top drawn. A pointed top requires an XYZ coordinate for the apex. A ridged top requires two coordinates for the ends. A truncated top requires three or four base points for the top surface. You can enter coordinates or use the cursor to specify point locations.

3D Wedge Construction

A wedge looks like a doorstop. The dimensions are similar to those required for a box. However, a wedge has an angled surface. A wedge can usually be drawn as a surface model or true solid. After entering the appropriate 3D drawing command and option, you must specify a starting location for the base. Then, the length, width, and height are specified. The wedge is drawn with the base sitting on the current drawing plane. In some programs, you are also asked for the rotation about the Z axis.

3D Mesh Construction

A 3D mesh is a type of mesh surface that can be created with irregular elevation. In other words, the faces making up the mesh do not have to be in the same plane. Refer to **Figure 8-18**. A mesh is a surface-modeled object. It is developed by identifying the first, second, third, and fourth corners of a plane. Once the fourth corner is identified, the computer automatically closes the perimeter of the plane. You must then specify the number of horizontal and vertical vertices used for generating line segments in each direction. The resulting grid lines define the rows and columns of the mesh. If the original perimeter lines do not lie in the same plane, the mesh lines are distributed equally to fill the distance between the opposing sides of the mesh.

Referring to **Figure 8-18**, the mesh was originally developed as a flat grid and then modified to assign different elevation values for vertices along the horizontal and vertical mesh lines. The **Stretch** command was used to drag line intersections to the desired X, Y, or Z direction. The result gives the appearance of terrain.

3D Dish Construction

A 3D dish looks like a bowl. Unlike other 3D shapes, a dish is drawn downward from the current elevation. See **Figure 8-21**. A dish is usually developed as a surface-modeled object. It resembles a surface-modeled sphere cut in half. To draw a dish, you must first locate the center point. If you do not specify a Z coordinate, the center point lies on the current XY drawing plane. If you want the dish to sit on a specific elevation plane, you need to specify a Z coordinate equal to the radius of the dish for the center or set a new elevation before drawing the dish.

After the center is located, the diameter or radius of the shape is specified. With some programs, you are also asked to specify the number of longitudinal and lateral segments. The longitudinal segments are lines that run north to south from the top to the bottom of the bowl. The lateral line segments run east to west and circle the shape. Refer to **Figure 8-21**. A dish appears more realistic if the number of longitudinal segments is greater than the number of lateral segments.

3D Dome Construction

A 3D dome is the opposite of a dish. It looks like a bowl turned upside down. However, the dimensions you provide are identical to those of a dish. You must provide the center point and a diameter or radius. In some programs, the number of longitudinal and lateral segments is specified. Unlike a dish, a 3D dome extends above the current drawing plane. Refer to **Figure 8-21**. The center point lies on the current drawing plane. A dome is usually developed as a surface-modeled object.

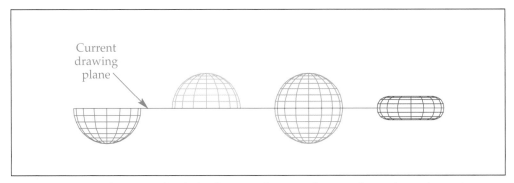

Current drawing plane

Figure 8-21. A side view of a dish, dome, sphere, and torus shows how these shapes are constructed in relation to the current XY drawing plane. Compare this view to the views in Figures 8-18 and 8-19. Notice the number of line segments assigned to each shape.

Both the dish and the dome are among the easiest of 3D objects to draw, but they are often difficult to edit. Both types of objects are normally used in conjunction with other 3D objects or in applications that require spherical constructions.

3D Sphere Construction

A sphere is created by specifying the same dimensions used for a dish and a dome. In some programs, the number of longitudinal and lateral segments is also specified. The center point of the sphere is located on the current drawing plane. Half of the sphere projects above the drawing plane and half projects below. Refer to **Figure 8-21**. If you want the sphere to sit on a specific elevation plane, you need to specify a Z coordinate equal to the radius of the sphere for the center or set a new elevation before drawing the sphere. A sphere can typically be created as a surface-modeled object or as a solid primitive.

3D Torus Construction

A 3D torus is shaped like a donut. It can usually be created as a surface-modeled object or as a solid primitive. To draw a torus, you must first specify the center point. The center point is located on the current drawing plane. Next, you must specify the overall diameter or radius of the torus. You must then provide the diameter or radius of the tube of the torus. With some programs, you must also specify the number of segments around the *tube* circumference and the number of segments around the *torus* circumference. The torus in **Figure 8-22** has 16 segments around the tube circumference and 10 segments around the torus circumference. As you assign more segments to the tube circumference, the arc curvature becomes smoother. However, as with all 3D objects, the size of the drawing database increases as well. You should always use the lowest number of line segments possible while still achieving a satisfactory appearance of the object.

Solid Modeling Applications

Solid modeling is very similar to surface modeling. The primary difference is the way in which shapes are defined. Whereas a surface model can be thought of as a representation with "skin," a *solid model* is a true solid representing the entire mass of an object. Solids created in a CAD system can be calculated for

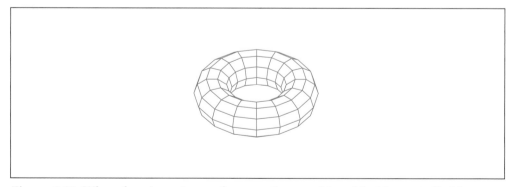

Figure 8-22. When drawing a torus, the smoothness of the object is controlled by the number of line segments specified for the tube circumference and the torus circumference.

mass and volume. This makes solid modeling very useful for manufacturing applications. Solids can also be shaded and rendered for presentation purposes, and they are typically easier to edit than surface models.

There are four basic ways to create solid models. Solids can be constructed using solid primitives, extrusions, revolved shapes, and editing operations. *Solid primitives* are very similar to the predefined surface-modeled shapes previously discussed. The solid primitives available in a typical CAD system include the box, sphere, cylinder, cone, wedge, and torus. As with surface modeling, these basic shapes can be used as building blocks. They are created by specifying parameters such as a center point or corner location and dimensions specific to the shape.

Extrusions are solids created from basic 2D geometric shapes. For example, a circle can be extruded to create a cylinder. As with surface models, solids can also be created from revolved shapes.

One of the most useful solid modeling methods involves the use of union, subtraction, and intersection operations with primitives and other solids. This makes it easy to construct composite models from a variety of solid shapes. Solid modeling methods are discussed in the following sections.

Creating Extrusions and Revolved Solids

There are two primary ways to construct solids from two-dimensional shapes. Shapes can be extruded or revolved. In a typical CAD system, extruded solids are created by first drawing a 2D shape as a closed polyline. A *polyline* is a continuous line made up of one or more segments. The segments can be straight or curved. Polylines and other closed shapes used for extrusions include rectangles, polygons, circles, ellipses, and splines.

After the 2D polyline is drawn, the shape is extruded into a solid. See **Figure 8-23**. In this example, the shape for the star is drawn in 2D. Then, the shape is selected and the thickness or "height" for the extrusion is specified. This is typically accomplished with the **Extrude** command. A two-dimensional shape can also be extruded along a path curve. In this instance, the path selected determines the direction and "height" of the extrusion. The path curve can be a polyline, line, circle, arc, ellipse, or spline.

Revolved solids are useful in a variety of modeling applications. A revolution is created by rotating a profile shape about a centerline axis. See **Figure 8-24**. The object to be revolved must be closed and can be a polyline, circle, ellipse,

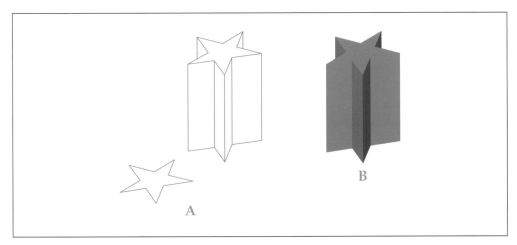

Figure 8-23. Creating an extruded solid from a closed two-dimensional shape. A—After the shape is drawn, the height for the extrusion is specified. B—The model after rendering.

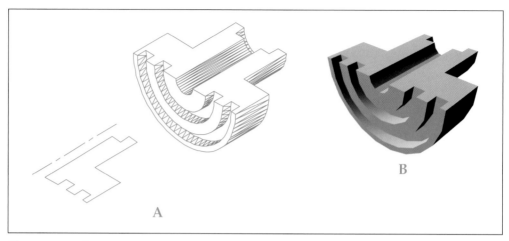

Figure 8-24. Creating a revolved solid. A—A closed shape and centerline axis are drawn for the revolution operation. A partial rotation of 180° creates the hub section. B—The model after rendering.

or spline. Revolved solids are typically created with the **Revolve** command. First, the 2D shape and axis of revolution are drawn. The axis can be a line, two pick points, or the X or Y axis of the current coordinate system. After the shape and axis line are selected, the angle of rotation is specified. This provides a great deal of flexibility. For example, you can create a section by specifying a partial rotation. In **Figure 8-24**, the polyline shape of the hub is revolved 180°. Had the shape been revolved a full 360°, the entire hub would have been generated. The revolve function is a powerful tool for creating complex solid models with a symmetrical shape.

Creating Composite Solids

Composite solids are combinations of solids created with solid editing operations. Models created in this manner are often constructed with the union and subtraction functions. The intersection editing operation can also be used to create a new model from existing solids. When creating composite solids, the geometric shapes are typically first drawn as primitives or extrusions. Then, they are modified with editing operations. For example, you can create several solid primitives with approximate shapes for a modeling application, orient them as needed, and then combine them to complete the construction. Advanced models can be constructed with solid editing operations. Although software programs vary, the union, subtraction, and intersection operations are most commonly used to create composite solids.

A *union* is created when two solids are joined to form a new solid. This can be thought of as welding two or more basic shapes together to create a more complex object. A *subtraction* occurs when the volume of one solid is removed from another solid to take away material. For example, solid cylinders are commonly used in subtraction operations to create hole features. An *intersection* is created when the volume shared by two solids is used to form a new solid.

A union operation with solid primitives is shown in **Figure 8-25**. These two primitives have the same radius value. The procedures for developing the cone and cylinder are discussed in the *Constructing 3D Objects* section of this chapter. In this example, the cone has a pointed top. To draw a solid cone with a truncated top, an editing operation must be used. After drawing the cone and cylinder, object snaps are used to move the cone to the top of the cylinder. Finally, the two primitives are joined into a single entity. Unioned objects are

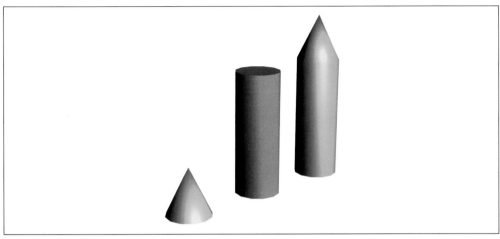

Figure 8-25. Creating a union from two primitives. After moving the cone to the top of the cylinder and selecting both objects with the **Union** command, the composite solid is generated automatically.

normally created with the **Union** command. When using this command, the primitives to be joined are selected. The new solid is then generated automatically.

As previously discussed, subtraction operations are useful for creating holes and other similar features. Subtractions are typically created with the **Subtract** command. After creating two primitives and orienting them as desired, one solid is subtracted from the other to create a new solid. See **Figure 8-26**. In this example, two box primitives are used to construct a box with a rectangular opening. The procedures for developing a box are discussed in the *Constructing 3D Objects* section of this chapter. The taller box has smaller width and length dimensions and is used as the drill object. The smaller dimensions are used to create the resulting wall thickness in the final object. After both boxes are created, the taller box is moved to the desired position inside the box where it will be subtracted. It is located at a higher base elevation to provide the desired base thickness. When using the **Subtract** command, you must first specify the object that will have material removed. The drill object is then selected, and the new solid is generated automatically. The drill object and the subtracted material are both removed from the view.

A very common type of subtraction operation is shown in **Figure 8-27**. In this example, one cylinder is subtracted from the other to create a hole feature. The cylinder shown in red is used as the drill object. It is drawn with a diameter

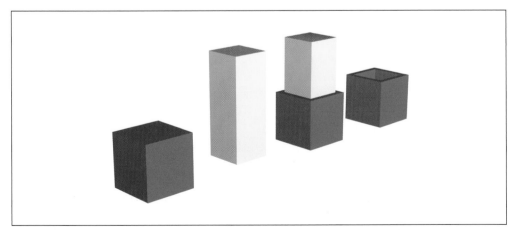

Figure 8-26. Creating a composite solid from a subtraction with two primitives. The taller box is subtracted from the original box to create the box at right.

that will result in the hole dimension. The cylinder is then moved to the desired location where it will be subtracted. Notice that the drill object is located so that the base extends through the bottom. This ensures that the hole will extend through the object. In applications where material will be left inside the object to be subtracted from, the drill object must be located at the correct elevation.

A subtraction operation using two composite solids and two wedges is shown in **Figure 8-28**. The composite solids shown in blue are unioned objects. Each object is constructed from a cylinder and a box. This shape will produce the rounded slot in the final object. Notice in this example that the drill objects are considerably larger than the wedges. As with the previous example in **Figure 8-27**, drill objects are usually created "taller" than the objects to be subtracted from to ensure that the desired material will be removed. The procedures for developing the wedges are discussed in the *Constructing 3D Objects* section of this chapter. After the composite solids are located inside the wedges, they are subtracted to create the final objects.

In some cases, you may find it useful to create a composite solid from the material shared by different solid objects. This is known as an intersection. See **Figure 8-29**. In this example, the cylinder and sphere are created as primitives

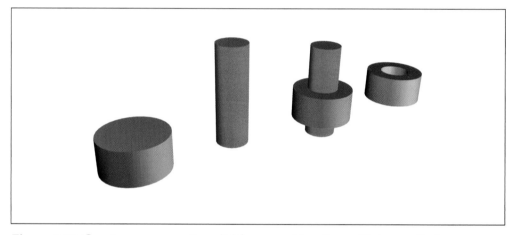

Figure 8-27. Creating a composite solid from a subtraction with two cylinders. Cylinders are commonly used to drill holes in solid modeling operations. Extending the drill object through the original cylinder ensures that the hole will extend through the final object shown at right.

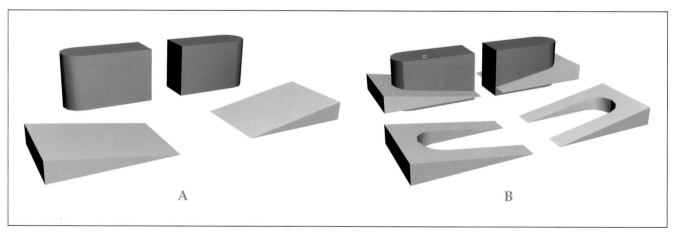

Figure 8-28. Creating a composite solid using other composite solids and wedge primitives. A—Each composite solid to be subtracted consists of a unioned cylinder and box. B—After the drill objects are located correctly, subtraction operations are performed to create the final objects.

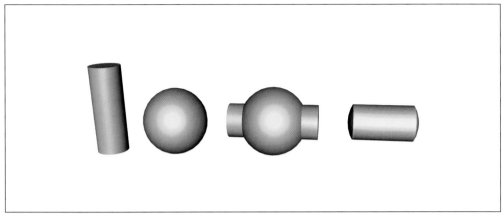

Figure 8-29. Creating an intersection from the shared volume of two primitives. After the basic shapes are constructed, they are overlapped to establish the intersection. They are then selected with the **Intersect** command to create the composite solid shown at right.

and overlapped. The common volume of the objects is then used to create a new cylinder with rounded ends. An intersection can typically be created with the **Intersect** command. After selecting the objects with common volume, the new solid is generated automatically.

The examples in this section are intended to give you a basic introduction to solid modeling methods. This overview does not address all of the modeling and editing tools available in some CAD systems. With practice, you will find that the methods discussed provide many different ways to create objects in 3D. When constructing a model, always consider which method will work best for your application.

Summary

It is important to understand the differences in the development of two-dimensional drawings, simulated three-dimensional drawings, surface models, and solid models. The major types of two-dimensional pictorials are axonometric, oblique, and perspective drawings. When these types of drawings are developed on a computer, they are simply 2D pictorials that give the illusion of three dimensions. They are developed using only X and Y coordinates. If viewed from any viewpoint other than the plan view, they do not project above or below the current elevation plane.

Simulated 3D drawings are created in a CAD system by adding a thickness or depth measure to objects in a two-dimensional view. When drawings are constructed in this manner, the third dimension can be seen after changing the viewpoint. Simulated 3D drawings are easier to visualize than 2D pictorials. However, they are not as easy to modify as surface-modeled or solid objects. Although they are adequate for representing relatively simple objects, simulated 3D drawings require considerable work to develop inclined features or views where hidden lines must be removed.

When constructing 3D drawings as simulated 3D drawings, surface models, or solid models, it is often necessary to use relative coordinate systems and special viewing functions. Changing from the default world coordinate system allows you to orient the drawing plane to specific features on an object. In many CAD systems, there are various viewing tools available to establish different viewpoints in a 3D drawing. These range from predefined orthographic and isometric views to viewpoint commands and dynamic viewing functions.

Surface modeling is a process of developing 3D objects by defining their exterior surfaces with mesh surfaces. There are several ways to construct surface models. Many CAD systems allow you to generate mesh patterns using edge defined surfaces, revolved surfaces, ruled surfaces, tabulated surfaces, and flat mesh surfaces. In addition, predefined shapes provided by software programs are commonly used to develop surface models. These objects can be quickly drawn by providing basic dimensions. Objects such as boxes, cylinders, cones, and spherical shapes can be drawn, located correctly, and combined into one model. The resulting surface model can then be shaded or rendered.

As with surface modeling, solid modeling methods allow you to construct 3D objects for shading and rendering purposes. However, a solid model is a true solid composed of actual material rather than "skin." The major solid modeling methods involve solid primitives, extrusions, revolved solids, and composite solids. Solid primitives are similar to the basic geometric shapes used to create surface models. They can be used as a starting point in the modeling process. Extruded and revolved solids are created from 2D-based shapes. Composite solids are combinations of solids constructed from common editing operations. These include unions, subtractions, and intersections.

When developing three-dimensional drawings with a CAD system, use the methods that best serve the end result of the application. It is important to use the tools that are available to you. While simulated 3D drawings can be developed quickly with basic drawing functions, primitive shapes and modeling tools offer many advantages to the illustrator. As you plan your drawings, always consider the tools and methods that will be the most effective in helping you complete each project.

1. Explain how a two-dimensional drawing differs from a three-dimensional model.
2. What are two limitations in relation to simulated 3D drawings created from objects with thickness?
3. What is the purpose of the **Hide** command?
4. Define *world coordinate system*.
5. What is the purpose of a relative coordinate system?
6. Describe how the coordinate system icon appears in a 2D view when the world coordinate system is active. How does the icon change when a 3D view is selected?
7. List the views that are typically available as preset views in a CAD system.
8. What is a *viewport*?
9. When establishing a viewpoint with the **Viewpoint** command, what two angles must be entered to define the viewing position when using the rotation method?
10. Define *orbit viewing*.
11. What is a *surface model* and how does it differ from a solid model?
12. How many edges must be used to define the boundaries when constructing an edge defined surface?
13. What two objects are required to create a revolved surface?
14. What is a *ruled surface*?
15. What type of surface is generated from a surface outline along a direction path?
16. What three specifications are required to draw a surface-modeled or solid primitive cylinder?
17. How can you draw a surface-modeled or solid primitive cone with a pointed top?
18. What is a *tetrahedron*?
19. Describe the procedure used to create a 3D mesh.
20. By default, where is the center point located when a 3D sphere is created?
21. What specifications are required to draw a surface-modeled or solid primitive torus?
22. List four basic ways to construct a solid model.
23. What two shapes are required when using the **Revolve** command?
24. What are *composite solids*?
25. What is a *subtraction* and how is one created?
26. What type of object is created from the volume shared by two solids?

Drawing Problems

Using the techniques discussed in this chapter and the orthographic views shown, develop the following problems as three-dimensional models. Use surface modeling or solid modeling methods and primitive shapes when suitable. For each problem, use the dimensions provided and orient the object to create the most realistic representation. Use viewing tools, viewports, and user-defined coordinate systems to your advantage. Select the best modeling method depending on the object and the time available.

1.
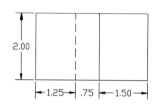

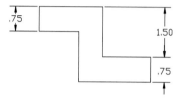

2.
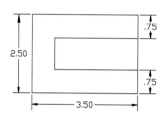

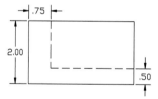

3.
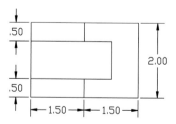

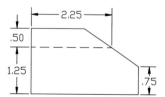

4.

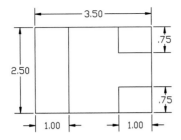

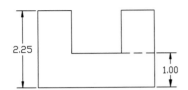

5.

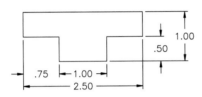

6.

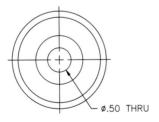

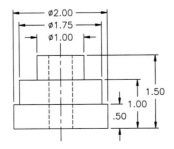

7.

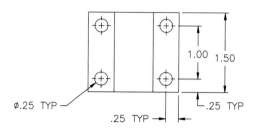

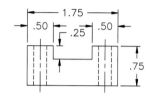

8.

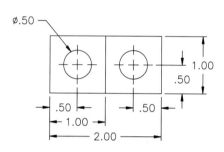

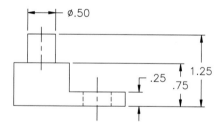

9.

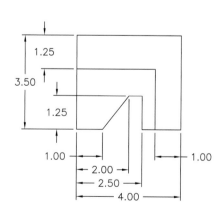

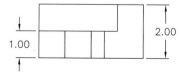

10.

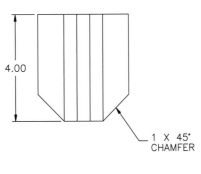

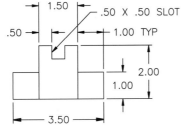

11.

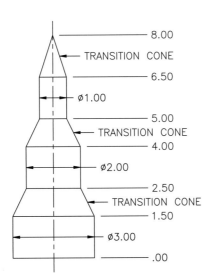

12.

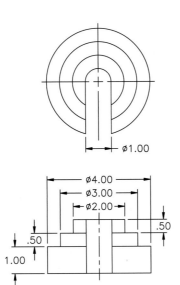

Illustration Enhancement

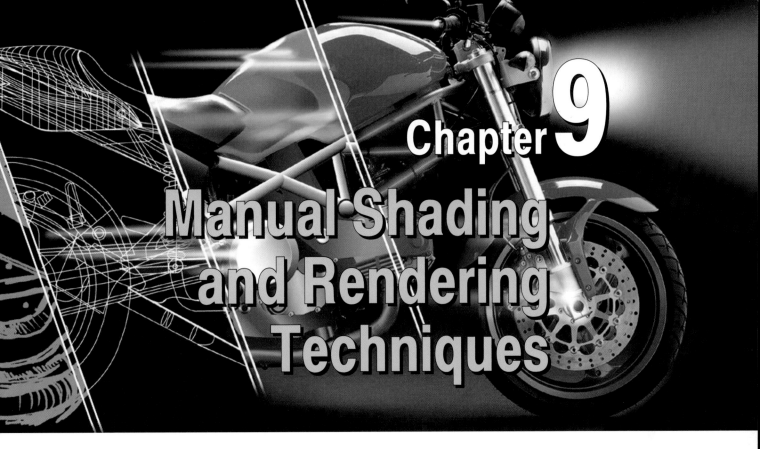

Chapter 9

Manual Shading and Rendering Techniques

Learning Objectives

At the conclusion of this chapter, you will be able to:

☐ Identify the primary applications of manually shaded illustrations and renderings.

☐ Describe the processes of line contrast shading, line separation shading, smudge shading, stipple shading, line surface shading, and appliqué shading.

☐ Identify the principal image-producing materials and media used in shading.

☐ Develop accurately drawn shadow effects appropriate to different light sources and object shapes.

☐ Describe the steps of procedure typically applied in developing a manual rendering of an object.

Introduction

Most pictorial drawings without shading or rendering are not realistic enough for the general viewer to understand. Shading and rendering add substance to surface features of an object to provide the illusion that the surface features can be seen.

This chapter focuses on manual shading techniques. The common materials, techniques, and principles used to develop shading are discussed. The primary applications, advantages, and disadvantages of manual shading techniques are also covered. In addition, an overview of the rendering process is presented. Chapter 10 introduces computer shading and rendering, which builds upon the principles presented in this chapter.

Shading Techniques

Shading is the process of adding darkness or color to an area of a drawing to produce a visual representation of a surface. Simply adding a basic surface color and a shadow effect can cause the surfaces of an object to appear much more three-dimensional than an unshaded pictorial, **Figure 9-1**. A theoretical light source is focused on the object and surface textures are added to depict areas of high light intensity, diffused light, shade, and shadow.

A *rendering* is a drawing or graphic representation that has the highest degree of realism possible. The term is also applied to the process of creating a rendering. A rendering is typically produced in color and normally includes background scenery to enhance realism and add a sense of scale to the object. Due to their complexity, manual renderings can become very time-consuming and expensive to produce. Therefore, manual renderings are reserved for special full-color applications.

There are two main types of manual shading. These are line shading and surface shading. Each type can be further divided into several shading techniques. Each technique has its own advantages and disadvantages. Line shading and surface shading methods are discussed in the following sections.

Line Shading Techniques

Line shading is a special treatment for the object lines of a pictorial to produce shadows and help define shapes. Line shading gives the drawing a

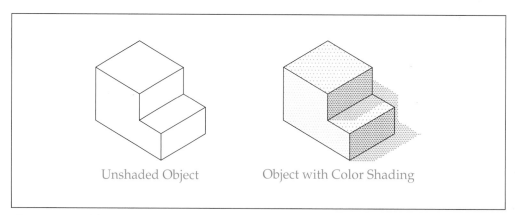

Unshaded Object   Object with Color Shading

Figure 9-1. Adding basic color and shadow to a drawing creates a far more interesting image and a greater three-dimensional effect.

more three-dimensional look than a standard line drawing. Although not as effective as most surface shading techniques, line shading is the quickest and easiest of the manual shading techniques.

There are two different techniques classified as line shading—line contrast shading and line separation shading. Complex objects should have a combination of both of these shading techniques.

Line contrast shading

Line contrast shading uses object lines of varying thickness to create an impression of different surfaces. The further an object line is from a light source, the thicker it is drawn, **Figure 9-2**.

The traditional location for a light source in illustration is above and to the left of an object. Light rays representing solar illumination are normally considered to be parallel to each other and strike an object at a 45° angle. Line contrast shading on an object with a light source in the traditional position is shown in **Figure 9-3**. Notice that three different line widths are used to provide contrast. These widths are thin, medium, and thick. Interior object boundaries are generally drawn with thin lines. Exterior boundaries closest to the light source are drawn with medium width lines. Thick lines are used for exterior edges on the shaded sides of the object. Areas inside the holes are also shaded to enhance the effect of the overhead light source.

Different size leads in mechanical pencils, drafting pencils sharpened to different points, or different size technical pens are used to create the varying line thickness. Since printing methods vary for technical illustrations, drawings must have good line quality on a clean, high-contrast drawing media. Therefore, using technical pens and plastic film or vellum is a good way to create line contrast drawings.

Many cylindrical and curved shapes are difficult to shade using line contrast shading with three different line widths. Unlike normal rectangular and inclined surfaces, curved objects often use a continuous curved line to represent interior and exterior boundaries and surface limits. You would need to start an exterior surface limit with a thick line width and then carefully transition to a thin width as the curved line evolves to an interior line. This is especially difficult when using technical pens and pencils to develop the drawing. Regardless of this difficulty when using line contrast shading on a cylindrical or curved surface, the curvature of the cylinder or curve should be exaggerated in some manner. This helps viewers know that they are not looking at a flat surface.

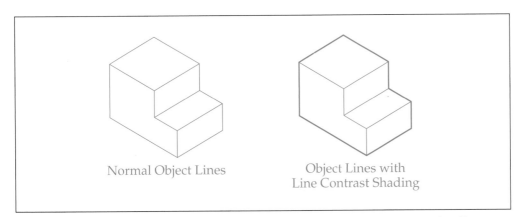

Normal Object Lines

Object Lines with
Line Contrast Shading

Figure 9-2. When using line contrast shading, line thickness increases as the distance from the light source increases.

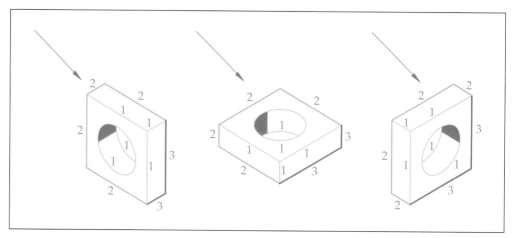

Figure 9-3. Examples of line contrast shading. The line widths are numbered to indicate line thickness. Light is traditionally located above and to the left of the object, as indicated by the arrows.

A series of objects containing cylindrical shapes with line shading is shown in **Figure 9-4**. There are two types of cylindrical shapes. Positive cylindrical shapes project out of the object, while negative cylindrical shapes project into the object. Positive cylinders have a wide shadow line at the darkest portion of the curvature, a narrow unshaded strip to represent reflective light, and then a thinner band of shading or a series of lines with decreasing width. The three objects in **Figure 9-4** have positive cylinder shading. Shading lines generally should follow the major axis of the cylindrical shape unless the light source is established at a location that changes shadow locations.

For negative cylinders, approximately one-fourth to one-third of the cylinder is darkened on the shaded side of the feature. In **Figure 9-4A**, this type of shading is used for the hole feature. In **Figure 9-4B**, the recessed feature is shaded in this manner.

Line shading is also commonly used to give threaded parts a three-dimensional appearance. The object in **Figure 9-4C** uses curved lines projecting "into" the cylinder to illustrate the threads. Notice the slight sawtooth edge added to the lower shading. This is a common effect used to help enhance the appearance of threads. Erasing a strip along the middle area of the threads also gives the illusion of reflected light.

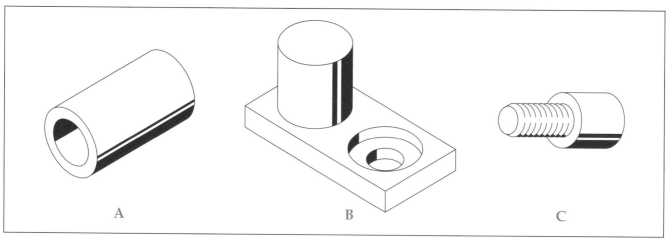

Figure 9-4. Examples of line shading for cylindrical objects. A—Positive cylinder shading is used for the exterior surface. Negative cylinder shading is used for the interior. B—Narrow shading strips representing negative cylinder shading are used to define the recessed feature. C—Line shading is used to enhance the appearance of the threads.

Line separation shading

Line separation shading uses slight gaps at the end of object lines to define features that lie behind other features in order to show depth. See **Figure 9-5**. Line separation shading can be used with numerous other shading techniques to transform two-dimensional pictorials into images that appear three-dimensional. However, line separation shading alone does not provide a complete 3D appearance. It is normally combined with other techniques to add realism. Line separation shading is easy to learn. It does not require any additional drawing time, and no additional illustration materials are required. This technique is primarily used for objects with complex shapes.

Surface Shading Techniques

Surface shading is the application of special highlights or textures to an object's surfaces. There are four major surface shading techniques used in manual illustration. These are smudge shading, stipple shading, line surface shading, and appliqué (transfer) shading. Surface shading differs from line shading in that shading effects are applied to entire surfaces. The advantage of surface shading techniques over line shading techniques is the added realism. Surface shading makes it easy to visualize an object as three-dimensional. Varying the density of surface shading can create the visual effects of highlights, shadows, and reflections.

Always start with a quick sketch to visualize how the object will look with surface shading. This sketch may identify problems with the style of shading chosen. It may also show that a combination of shading techniques is the best presentation. A sketch allows you to try a variety of ideas before starting on the actual illustration.

When applying shading, always start light. It is easier to add more shading later than to remove too much shading. As you continue to touch up surfaces, the overall illustration tends to get darker. Also, it is very difficult to avoid running the smudged material over exterior object lines as you try to blend the material into smooth transitions of density. Therefore, use an erasing shield and a soft eraser to clean up edges and create crisp definitions before final darkening of

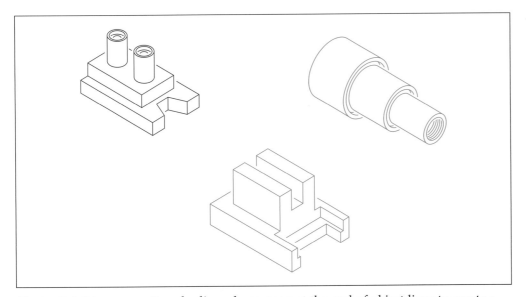

Figure 9-5. Line separation shading places gaps at the end of object lines to create a three-dimensional effect.

object lines. Art gum erasers work well for graphite and plastic lead. However, they have a tendency to streak with chalk or charcoal. Therefore, a kneaded rubber eraser should be used for chalk or charcoal.

Always be careful to keep your drawing hand from touching and streaking drawing material. Many drafters and illustrators tape a protective cover sheet over completed areas. When using ink, it must be allowed to dry completely before placing a cover sheet over it. Also, never spray a fixative or lacquer protective coating over chalk or charcoal shading until you are absolutely certain the illustration is finished. Once the protective spray has been applied, you will not be able to touch up any imperfections in the shading or make modifications to the drawing.

Smudge shading

Smudge shading is applying a drawing material to the media (paper) and smearing it to create a softly shaded or textured effect. Practice on scrap paper first. Once you have mastered the technique, you may want to use chamois to smudge the shading material. A *chamois* is a piece of very soft leather. Another blending aid useful in smudge shading is a *blending stump.* Blending stumps are double-ended soft felt pads. Their edges can be cleaned and sharpened with a sandpaper pad. Blending stumps come in small, medium, and large sizes.

To create smudge shading, use drawing strokes parallel to a horizontal or vertical axis to apply the shading material. If the object is cylindrical, apply the drawing material in strokes parallel to the centerline of the cylinder. In the event the surface has a compound curve such as a sphere, dome, or similar shape, it is preferable to make the application strokes in a series of curves parallel to the surface boundary.

After the shading material is applied, individual lines of the material are smoothed and blended. Use tissue, cotton, or another type of blending pad to make light strokes running parallel to the original strokes of the shading material. After a few passes, the smudging pad will pick up enough of the material to fill in lighter areas between the original application strokes.

Once the area is blended to a smooth or uniform density, any areas that should be darker can have a slight touch of drawing material added with the smudging pad. Applying the shading material directly tends to make the area too dark. Using the smudge pad applies less material and allows better control of the density and blending during final touch-up.

An illustration with smudge shading applied is shown in **Figure 9-6**. Notice that this type of shading can create a three-dimensional effect *and* add texture in the same application. Rough drawing paper or illustration board can also give the impression of surface texture on an object. The direction of the strokes when applying the shading material can also suggest shape. For example, a sphere is drawn with darker strokes near the surface boundaries. Then, lighter and lighter drawing strokes are used to taper the shading density until a nonshaded area is created. This is the spot of highest light intensity.

Stipple shading

Stipple shading is applying a pattern of dots to a drawing to create shadows or surface textures. Dot patterns give the illusion of shadows created from tiny bumps and depressions in an object's surface. Stipple can also represent an area shaded from direct illumination. As a shadow changes from light shading to absolute darkness across the face of a surface, the stipple pattern should increase in density. Stipple patterns are normally added to a drawing

Figure 9-6. In smudge shading, material is applied to emphasize smooth transitions between areas of density and highlight.

before final darkening of object lines. In some cases, stipple patterns may replace the object lines on rounded edges or in areas of high reflectivity. See **Figure 9-7**.

Stipple patterns are usually produced with a pencil or fine-point technical pen. Pencils should be sharpened to a slightly rounded conic point. A medium hard lead between H and 3H is recommended. Inks are available in a variety of colors and levels of opacity. Black inks tend to be completely opaque and provide excellent reproductions. In contrast, many of the colored inks are translucent and produce soft, faded images that may not reproduce well. Technical pen points produce both consistent size and density in the ink dots. For this reason, some illustrators prefer to use pencil so they can intentionally vary the size and density of the stipple dots. Varying the density of the pattern can replace normal object lines creating object boundaries, represent surfaces in heavy to light shade, or generate the loss of surface edge visibility when the edge is supposed to be reflecting from a high-intensity light source.

When creating stipple shading, sit in a relaxed and natural drawing posture. The pencil or technical pen should be held in a vertical position as you lightly tap the point on the drawing surface. Stipple shading is an effective, but very

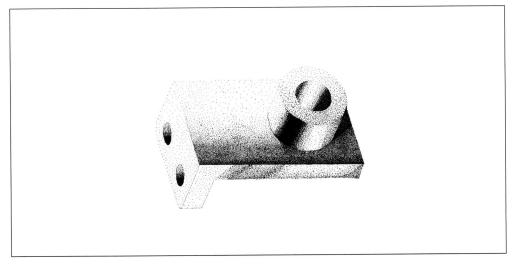

Figure 9-7. In stipple shading, patterns of dots are applied to create various shading effects. In this example, stipple dots are used for object lines.

tedious, process. Inconsistency of dot patterns is one of the major problems. As your wrist begins to tire, or when the process becomes boring, you will probably see either the pattern getting too dense or pockets of fading appearing in the pattern. The best remedy for this problem is to only stipple shade for 15 to 20 minutes at a time. Then, do some other less repetitious drawing task, exercise your hands and wrists, or find another productive activity for at least five minutes. Also, avoid developing either a steady rhythmic or musical effect when stippling. The tapping sound of stippling is monotonous, but your concentration and pattern consistency will be improved if you force yourself to tap the stipple pattern to the profiles and density changes of the surfaces rather than to a musical beat. Remember, you should develop a stipple-shaded drawing with a light touch. Then, you can add more stippling to develop more contrast among the various surfaces of the object.

Line surface shading

Line surface shading is drawing lines, or line patterns, on object surfaces to create the appearance they are made of a solid material. Some line surface shading is done with crosshatch patterns. Therefore, this method is sometimes called *crosshatch shading* in industry. The process is relatively simple and has considerable flexibility. Surfaces can be made either plain or detailed.

A variety of line types, line patterns, line angles, and pattern densities can be used to simulate surfaces. See **Figure 9-8**. Notice that just a little shading can effectively improve the visualization of an object. Lines drawn on the surfaces are generally parallel to the major axis of the surface. Large areas can be effectively line shaded using an Ames or Braddock lettering guide to establish uniform line spacing. These devices are also useful when you want to skip a line in an area of highlight or reflection.

You can also approximate object features and profiles using line surface shading. For example, fillets and rounds are shaded in several ways in **Figure 9-9**. Fillets and rounds with small radii should have a large number of ellipse

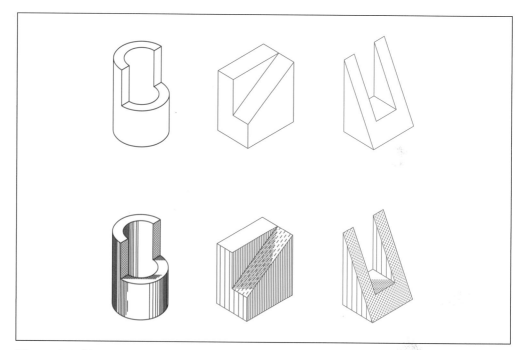

Figure 9-8. Line patterns and densities are varied to create different shading effects in line surface shading.

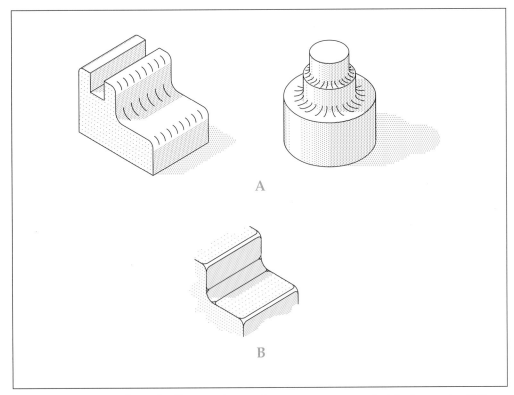

Figure 9-9. Line surface shading can be used to represent rounded features. A—Ellipse segments are drawn to represent the different size radii for fillets. B—An alternate method uses lines rounded at the ends and gaps to define the filleted areas. This is a quicker method, but the results are less realistic.

segments representing the curve. Fillets and rounds with large radii should have just a few ellipse segments to show the curve. The segments are drawn one-fourth in size. Refer to **Figure 9-9A**. An alternate method of using line surface shading to represent fillets and rounds is shown in **Figure 9-9B**. This method is quicker than using the radius method with ellipse segments. However, it is not quite as realistic in appearance.

Appliqué shading

Appliqué shading or *transfer shading* is adhering images from a printed transfer sheet to a surface in the drawing. *Appliqué sheets* are commercially prepared, transparent sheets with an adhesive backing. When shading images are removed from a sheet and applied to the drawing, the adhesive side sticks to the drawing sheet. Appliqué sheets contain printed lines, patterns, or colors. Common objects used in illustrations, such as nuts and bolts, are also available on appliqué sheets. A pictorial with appliqué shading is shown in **Figure 9-10**.

Another form of shading very similar to appliqué shading is accomplished using transfer sheets. *Transfer sheets* have shading patterns, lettering, or artwork mounted on the bottom of a transparent sheet. Each image has a wax coating on the bottom. The wax coating adheres to the drawing sheet when the image is transferred. To use a transfer sheet, the image is positioned in the desired location. Then, a burnishing tool is used to rub the image onto the drawing. As the top of the sheet is rubbed, the wax on the bottom adheres to the drawing sheet and the image is transferred. The term *transfer shading* is sometimes used to describe appliqué shading. Although true transfer shading is still sometimes used in illustration, most transfer shading is done with appliqué sheets.

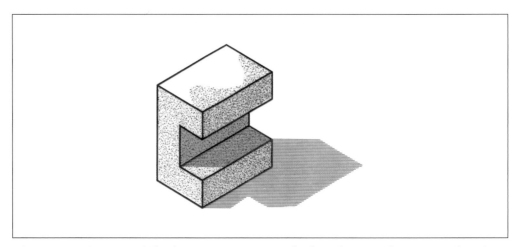

Figure 9-10. Preprinted shading patterns are applied to object surfaces in appliqué shading.

Draw all object lines in their final density before using appliqué shading. Once the appliqué is adhered to the drawing surface, use a razor knife to trace along the object lines. This cuts out a piece of the sheet the exact shape of the area to be shaded. Use light pressure when cutting so that you do not cut through both the appliqué sheet and the drawing sheet. A straight blade should be used for straight object lines. For curved object lines, use a swivel knife. A *swivel knife* is a razor knife with a blade that can be rotated in the knife handle. Be sure to follow the correct cutting line on the first pass of the knife. Trying to repair a miscut often leaves a visible blemish on the drawing. After the contour is traced with the razor knife, the excess appliqué sheet is peeled away. Then, the remaining sheet is gently rubbed down to firmly attach it to the drawing. Rub the sheet from the center outward to remove any air bubbles under the sheet. Use gentle pressure to prevent the appliqué sheet from shifting or stretching.

Stipple, line, and crosshatch patterns are available on appliqué sheets in many different sizes and densities. If one appliqué does not produce a desired effect, you may need to use additional layers of appliqué sheet on top of other layers to customize the shading. Also, there may be instances when it is necessary for one pattern to stop and another to immediately begin. If one edge is simply butted against the edge of another sheet, gaps that are left may be visible on the finished drawing. In this situation, gently adhere one of the appliqué images over the other with approximately 1/8″ to 1/4″ of overlap. Then, cut through both sheets along the middle of the overlap area using a straight razor knife and a steel straightedge. Peel back the cut edge of the overlapping sheet to remove the two strips of scrap. The overlapping sheet should then roll into place on the drawing and create an invisible butt joint with the other sheet.

Shading Media

There are many common materials used for surface shading. These include plastic and graphite-based pencil leads, colored chalks (pastels), charcoal, inks and dyes, watercolors and washes, paints, and appliqué transfers. Most of these materials have specific surface shading applications. For example, colored chalks and charcoal are used almost exclusively for smudge shading. Inks and dyes, watercolors and washes, and paints are primarily used to produce color renderings. There are also many different types of drawing sheets available. These include vellum, plastic film, paper stock, illustration board, and canvas. The type of drawing sheet used depends on the type of shading material used. An overview of these materials is presented in Chapter 2.

Shading Materials

Graphite pencils are inexpensive and readily available. They erase easy and blend smoothly for smudge shading. See **Figure 9-11**. Also, the different grades of hardness provide many line characteristics and densities. However, graphite-based lead may not be opaque enough and may therefore reproduce poorly. Also, soft and medium-grade graphite leads can smear and streak. Colored plastic leads have these same advantages and disadvantages. In addition, except on plastic drawing film, completely erasing a line may be difficult.

Charcoal and colored chalks blend easily and produce excellent smudge shadings. A single heavy stroke of charcoal or chalk can be blended from a solid, opaque density to a very light shading with just a few smudge strokes. The small particle size leaves a fine grain coating that can appear extremely smooth on slick finish paper, or as a coarse texture on a rough textured illustration board. Darker tints of charcoal tend to provide fairly good reproduction quality. However, lighter charcoal tints and some pastel-colored chalks do not have the intensity to photograph and reproduce well. Also, these materials smear easily.

Black waterproof ink on a white background can produce outstanding reproduction quality. See **Figure 9-12**. Once dry, waterproof inks on paper media are permanent and will not streak or smear. However, they are also very difficult to modify. Water soluble inks, dyes, and watercolors can be applied to a drawing with varying density and color intensity. Their application can range from opaque to transparent. If needed, water can be applied with a brush to soften and blend distinct brush strokes using a wash technique. Water soluble shading materials can usually be given a finishing touch-up even after their initial drying.

Paint is normally used to obtain a very specific color, provide opaque values not attainable with other colored drawing materials, or to give the illustration an elegant artistic image for presentation to an audience. Most renderings and illustrations with paints that are not water soluble are used for architectural applications or for sales brochure artwork where realistic color is more important than time and cost.

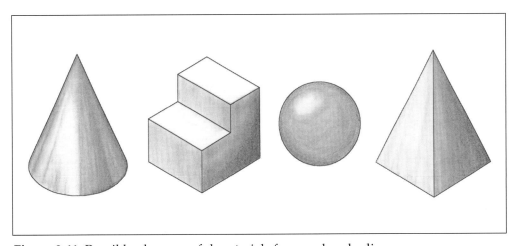

Figure 9-11. Pencil leads are useful materials for smudge shading.

Figure 9-12. An example of high-quality manual illustration work drawn with ink. The exceptional shading detail in this black ink rendering produces a photorealistic effect.

Drawing Sheets for Shading

There is an extremely wide range to select from when using drawing sheets for shading and rendering. Vellum and plastic film are used for line shading. Surface shading is generally done on a specific type of sheet stock designed to accept the drawing material. For example, a textured 100% rag content paper can accept charcoal and chalks quite well. However, if ink is applied to this paper, it is absorbed by the rough texture and a blurred image is created. Special paper designed to accept watercolor is available with a smooth finish or rough texture. Illustration board can accept most shading materials if the texture and finish are matched to the material. Inks, ink washes, and paints are applied to canvas, illustration board, and some types of paper sheets heavier in weight. Drawing sheets are available in weights ranging from 140-lb. (light) to 300-lb. (heavy). In general, the *weight* of a drawing sheet refers to its relative strength.

Shadow Techniques

Shading and shadowing are really the same thing. However, shading is used to describe a visual treatment applied to the surface of an object to distinguish surfaces. *Shadowing* refers to darkening an area of an object that has blocked illumination from the light source.

It is important to distinguish whether the illumination of an object is from solar light or an artificial light source such as an electric bulb. Solar light rays are considered to be parallel. Shadows created by an object blocking parallel light rays recede from the object along a parallel axis. Artificial light sources, however, project light rays in a cone of illumination. This cone causes shadows to recede from the object at nonparallel angles, **Figure 9-13**. The majority of shadows developed in manual technical illustration are based on solar illumination.

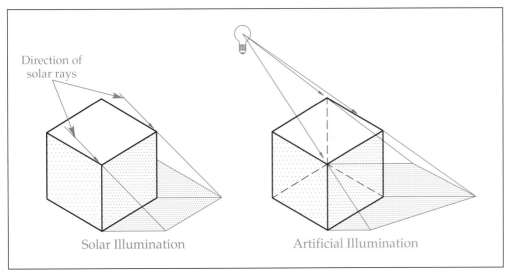

Figure 9-13. Shadows created from natural solar illumination project along parallel axis lines. Shadows produced from artificial sources of illumination project along nonparallel axis lines.

Shadows from Solar Illumination

Typically, solar rays are drawn passing across the top front edge of the object at a 45° angle. Any vertical or steeply inclined surfaces of the object below that edge are shadowed. The surface the object is sitting on also has a shadow created by the object blocking rays from the light source. Notice in **Figure 9-13** that shadow corners on the solar-illuminated object are located at the intersections of a horizontal line from each base corner and lines drawn at 45° angles from each of the nearest top corners. Always start laying out shadows by drawing a horizontal line from the bottom nearest corner corresponding to a top edge. Draw this line to the side of the object opposite the light source.

A general rule is that vertical edges on an object cast horizontal shadow edges when they block solar illumination radiating at a 45° angle. Referring to **Figure 9-13**, after drawing the horizontal line from the bottom corner, a line is drawn at a 45° angle down from the nearest corner above the origin of the horizontal line. The intersection of the two lines represents the shadow location for the upper corner where the 45° projection line began. This procedure is repeated for the top-right corner of the surface. Then, a line is drawn from the first shadow corner to the second shadow corner. This is the shadow cast by the top edge. Finally, the shadow cast from the upper-rear corner is located. This point should be along a line that is drawn from the last shadow corner located and parallel to the top edge that creates the shadow.

Shadow techniques for cylinders are identical to those for other objects, except that centerline layout is used to locate object features. Examples of locating shadows from a solar light source are shown in **Figure 9-14**. Principal layout points are labeled with numbers or letters and the corresponding shadow locations are labeled with a letter *S* subscript. The top edges and bottom corners of the objects determine the maximum projection of shadows. Horizontal lines from the bottom corners determine how low or high the shadow lies. The light rays passing over the top determine how far out from the object the shadow will project.

As natural light rays travel in parallel paths, shadows are cast parallel to the features that create them. In other words, shadows will have the same basic

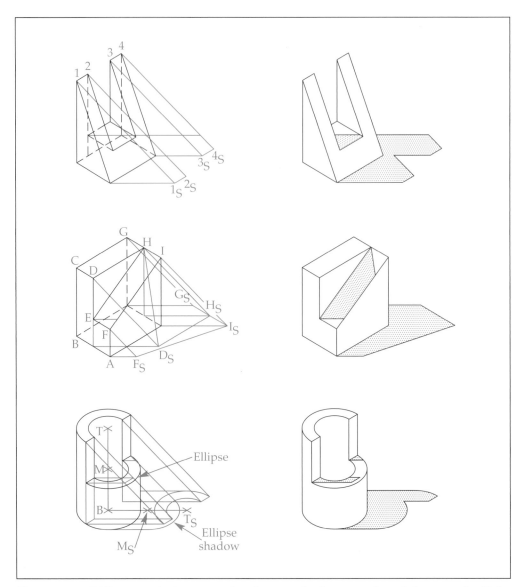

Figure 9-14. Different techniques used to lay out shadows produced from a solar light source.

shape as the features that are blocking the light rays. The object at the bottom of **Figure 9-14** creates a duplicate of itself in shadow form except for the portion of the shadow that is blocked by another shadow. Although the C-shaped lip at the top of the tube creates a shadow, the elliptical shape at the middle of the tube projects further onto the base shadow field. As a result, the C-shaped shadow is mostly hidden by the ellipse shadow.

Shadows from Artificial Light Sources

As previously discussed, light rays from an artificial light source are projected in a cone rather than along parallel lines. As a result, shadows project along nonparallel lines. The first step in creating a shadow from an artificial light source is to locate the light. The closer the light source is to an object, the wider the cone of the shadow. However, if the light source is too far from the object, it essentially creates solar illumination. Also, the size of the object, scale of the drawing, and size of the drawing board impact the location of the light source.

The shadowing procedure outlined in this section creates an approximation that works in most situations. Shadows can be located precisely using descriptive geometry or trigonometry techniques. However, these procedures are not covered in this text. When locating the light source, the light is typically oriented along a line drawn at a 30° angle through the center of the base of the object. This allows you to use an isometric ellipse template to transfer height measurements to the shadow projection.

After locating the light source, draw lines from the center of the light through each of the base corners nearest to the light. See **Figure 9-15**. If the object is cylindrical, draw the lines from the center of the light through the points of tangency on the base ellipse. These lines establish the cone of the shadow. Next, establish the length and final shape of the shadow. Usually, the length of the shadow is equal to the height of the object. Length can be established by direct measurement, by projecting a 45° line from the top edge to the surface the object is sitting on, or by using an isometric ellipse template to measure. Direct measurement should be used on a projection line from the center of the light to the center of the object. A 45° projection line should intersect the same center-line projector. If you are using an isometric ellipse template, begin by locating the center of the ellipse on the intersection of the 30° line from the light source and the nearest base line. Refer to the rectangular objects in **Figure 9-15A**. In **Figure 9-15B**, the center of the ellipse is aligned with the center of the base ellipse. Select an ellipse size nearest to the vertical measurement of the object. Then, mark where the ellipse intersects the 30° line on the opposite side of the object from the light source. The final step in creating the shadow is to draw the rear edge. This edge typically appears as an oversize outline identical in shape to the tallest surface edges on the object. In **Figure 9-15B**, the rear intersection of

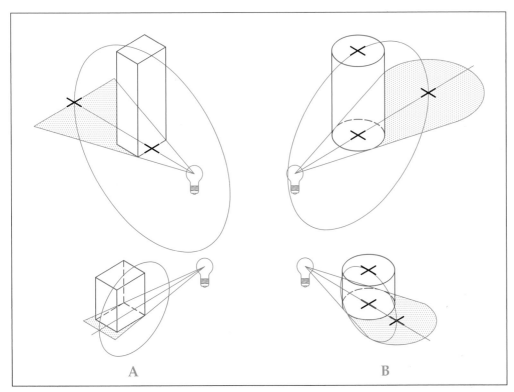

A   B

Figure 9-15. Drawing shadows for objects illuminated by artificial light sources. A—Projection lines and isometric ellipses are used to lay out the boundaries of the shadow. Object height determines the length of the shadow. B—Shadow layout for cylindrical objects. The arc outline forming the rear edge of the shadow is created by locating the center point and drawing arcs from the edges of the lines forming the cone.

the ellipse with the 30° line from the light source is used to locate the center of the arc outline. The shadow for each object is completed by using the measurement from the center mark to the outer projection lines to draw the arcs.

Be sure to verify whether hidden corners of the object create visible shadow lines. Referring to **Figure 9-13**, for example, the rear corner of the solar-illuminated object creates a visible shadow. In **Figure 9-15**, shadows are created by the artificial light source located in front of each object. Notice how the height of each object affects its shadow.

Other Shadow Elements

Additional shadowing techniques may be used to add more realism to illustrations when necessary. Two techniques used for added effect when creating shadows are shadow fill and shadow softening. See **Figure 9-16**.

Shadow fill is using a solid fill or darkening effect on surfaces not illuminated by either direct or reflected light. Recessed cavities, holes, and other surface features that are not illuminated should be shadow filled as the darkest surfaces on the object. Then, the remainder of the object is rendered with decreasing densities of shadows. Refer to **Figure 9-16A**. The vertical surface within the slot is darker than the surfaces closer to the viewer. Also notice that the shadow fill is even darker than the shadow cast by the object.

Shadow softening produces shadows that transition smoothly from dark shading to lighter shading. This produces a very realistic shadow effect. Notice in **Figure 9-16A** that the density of the stipple pattern decreases across the horizontal and vertical surfaces as the stipple dots approach the light source. This is an example of shadow softening. In **Figure 9-16B**, the shaded cylinder uses line surface shading to achieve shadow fill and shadow softening. Partial shadow fill is first drawn on the area inside the cylinder. Then, shadow softening is applied by gradually widening the gaps between the vertical surface shading lines. Gaps are left intentionally to represent highlight areas from reflected light.

Surface shading should incorporate shadow softening wherever possible to add realism to the object. However, too much uniformity in the transition from dark to light can detract from the overall visual effect. For best results, incorporate these slight shading irregularities in areas near the transition zones where little shading dissolves into areas of no shading.

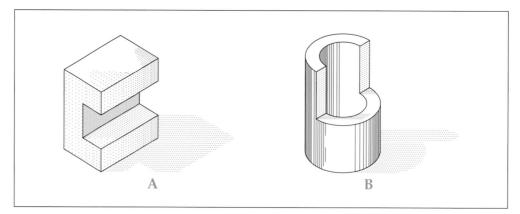

Figure 9-16. Applying shadow fill and shadow softening to add realism to the drawing. A—The vertical slot is shadow filled to highlight the area with less illumination, while softening is used to decrease the density of the stipple dot pattern closer to the light source. B—Shadow fill and shadow softening are used with line surface shading to highlight the interior portion of the object.

Introduction to Rendering

A rendering is a color illustration in pictorial form with a variety of effects applied to show a finished product. The term *rendering* is also used to describe the rendering process. Most manual renderings are produced by artists and illustrators who specialize in painted or inked media. Rendered illustrations are among the most expensive of shaded drawings because of the time required. An illustrator developing a rendering must draw the pictorial, paint or ink the background, the object, and the foreground, and finish the process by mounting and framing the rendering. Although renderings are used in civil and mechanical design applications, they are most often used in architecture and interior design. If an illustration is produced to aid in simple color evaluation, an artistic watercolor, inked painting, or airbrushed painting may provide an adequate and less expensive color comparison. However, when preparing a very important illustration of an expensive product for presentation to clients or potential customers, a complete color rendering may need to be developed. In many cases, renderings are now produced with the use of computer-aided drafting software. Computer rendering techniques are discussed in Chapter 10.

Rendering Applications

The most common types of renderings are those used by architectural firms to show interior or exterior building designs. See **Figure 9-17**. Architectural renderings are usually done to show clients what the final construction will look like. Gaining acceptance of a visual concept by an audience is often the most important function of a rendering. Renderings may also be used to advertise. These usually show what a product under development will look like when produced and on the market. Some engineering decisions are made based on the appearance of a few equally acceptable solutions. Renderings can

©2002-Engle Homes; Phoenix, AZ

Figure 9-17. Renderings are designed to show the detail of a final product and are typically used in architectural applications. (Engle Homes)

show how possible solutions differ visually and can assist in the decision-making process.

Manual Rendering Techniques

Although some manual renderings are done with paint on canvas, most are created with colored paint, watercolor, or ink applied to illustration board. The illustration board usually has a smooth or slightly textured finish.

Painted renderings tend to have very realistic colors and an excellent three-dimensional appearance. Layers of paint help "lift" surfaces away from other surfaces painted with a single layer. Another advantage of using paint is the ability to mix colors and create an almost infinite range of colors and shades. Texture differences and subtle changes in color can combine to enhance the illusion of a three-dimensional shape.

Watercolor and inked renderings generally do not produce the high contrasts found in painted renderings. They are often used when the goal of the rendering is to produce a visual effect of soft tones and low contrast. A watercolor or inked rendering may also be used to develop an artistic conception. Painted renderings can almost appear as realistic as color photographs.

The importance of the rendering serving the visualization needs of the audience must be balanced with the goals of the individuals having the rendering produced. Therefore, all aspects and needs of the project should be carefully analyzed before developing the rendering. For instance, an audience may accept a plain building exterior better if the drawing scale is changed to include more of the background and foreground. For an attractive building in an unattractive location, use a larger drawing scale to reduce the background and foreground. Changing the elevation of the ground line in the perspective view can also reduce distracting surroundings in the rendering. Artistic license can even be used to omit or alter the appearance of the surroundings. Using colored illustration board and inks or watercolors that are pleasing to the eye—but not the actual building colors—can help focus the audience on the shape and design, rather than the exact color and texture characteristics.

The same general steps are used whether creating a painted, watercolor, or inked rendering. These steps are as follows:

1. Sketch and rough shade numerous views of the design to determine the approximate layout of the perspective. This will help identify alternative viewing directions and elevations, and provide a trial evaluation of shading and shadow effects.
2. Evaluate the sketches and select the best one for the project. Be careful to verify that shading and shadows are correctly located based on the orientation of the object or building and the illumination. Also be sure that all shadows from a light source project in the same direction.
3. Identify whether the project requires a rendering with high contrast and a realistic appearance or a softer watercolor or inked representation. Consider whether there are special circumstances that the rendering must address, such as the use of color, texture, scale, or other effects.
4. Draw the basic shapes on the selected media using light construction lines.
5. Apply the background to the media. Work from the top downward to avoid smearing the rendering. Complete the background so there is a slight overlap with the object lines.
6. Create the basic shape of the design. Overlap the edge of the background. Add details such as windows, doors, trim, and roofing treatments.

7. Apply the foreground to overlap hidden features. Carefully blend the background and foreground transition areas.
8. Double-check all details of the rendering for accuracy.
9. Allow the rendering to completely dry. For renderings that will not be frame mounted under glass, apply a thin spray coating of fixative to protect the rendering from smearing.
10. Mount and frame the rendering.

Summary

Manual illustration shading encompasses the two distinct categories of line shading and surface shading. Line shading does not actually place a shading effect on the surfaces of an object. Line shading uses the line contrast method, the line separation method, or a combination of both. Line contrast shading is using thin, medium, and thick object lines to give depth to an object. With line separation shading, small gaps are left at the end of object lines for features that recede behind other features. These two methods are the easiest and quickest of all shading techniques.

Surface shading is the application of special treatments to entire surfaces to produce highlights or textures. Surface shading methods include smudge shading, stipple shading, line surface shading, and appliqué shading. Smudge shading uses materials such as graphite, plastic lead, charcoal, or chalk. These materials are applied to the drawing surface and smudged with a pad to smooth and blend their density. Stipple shading involves the use of a pencil or technical pen to apply a dot pattern on surfaces. The closer the dots are drawn, the darker the surface shading appears. Line surface shading consists of drawing lines, line patterns, or crosshatching. The closer the lines are drawn to one another, the darker the shading appears. In appliqué shading, or transfer shading, preprinted graphic images are applied to object surfaces.

The materials most frequently used to create manual shading effects include plastic and graphite leads, colored chalks (pastels), charcoal, inks and dyes, watercolors and washes, paints, and appliqué transfers. The primary shading applications of plastic and graphite lead include stipple shading and line surface shading. Ink is used in stipple shading, line surface shading, and color renderings. Other materials such as dyes, watercolors, watercolor washes, and paints are also used to produce renderings.

Shadows add a final touch of realism to illustrations. Two types of illumination are used to create shadows. These are solar and artificial light. Traditionally, solar light rays are projected on an object from above along 45° angles. Solar rays are assumed to be parallel. Artificial light creates a cone of illumination. Two common techniques used to add shadows are shadow fill and shadow softening. Shadow fill is created for areas of an object that are in a total shadow. These areas are the darkest on the drawing. Shadow softening is producing a smooth transition of shading from very dark areas to areas without shading. Shadow fill and shadow softening can be used in all types of surface shading.

Although renderings are used by all areas of industry, they are primarily used for architectural applications involving interior and exterior views of buildings. Rendering is the process of developing a highly realistic color illustration depicting an object within a background and foreground. Manual renderings are expensive because of the time required to produce them. However, they can be used to establish a design concept or illustrate different visual solutions for a final product.

Review Questions

1. Identify the two general types of manual shading and explain how they differ.
2. Define *line contrast shading*. What types of lines are used to represent exterior edges on the shaded sides of an object?
3. Explain how line separation shading is used to show depth in an object.
4. Describe the smudge shading process.
5. What is a *blending stump*?
6. Define *stipple shading*. How should a stipple pattern appear as areas of light shading change to areas of darker shading?
7. Compare the advantages and disadvantages between using a pencil and a technical pen for stipple shading.
8. What is *line surface shading*?
9. Briefly explain how line surface shading can be used to approximate rounded features in an object.
10. Define *appliqué surface shading* and briefly describe the steps used to apply an appliqué image.
11. Name four common materials used in surface shading.
12. How does *shadowing* differ from *shading*?
13. Briefly describe the main differences between solar illumination and artificial light illumination.
14. Explain how to develop the shadow for a rectangular object receiving solar illumination.
15. How is an artificial light source typically oriented for an object when projecting shadows?
16. Define *shadow fill*.
17. What shadowing technique is specifically used to produce shadows that transition smoothly from dark shading to lighter shading?
18. What is a *rendering*?
19. What is the most common application for manual renderings?
20. List the common steps used to develop a manual rendering.

Drawing Problems

Develop the following problems as three-dimensional models or draw them by hand. Use your own dimensions. For each problem, discuss the appropriate shading techniques with your instructor and develop a shaded illustration as directed. Use combinations of techniques where necessary. Orient the object and shading to create the most realistic representation.

1.

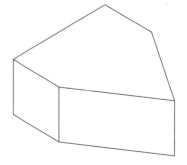

2.

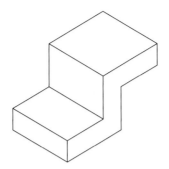

3.

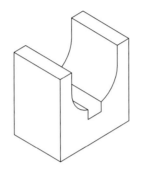

4.

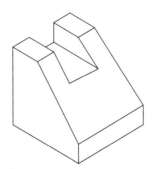

5.

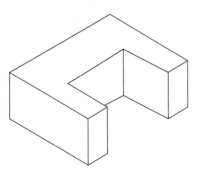

6.

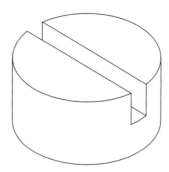

7.

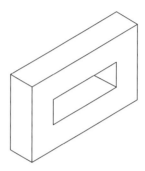

8.

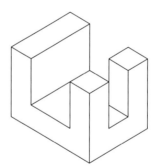

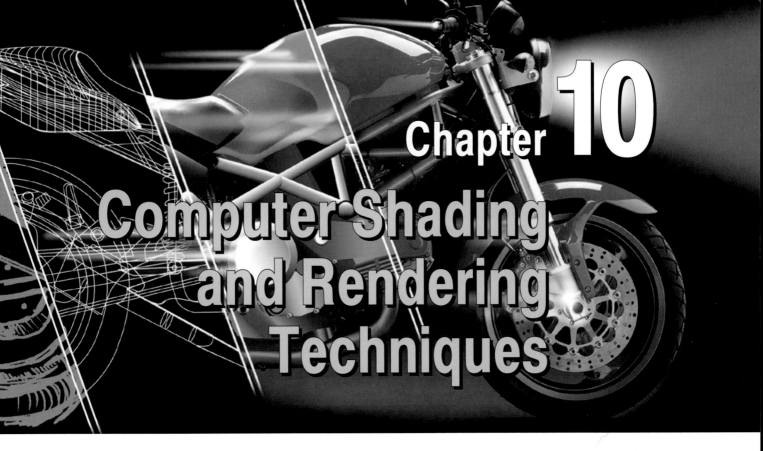

Chapter 10

Computer Shading and Rendering Techniques

Learning Objectives

At the conclusion of this chapter, you will be able to:

☐ Describe the common functions of paint programs and image editors and explain their applications in technical illustration.

☐ Explain the computer-based techniques used to produce line shading and surface shading.

☐ Identify the applications of automatic computer-generated shading and rendering.

☐ Understand the basic terminology related to computer-generated shading and rendering.

☐ List and describe the types of objects used to define a scene and prepare it for rendering.

Introduction

Many of the shading and rendering techniques discussed in Chapter 9 can be performed using computer software. Often, shaded drawings and renderings are created using the automatic functions provided by a computer-aided drafting program. Line shading and surface shading are also quite easy to produce using CAD systems or other software. For example, one way to use a computer to generate smudge or appliqué shading is to first create a line drawing on a CAD system with the necessary highlights. Then, smudge shading or appliqué shading effects can be automatically applied using the appropriate tools in the CAD program or in a separate program.

An important aspect of computer graphics is how easy color can be applied to objects. Basic CAD systems can display objects in red, yellow, green, cyan, blue, magenta, black, white, and varying shades of gray. Most midrange CAD systems can display at least 256 different colors. An infinite number of photorealistic colors can be displayed on high-level systems. The output device you are using controls how many different colors you can actually print. For example, an eight-pen plotter can only produce eight colors on a drawing. However, color laser and inkjet printers can produce a much wider variety of colors and are capable of outputting high-quality graphics.

Some software systems allow you to shade and render objects with specific light and reflectivity conditions. These systems create color renderings based on light sources, surface finishes, and viewing orientations. Very realistic illustrations can be produced with this type of software.

Overview of Computer-Generated Shading and Rendering

There are many different ways to produce shaded or rendered illustrations using computer software. If the model is created in a program with advanced rendering capability, you can produce a realistic full-color rendering with various treatments applied, such as lighting and surface finishes. See **Figure 10-1**. If the existing program is not capable of generating high-quality color renderings or images, you can export the image to a higher-end program or to an image editor. If the image is created in a vector-based program, you can also export the image to a bitmap editing program to apply color or shaded graphics in the bitmap format. Likewise, a bitmap image can be applied to a portion of a vector-based model in order to create a photorealistic image. Some software programs permit the conversion of bitmap images to vector images. This makes some images easier to modify because vector-based object features are defined with X and Y coordinates, rather than with individual pixels.

You can also scan existing images into a software program and apply various drawing tools to create visual effects. The image files can be converted from one format to another depending on the needs of the application. This chapter discusses the various tools and techniques used with computer software to create full-color shaded and rendered illustrations.

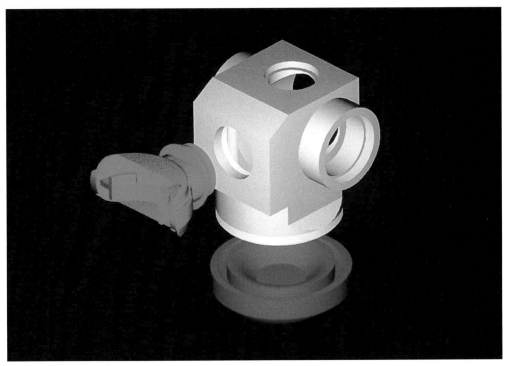

Figure 10-1. Realistic illustrations can be created from three-dimensional models using software with rendering capability. This model has basic lighting and surface finishes applied. (Autodesk, Inc.)

Computer-Painted Illustrations

As discussed in Chapter 8, three-dimensional (3D) models are typically created in CAD programs as surface models or solid models. The surfaces are then shaded or rendered to create the desired representation. In the event that you do not have a true 3D drawing of an object and a more photorealistic color representation is needed, you may want to utilize a paint program or an image editor. *Paint programs* create images in the bitmap format. CAD programs, by contrast, are vector-based programs classified as *draw programs*. *Image editors* are special graphics programs that provide advanced bitmap or vector-based editing functions.

Paint programs and image editors allow you to apply color to a drawing in a manner similar to how an artist paints a picture. Various tools in the program can be used to add effects such as simulated light, shading, reflection, color transitions, and font or text effects. This is another option to choose from when creating the original two-dimensional (2D) drawing. For example, you may want to draw the original object using a CAD system. You can then save the drawing in a bitmap file format so that it can be exported to the paint program or image editor. Bitmap files can be typically saved as *Tagged Image File Format* files (referred to as TIFF or TIF files), *Encapsulated PostScript* files (EPS files), or as Macintosh-based PICT files. The resulting file can then be exported and edited as desired.

Images brought into editing programs do not have to be developed on a computer. In many cases, photographs or manually drawn images are scanned and converted into bitmap files. Several different images can even be merged into a composite bitmap image and edited to produce a desired effect.

Using Paint Functions

The majority of paint programs and image editors permit you to apply specific colors to individual areas one at a time until the drawing is fully color illustrated. Paint programs do not have the advanced capabilities of image editors. However, nearly all paint programs are alike in that they allow the illustrator to apply the "paint," or color, to an image by editing the individual pixels defining the bitmap. Each pixel may be assigned any specific color available to produce the painted image.

If you want to apply advanced bitmap editing functions, it is best to use an image editor. See **Figure 10-2**. Tools are provided for drawing basic objects, modifying objects, cropping images, painting, and enhancing color. Images may be created from drawing commands within the software or from imported files.

Applying paint

Paint programs and image editors offer many different control functions for applying paint. When working with composite images, for example, it is often useful to mask certain areas to protect them from accidental painting. A variety of brush shapes are usually available to apply the paint. In addition, special "airbrush" functions may be used to vary the spray pattern, pattern shape, flow rate, spray area, or dot size. After an area is painted, it may be masked to protect it from overspray, or its edges may be oversprayed intentionally. Smudge tools can be used to soften the transition of colors at edges. If overspray is not desired, the painted areas may be masked for protection, and a sharp transition will result between colors.

As with manually prepared drawings, colors can be applied using instant transitions or gradual transitions called *gradients*. This is very useful when working with composite images. See **Figure 10-3**. Additionally, a typical software program with paint functions offers a wide variety of textures to produce

Figure 10-2. Image editors provide a wide variety of functions to create and edit bitmap images. Numerous drawing and editing tools were used with separate images in the development of this illustration. (Adobe Systems, Inc.)

distinctly different surface characteristics where color treatment alone is not sufficient. In the examples shown in **Figure 10-3**, image enhancement options in the software were used to vary the brightness and contrast of colors and color balance. Using this type of software, colored areas may be smudged, diffused, lightened, darkened, and even reversed or inverted.

Of particular importance in applying paint is the ability to zoom in to a highly magnified view so that color can be assigned to specific pixels. Although working at the individual pixel level is tedious, it provides the maximum feature control possible in color illustration. However, many illustrations do not require such a fine degree of paint treatment. As an option, boundaries can be established for color application so that different textures and gradient fills can be applied to vary the appearance.

Creating image layouts

Referring to the composite illustrations in **Figure 10-3**, take special notice of the spatial techniques used to give the illusion of a three-dimensional layout. Sharp contrast and separation between features combine to define a sense of depth. In **Figure 10-3A**, the image of the vintage car creates an overlay upon the sunset and other images in the background. Also, the portion of the gear teeth in color on the left is overlaid by the portion in black and white and the other mechanical parts.

In **Figure 10-3B**, the horizon and sky were developed to establish the background. Then, the limes and umbrella were drawn over the sky. Finally, the interior room and the exterior details were drawn. Each of the elements is applied on a separate layer. This is a common way to orient a painted image. Placing the background on a background layer and the foreground objects on successive layers achieves the effect of orienting objects in the foreground closer to the viewer. Using layers also allows the individual elements in the image to be edited separately. Other techniques are used to create a sense of depth in **Figure 10-3B**. Notice the overlay of the curtain in the foreground. This helps attract visual attention and emphasizes the spatial relationships of the various objects.

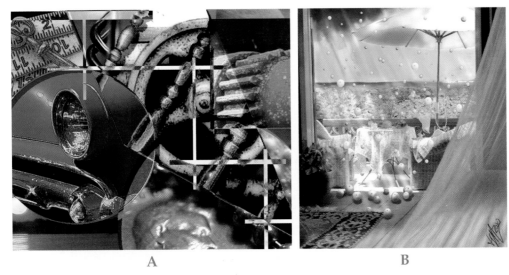

A B

Figure 10-3. Paint functions and imaging tools offer many ways to apply distinct transitions and other visual effects in composite images. A—Color gradients and shadows enhance the overlapping images and help define a sense of depth. (Jason Weiesnbach) B—Cut and paste functions were used with layers to insert the individual elements in the lime shower image. Contrasts in color establish spatial relationships between the various objects. (Jo-Anne Mason/www.joannemason.com)

Similar techniques were used in the development of the illustration in **Figure 10-2**. The overlay of the bench and trees with the metallic ball in front of the background elements is enhanced by both reflection and shadow. In this example, cut and paste tools were used with layers to arrange the objects in a three-dimensional layout.

Assigning color specifications

The color production capabilities found in a higher-level paint program or image editor provide powerful controls in creating customized colors. In more advanced programs, any of the standardized Pantone® colors may be used, or customized colors may be developed using different systems based on the RGB, CMYK, and HSL color models. Color models are discussed in greater detail in Chapter 11.

Some programs allow colors to be assigned using RGB specifications, while others permit color selection using either CMYK or HSL specifications. When using RGB color formation, colors are identified by specifying the intensity of red, green, or blue. Most color monitors can display as many as 256 different intensities of red, 256 intensities of green, and 256 intensities of blue. Monitors with this capability can display almost 17 million possible combinations of RGB color generation. When using software with CMYK color formation capability, colors are specified using values of cyan, magenta, yellow, and black. An example of CMYK color editing is shown in **Figure 10-4**. In this illustration, CMYK color scans are merged to produce a full-color image. As shown in the color menu, the separate layers of cyan, magenta, yellow, and black combine to produce the four-color image. The illustrator can adjust the intensity of any of the four color components to vary the appearance of the image.

When making color specifications with HSL color formation, colors are defined based on hue (color), saturation (intensity), and lightness. In **Figure 10-5**, the hue and saturation components of the image are edited to modify individual colors. For example, it is possible to adjust the hue in the red flower without affecting the appearance of the small yellow petals in the center.

Figure 10-4. In CMYK color editing, colors are specified using values of cyan, magenta, yellow, and black. (Adobe Systems, Inc.)

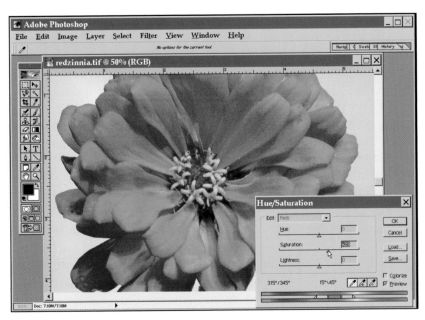

Figure 10-5. An example of HSL color editing. The menu palette shown is used to make hue and saturation adjustments to individual elements in the image. (Adobe Systems, Inc.)

Some software programs have color control functions based on manual application methods. For example, paint may be applied using a drawing tablet and stylus. See **Figure 10-6**. By varying the stylus pressure on the tablet surface, the user can apply more or less paint or thicker or thinner brush strokes. This is a way to develop a screen-printed effect.

When using any software program or color specification system to produce color images, it is very important to understand that the generation of color by a monitor is often different from the generation of color in the final illustration. Actual output color in many applications does not match what you see on the monitor. The primary reason for this is that different output devices use different

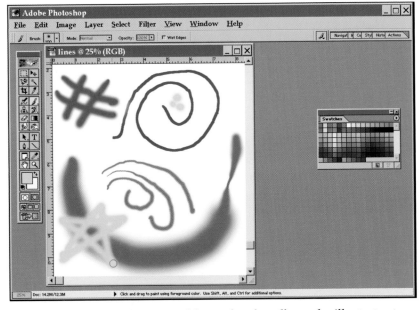

Figure 10-6. Using a drawing tablet and stylus allows the illustrator to vary applications of color and painting effects. (Adobe Systems, Inc.)

means to produce color. For example, monitors typically use RGB color formation to produce color, while printers typically use CMYK color formation.

Monitor display quality is normally described by the computer graphics acronym *WYSIWYG* (pronounced *wissy-wig*), which stands for "What You See Is What You Get." Most vector-based images on a monitor display do achieve accurate WYSIWYG output in terms of shape, proportion, and completeness of detail. Quite often, however, WYSIWYG output does not apply to color output. What you see on the monitor display is *not* what you get from an output device for most custom colors.

One accurate method of color definition is to attach database attributes to screen colors corresponding to Pantone color codes. The technical illustrator can select standard colors from a Pantone color chart and render the drawing with the specified colors. Then, production personnel can match printing ink to the Pantone colors with some assurance that the colors that are printed are the colors desired by the illustrator. A similar method is to use a software program that combines Pantone color formation with CMYK printing support. See **Figure 10-7**. Using a color palette menu, colors in the Pantone system can be assigned as needed. The equivalent CMYK colors can then be extracted from the user-defined Pantone colors for four-color printing purposes.

Detailed color specifications made with a paint program or image editor will not always ensure absolute color accuracy. There are several reasons for this. One is that the paper or media the illustration is printed on will often draw from or add to certain color intensities. In addition, in many cases, the different color conversion processes that take place as graphics are converted from one format to another may cause the graphics to appear differently. For example, colors may change in appearance when film is made from color proofs. Therefore, more advanced software programs with color management systems have been developed to provide a greater degree of control over color specifications. A color management system helps ensure that color remains consistent throughout the production process by making adjustments between the different input and output systems, including scanners, monitors, and printers. When images are converted from one form to another, controls are applied to each device so that color is rendered properly.

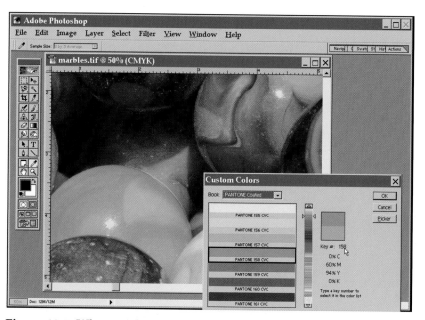

Figure 10-7. When applying Pantone colors to an image, special editing functions can be used to extract the equivalent CMYK colors for printing purposes. (Adobe Systems, Inc.)

Using editing tools

As previously discussed, image editors are typically used to modify scanned or imported images in order to create photorealistic illustrations. Some programs offer specialized tools for operations such as editing portions of scanned bitmap images and merging them into other images. This is often useful when working with scanned photo and line images that are incorporated into other applications. For example, objects in a scanned image may be edited, copied, and then pasted as a bitmap image into a desktop publishing program for page layout. See **Figure 10-8**.

Images brought into desktop publishing programs are commonly imported as EPS files. This is because the EPS format can display bitmap or vector images and is capable of supporting a number of color formation systems. In addition, vector-based EPS images can be scaled to a larger or smaller size without a significant loss of shape or proportion. In **Figure 10-8A**, the jukebox image is first selected by tracing its specific outline with a curve definition tool called a Bézier curve tool. A *Bézier curve* has control points that allow you to make precise adjustments to the shape of the curve. Once selected, the outline and images within it may be edited, saved, and exported. In this example, the image is exported as an EPS file. The resulting image may be scaled as needed. In **Figure 10-8B**, the image is pasted onto a page layout. The Bézier curve defining the shape of the jukebox is also used to determine the shape of the page layout so that text wraps around the image in the shape of the curve.

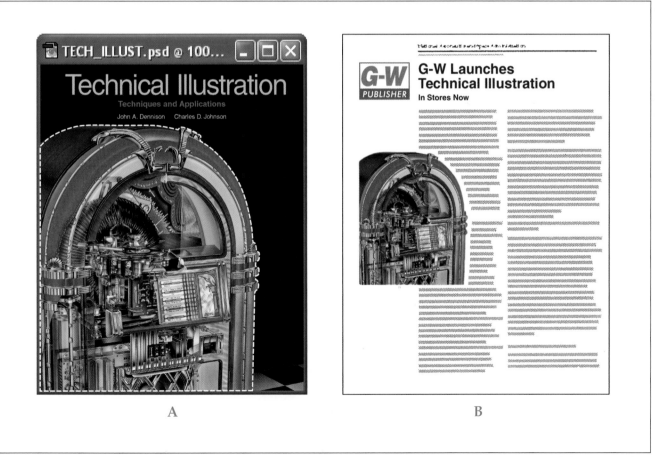

Figure 10-8. Portions of bitmap images are often selected for use in other applications. A—A curve definition tool is used to trace the outline of the jukebox. B—The image is imported into a desktop publishing program as an EPS file and resized to fit the layout of the text.

When creating composite images, it is often useful to edit several different image files within the same software interface. This capability is typically provided in image editors and in more advanced paint programs. See **Figure 10-9**. In the example shown, scanned images are combined with painted images during the editing process. Different editing operations are then performed on the separate images in the same file. This allows the illustrator to follow a more natural workflow. Systems without these capabilities typically require you to edit each image in a separate file until all images have been edited. This creates difficulty because many complex illustrations are easier to develop by editing a number of different images in a single file. This is especially true when blending one image into another or applying special color treatments to the images.

Combining paint and draw functions

In many cases, images are created as either bitmap or vector images without combining the two formats. However, some software programs provide tools that permit the creation of bitmap and vector graphics within the same image. These types of programs allow you to switch drawing modes when applying different paint and draw tools. In some cases, it may be useful to use a draw program rather than a paint program when creating a color illustration. For example, it may be desirable to convert bitmap images into vector images and modify them using vector-based methods. Some draw programs have tools that allow you to apply painted effects such as brush strokes and spray patterns. Tools used to create transparent images and reflections may also be available. See **Figure 10-10**. The illustration shown is an example of a photorealistic vector-based image. The variances in color were produced by using custom colors and other tools designed to produce reflections and gradient effects. Notice that both the graduated softening of colors along irregular arcs and the highlights on the glass lenses aid the illusion of curvature and close proximity of an intense light source.

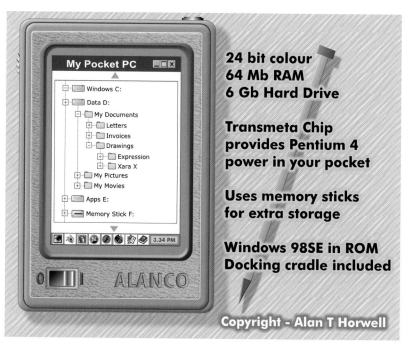

Figure 10-9. Screen capture, text, and painted images were combined for editing in this illustration. The images can then be modified in the same file. (Alan T Horwell—Alan's Image Factory/www.aifweb.com)

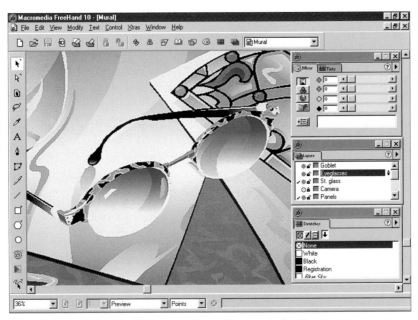

Figure 10-10. Photorealistic images can be created with draw programs. A variety of tools were used in this illustration to create the highlights, reflections, and gradient effects. (Macromedia FreeHand)

Advantages and disadvantages of paint-based programs

There are many advantages in using paint programs or image editors to create full-color illustrations. These include the quality of the finished illustration, the wide-range adaptability of many programs, the ease in learning how to use them, and their user-friendly nature. However, there are instances where other systems or drawing methods are more suitable for a particular application. This is usually because of the limitations of the program. For example, poor choices of color can create illustrations that appear artificial. In addition, the locations of highlighted areas, shading, and areas of shadow must be developed by the illustrator, rather than automatically generated by the computer. Complex object shapes may also be difficult to create when using the basic drawing tools available in a paint program or image editor.

Computer-Aided Drafting Illustrations

Computer-aided drafting programs provide a number of tools and methods to create high-quality illustrations. Basic 2D drawing tools can be used to create pictorial drawings that appear three-dimensional, such as isometric drawings. Simple shading effects, such as line surface shading, can be easily applied to these types of drawings using the same drawing functions. In more advanced CAD programs, true 3D drawings known as models can be created using XYZ coordinates and 3D construction methods. See **Figure 10-11**. When using a CAD program with rendering capability, surface models or solid models can be automatically shaded with lighting and material surface patterns applied.

When working with 2D-based drawings, applying line shading and surface shading with basic drawing commands is relatively quick and effective. This is the preferred method for isometric, dimetric, and trimetric drawings. The major reason that this type of shading—rather than automatic computer-generated

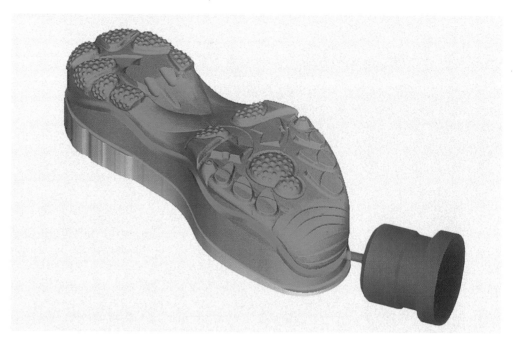

Figure 10-11. CAD programs with 3D drawing tools and rendering capabilities are used to create and automatically shade models with lights and materials. (Cimatron Ltd.)

shading—is applied to axonometric drawings is the inability of automatic shading and rendering programs to recognize the spatial relationships of 2D surfaces. A computer must have the information necessary to define the Z coordinate relationships of surfaces for the calculation of shadows, shading, and areas of high light intensity. For this reason, automatic computer-generated shading and rendering systems typically do not perform satisfactory shading of objects unless they have been drawn in true 3D form.

The following sections discuss common CAD drawing functions used to create shaded 2D illustrations. Automatic shading and rendering procedures used with 3D models are discussed later in this chapter.

Applying Line Shading to CAD Drawings

There are several ways to apply line shading to 2D-based CAD drawings. Most major CAD programs have a polyline function that can be used to create line shading. As discussed in Chapter 8, a *polyline* is a continuous line made up of one or more segments. Most polylines can be created with straight or curved segments. Most polyline functions also allow you to control the width of the segments. The polyline function can therefore be used to create line contrast shading. Draw the object using polylines with a single width. Then, use the polyline editing function to change the widths of the appropriate lines.

Line separation shading can be created using the **Break** command. First, draw the object with all object lines touching for ease of geometric construction. Then, use the break function to create slight gaps at the ends of any lines that pass behind other objects. The zoom function can be used to make it easier to create the gaps. However, an enlarged view can also distort your sense of proportion in relation to the relative size of the gaps you are creating. Before plotting or printing, be sure to zoom the view to a size that is approximately proportional to the actual drawing size. This will help you check to make sure all gaps are proportional.

Applying Surface Shading to CAD Drawings

A variety of CAD functions can be used to create drawings with surface shading. As with manual surface shading, different shading methods are often combined depending on the application. However, excessive mixing of line surface shading, crosshatch shading, and stipple shading on the same drawing may create an inconsistent appearance. The following sections detail the major surface shading techniques that can be used to develop shaded CAD drawings.

Line surface shading

The same concepts involved in manual line surface shading are used when applying line surface shading to a CAD drawing. See **Figure 10-12**. Straight lines and arcs are used to shade the surfaces of the object shown. When applying surface shading within holes, shading lines should always be aligned parallel to the centerline of the hole. Since holes in pictorial drawings are elliptical, you will need to align the shading lines with the *minor* axis of each hole located on a normal surface. Elliptical holes drawn on inclined or skewed surfaces may create some problems if the light source is at a nontraditional location. Normally, nonisometric axis holes should have shading lines drawn parallel to an isometric axis that is closest to perpendicular to a line from the light source.

In **Figure 10-12**, notice that the exterior of the vertical cylindrical shape in the middle of the object has surface shading lines to give an illusion of higher intensity light near the middle portion and darker areas at the extreme left and right. This darkening is a result of curvature that does not reflect light directly to the viewer. In addition, observe the subtle shading applied to the horizontal gap at each end of the yoke. The preciseness of this shading is more visible in the enlarged view, where the light source illuminates a small crescent-shaped

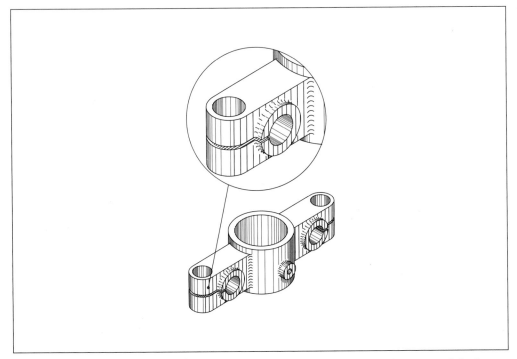

Figure 10-12. Line surface shading is used to establish highlighted areas and darker portions in this CAD drawing of a fork yoke. The enlarged view illustrates shading details and areas illuminated by light.

portion on the left rear of the gap. Attention to detail such as this is important in the development of shaded illustrations. A more obvious example of line surface shading in **Figure 10-12** is the absence of any surface shading on the top of the object. Omitting all traces of shading on the top while applying only slight shading on the nearest vertical surface provides the illusion of an overhead light source. Had some scattered low-density shading been used on the top surface, it would have been necessary to add considerably more dense shading to the remainder of the shaded surfaces to maintain the visual balance.

Notice that elliptical arcs are used to represent rounded features of the object. The development of these shading features is discussed in greater detail later in this chapter.

Line surface shading and crosshatch shading

A variety of line surface shading and crosshatch shading techniques are used to shade the adjustable wrench in **Figure 10-13**. The object has shading applied to straight vertical surfaces, irregularly curved surfaces, inclined surfaces, cylindrical surfaces, and a knurled thumbwheel. There are three in-line vertical shaded surfaces and two more slightly shaded vertical surfaces to the left of the knurled knob used to adjust the wrench. As all five of these surfaces are vertical, their surface shading lines have been drawn as vertical lines.

One method used to help the viewer visualize that the three slightly recessed surfaces are in alignment with each other is to apply line surface shading with identical spacing, width, and pattern contours. To visually separate the vertical surfaces for the wrench jaws from the other visible surfaces, a shorter shading pattern has been used. For these surfaces, the spacing between lines is also uniform. One way to produce evenly spaced shading lines is to use *crosshatch patterns*. These are typically available as predefined patterns in CAD programs. Custom patterns can also be created. The wrench jaws and recessed surfaces in **Figure 10-13** were shaded with a predefined crosshatch pattern

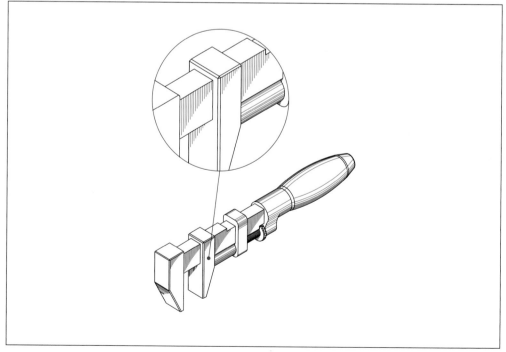

Figure 10-13. Line surface shading is used with crosshatch patterns to distinguish surfaces in this CAD drawing of an adjustable wrench.

normally used to represent cast iron or general purpose lines. The spacing for the patterns can be seen more clearly in the enlarged view. The patterns were easy to apply to the surfaces after some advance planning.

Notice the curvature used with each of the crosshatch patterns for the shading. Many CAD systems do not allow you to modify or edit the individual lines of a crosshatch pattern. Although the line pattern can be broken apart, or "exploded," into separate lines, this increases the size of the drawing database. Therefore, to establish the curvature of the patterns, two slightly arched lower boundary lines were drawn on a construction layer. One arc was used for the wrench jaws, and one was used for the recessed surfaces. The arcs were then copied to the other corresponding surfaces to produce an identical pattern shape for the bottom of the shading lines. To create the crosshatch patterns, the boundaries of the areas were selected, and the patterns were automatically drawn by the computer. By default, the crosshatch pattern used in this example is drawn with the line segments at a 45° angle. The vertical patterns were created by changing the angle of the pattern. After all of the patterns were completed, the construction layer containing the boundary arcs was turned off to remove it from the display.

The crosshatch pattern used for the shading on the inclined surface of the left wrench jaw was drawn in the same manner, except the angle used for the pattern was set at 150°. The knurling on the thumbwheel was also drawn using a crosshatch pattern. This pattern is a double crosshatch normally drawn with one set of lines at a 45° angle and the other set of lines at a 135° angle so that the lines are perpendicular. These default angles are sufficient for orthographic views. However, this pattern orientation will cause most pictorial surfaces to appear distorted. When working with pictorial drawings, it is recommended that the crosshatch patterns be rotated to align with an appropriate pictorial axis. Patterns with axis lines drawn at 30°, 150°, or 210° are typically used on isometric drawings.

Some dimetric or trimetric drawings will require you to generate your own knurling pattern. However, this process can be simplified by drawing a set of general purpose crosshatch lines in one angular direction and then another set at a different angle. The knurling pattern normally appears more realistic if one crosshatch angle is perpendicular to the right receding axis and the other crosshatch angle is perpendicular to the left receding axis. This alignment will give the viewer an additional cue to distinguish visually between pictorial surfaces. The principles outlined in this section should also be used to align crosshatching angles for stipple dot patterns when working with different pictorial axes.

Applying shading to rounded features

Elliptical arcs and straight lines can be used to apply line surface shading to a CAD drawing with rounded features. This type of shading is applied to the spoked hub in **Figure 10-14**. The shadow on the lower side of each spoke and hub was created with individual lines of varying length. As the lines approach more direct lighting, the spacing is increased to lighten the shading density. Areas beneath the cylindrical hub and the smaller cylindrical shape at the end of the left spoke have been shaded with 30° angled lines so dense that these areas are nearly filled solid.

The elliptical arcs shown in the two views were not produced with a special crosshatch pattern. Instead, ellipses aligned with the axis of each spoke surface were drawn and trimmed to produce the resulting arcs. For each spoke, a construction line was first drawn marking the theoretical edge of an intersection of the

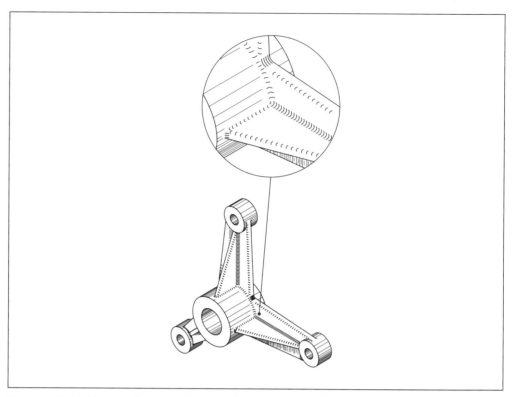

Figure 10-14. Line surface shading is achieved in this illustration of a spoked hub by using lines of varying length and density. Elliptical arcs are used to represent the filleted features.

horizontal and vertical surfaces on the spoke. Then, the **Divide** command was used to divide the line into 30 segments. When using this function, each resulting segment length is indicated by a small "X" automatically drawn by the computer. This is a particularly useful function in this example because the number of divisions for each spoke was determined by trial and error. Initially, the construction line was divided into 20 segments, each represented by a small "X." However, this spacing provided visual evidence that 20 divisions would not give adequate density to the pattern of arcs. After the first division operation was undone, the line was then divided again using 30 segments. When using the divide function, the resulting indicator points represented with "X" marks are identified as node (or point) object snap points.

Notice that for each spoke, the minor axes of the elliptical arcs align with the axis of the spoke. Arcs used to represent rounded features should align with the radial axis in this manner. To draw the ellipses for each spoke, a single small isometric ellipse was drawn with the minor axis in alignment with the spoke axis along the division construction line. Next, a construction line was drawn through the major axis, and half of the ellipse was trimmed away at the major axis line. Multiple copies of this isometric arc were then placed along the spoke axis at the points on the division line using the node object snap. This procedure was simplified by placing the first isometric arc at the first point on the division line. After the first copy was made, the two arcs were selected to make two more copies along the spoke axis. Then, the four arcs were selected to make four more copies, and so on until enough arcs were copied to locate on the remaining division points. In each copy operation, the node object snap was used to define the base point and destination point. Notice that the small fillet arcs at the intersections of the rib webs were created in the same manner. The only difference in their development is the smaller size and greater density.

Applying shading to curved surfaces

A number of techniques can be used to shade curved surfaces with combinations of rounded and planar features. The micrometer shown in **Figure 10-15** is an example of an object that has a transition from a cylindrical shape to a planar surface. Referring to the enlarged view, the barrel uses a combination of straight and curved lines to visually define the shape transition. Notice that the horizontal shading lines on the vertical surface end just before intersecting the curved lines denoting the fillet between the cylindrical and planar surfaces. Also notice that the elliptical fillet arcs are aligned with the axis of the fillet.

To draw the horizontal shading lines, a construction arc was first drawn with its curvature approximately parallel to the ends of the fillet arcs. The shading lines were then drawn individually across the plane of the vertical surface with variable gaps. Next, the inside arc defining the crescent-shaped contour of the vertical surface and the construction arc were used as boundary lines to trim excess length from the horizontal shading lines. Using a crosshatch pattern in this application would not have been practical because of the variable gaps between individual lines and the nonuniform line lengths achieved through trimming. The irregular gaps and line lengths are drawn intentionally to show the areas that receive more light in contrast to the more heavily shaded areas. Reflected light and shading are seldom found in very similar or identical patterns across entire object surfaces. Surface shading with too much regularity in line spacing, identical line lengths, and uniform pattern density will make the object appear artificial.

The double crosshatch pattern used for the knurling on the micrometer handle is drawn at a 30° angle so that it aligns with the isometric axis of the barrel. The visual image of this pattern will detract from the cylindrical shape if it is drawn at the default setting of 0° of rotation. This setting will draw the diagonal lines at 45° and 135° angles as previously discussed. Setting the angle of the crosshatch pattern to 30° will rotate the entire pattern 30°. The crosshatch

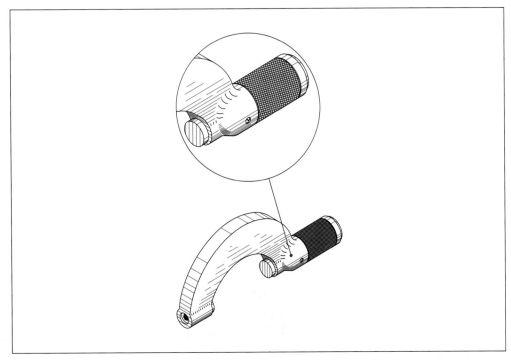

Figure 10-15. Variable line lengths and irregular gaps are used with shading lines in this illustration of a micrometer to define transitions between curved and planar surfaces.

lines are still drawn perpendicular to one another. Therefore, the actual angles of the lines are 75° (45° + 30°) and 165° (135° + 30°).

The vast majority of industrial knurling is used on cylindrical shapes to provide a better grip for the human hand. The visual combination of a cylindrical shape and knurling pattern can cause the illustrator drawing problems in some design situations. If you use a crosshatch pattern with very light density to represent the knurling, you will find that the cylindrical surface will begin to lose its curved appearance. The more open the knurling pattern density, the flatter the appearance of the surface. However, if you use a pattern that is too dense, the surface will begin to lose the appearance of knurling and take on the visual effect of heavy shading. Another concern in drawing knurling patterns is the area in which crosshatch lines are applied. A large crosshatched area will normally appear less realistic than a smaller crosshatched area with identical density. Experimentation and evaluation of the printed or plotted pattern is often needed to judge the proper pattern density in different illustration situations. It is often common, for example, to divide single surfaces into separate areas and apply patterns of varying density to create transitions from shadows to areas with more direct lighting.

Applying shading to spherical and conical surfaces

Special surface shading techniques are used to establish highlighted and shaded areas for spherical and conical features. Examples of line surface shading treatments applied to spheres, cones, and conic chamfers are shown in **Figure 10-16**. As with manual smudge shading and stipple shading methods, shading for spherical shapes should be developed to create the illusion of a bright spot of high reflectivity in relation to the light source. The highlighted

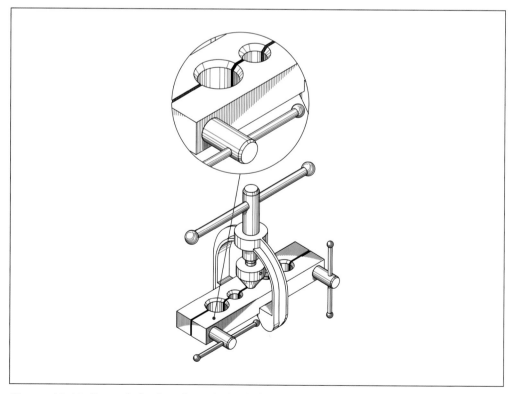

Figure 10-16. Curved shading lines help define the spherical balls on the ends of the cylindrical rods of this flaring tool. Shading lines representing conic features and chamfers theoretically intersect at the center of the shape.

area should transition to light shading and then to darker shading near the edges of the sphere. This transition is a little more difficult to develop in relation to other shading effects when using CAD drawing functions only, rather than computer-automated shading or rendering methods.

Arcs used to create shading lines on spherical surfaces can be drawn as simple arcs or as curved polylines. The advantage in using polylines is the ability to edit vertices defining the curve. Regardless of the method used to draw the curved lines, it is important that they appear circular until they intersect the edge of the sphere. Developing shading on conical shapes is much easier by comparison because all lines extend to a single point. Referring to the cone-shaped adapter of the tube flaring tool in **Figure 10-16**, the shading lines extend to an imaginary point on the cone even though the shape is truncated. Any excess shading lines are trimmed off at the boundary where the cone is truncated or cut off. This same concept applies to conical chamfers. The enlarged view in **Figure 10-16** shows that all of the shading lines on each chamfer theoretically intersect at the center of the ellipse. There are other shading treatments that add realism to the illustration, such as the higher-density shading patterns under the cylindrical shapes and within the gaps between the halves of the flaring tool.

Applying Shading Patterns and Fills

There are several ways to apply shading patterns and fills to 2D-based CAD drawings. Using a crosshatch pattern is one of the quickest methods. As previously discussed, crosshatch patterns can be used for a variety of shading applications. A sampling of predefined crosshatch patterns available in a typical CAD program is shown in **Figure 10-17**. Each pattern is drawn at a

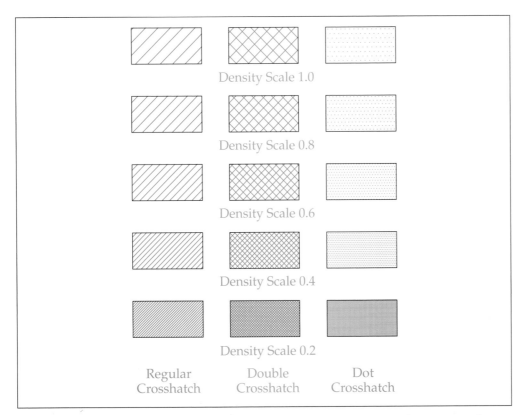

Figure 10-17. A comparison of sample crosshatch patterns. Pattern densities can be varied depending on the application.

progressively greater density. However, the line widths are identical. When applying crosshatch designs with greater pattern densities, the resulting image quality may be dependent on the printer or plotter you are using. Some output devices may not be able to reproduce patterns with extremely high density without creating unwanted phantom geometric patterns. Ask your instructor if you are unsure about what density scale to use for a particular crosshatch pattern in relation to the output devices available. Advance planning can save considerable revision work, wasted paper and ink, and most importantly, time.

Another way to create shading patterns in some CAD programs is to apply predefined fill patterns based on PostScript files. The drawing can then be exported to another software program as an EPS file. Fill patterns can be typically applied to objects made up of closed polylines. After entering the appropriate command, you are asked to select a polyline you want to fill. You are then asked to specify the desired fill pattern. Patterns based on grayscale shading are commonly used in surface shading applications. Grayscale shading allows you to apply shading in transitions of gray. For example, you may want to shade the interior of a closed polyline object using a percentage of gray. You may also want to create a shading transition that generates an area of highlight in contrast to darker areas. Several types of grayscale fill patterns and other fill patterns are available in some CAD programs, including color fills. See **Figure 10-18**.

When applying a grayscale fill pattern, the resulting shading is generated within the polyline boundary. The most basic grayscale fill pattern allows you to specify the percentage of fill density. The default value is 50%. Entering a lower value creates a light shading effect, while entering a larger value creates a darker percentage of shading. The linear grayscale fill pattern allows you to create a fill that transitions evenly from 100% shading to 0% shading. This fill pattern works especially well on normal object surfaces. The radial grayscale fill pattern generates a fill that transitions from 100% shading at the boundaries to a 0% shading highlight in the center of the boundary. The level of foreground and background shading density can be varied. This type of shading can be used to generate the center highlight on a spherical surface where areas closer to the boundaries in the background are darker. This is one of the advantages that makes radial grayscale shading more realistic than the other shading types.

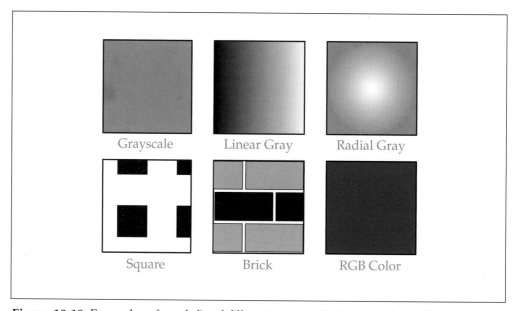

Figure 10-18. Examples of predefined fill patterns applied to closed polyline objects.

As previously discussed, color renderings produced in CAD programs are most often limited to drawings created as 3D surface models and solid models. However, shading patterns and other line work in 2D-based illustrations can be enhanced with the use of color. As with many software programs, CAD systems allow you to assign colors to lines and other 2D drawing shapes. This basic function can be used to add variety to an illustration. For example, color fills can be created by adding color to an object with line surface shading. In some applications, an object with colored line work or fills can generate a more realistic appearance than a single-color illustration. However, as with other treatments, color shading should be used in a manner that adds realism to the drawing. In most cases, color should be applied to create subtle effects. For example, many illustrations of manufactured products use only one or two basic surface colors and a few accent colors. Too much variety in surface shading should be avoided because excessive coloring may actually reduce the realism of a drawing. Generally, it is best to apply a single dominant color to object surfaces and use darker hues of the same color for shadowed areas. Areas receiving more light can be filled in with a lighter shade of the dominant color.

Automated Shading and Rendering

The variety of CAD software available provides an extremely wide range of automatic shading and rendering capabilities. Although some CAD programs are similar in terms of how renderings are produced, the presentation effects that can be added and the resulting output quality may vary widely. A basic rendering of a mechanical assembly is shown in **Figure 10-19**. The specific techniques used by a CAD system to shade and render models are usually detailed in the software user's manual. Generally speaking, there are a number of commands and functions that shading and rendering programs have in common. For example, these programs will typically only recognize objects or surfaces that have been

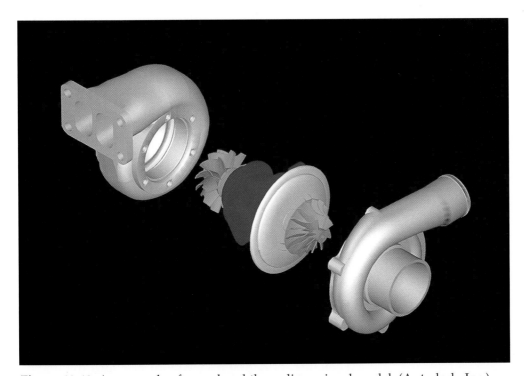

Figure 10-19. An example of a rendered three-dimensional model. (Autodesk, Inc.)

drawn using 3D faces, 3D meshes, primitive shapes, or extruded or joined solids. Therefore, the first step in creating a computer-generated rendering is to draw the object as a true 3D model using a standard CAD system.

Once the model has been completed, the illustrator may select a compatible shading or rendering method. Many CAD systems have shading and rendering functions integrated within the basic program. In some cases, drawings created within one CAD system are imported into another system with more powerful rendering capabilities. High-end rendering systems involve greater expense and require significant computer resources to operate properly. They also tend to require greater artistic and CAD drawing skills from the illustrator. Due to the wide range of CAD systems available, an overview of the typical functions and procedures found in the majority of the software should start with an introduction to the basic terminology used in shading and rendering applications.

Shading and Rendering Terminology

A few fundamental terms related to shading and rendering functions are listed below. While specific terminology may vary between different software programs, you should have a basic understanding of the following terms.

➤ *Scene.* A view of a 3D model specifically set up for rendering. A scene is typically made up of a 3D view of an object with materials and lights applied. A camera can be used to establish the view. Scenes can be rendered with backgrounds and a variety of lighting and environmental effects.

➤ *Lighting.* Any light sources used to illuminate a scene. Lights project light rays in all directions, along parallel paths, or in a cone of illumination. In many cases, several different lights are positioned in a scene to illuminate various focal points and object features.

➤ *Ambient light.* The evenly distributed background light that is cast throughout a scene. Ambient light is "soft" light that does not have an identifiable source. It does not create specific highlights or shadows.

➤ *Hotspot.* The area of projection from a light source where the greatest amount of illumination is generated.

➤ *Falloff.* A term that refers to the decrease in lighting intensity as the distance of the object from the light source increases.

➤ *Inverse linear.* A calculation for falloff that specifies the illumination intensity of a light source will decrease in inverse proportion to the distance from the light. An object three units away from a light source will receive one-third of the light intensity, while an object six units away from a light source will receive one-sixth of the light intensity.

➤ *Inverse square.* A calculation for falloff that specifies the illumination intensity will decrease in inverse proportion to the square of the distance from the light source. An object three units away from a light will receive $(1/3)^2$, or 1/9th, of the light intensity. An object six units away from a light will receive $(1/6)^2$, or 1/36th, of the light intensity.

➤ *Reflectivity.* A term that describes the relative amount of light that strikes the surface of an object and is then reflected. The degree of reflectivity is directly related to the intensity and angle at which the light strikes the object, the surface texture, and the viewing angle. Reflected light generally travels in a path that is perpendicular to the angle at which the light strikes the object. Therefore, orienting a viewpoint in which the line of sight is perpendicular to the rays of illumination will create the greatest amount of reflection, or highlight.

➤ *Diffuse color*. Color that is illuminated by the distribution of diffuse light as light strikes an object's surfaces. The amount of diffuse color can be controlled by both materials and lighting. Although diffuse light is typically produced by matte surfaces rather than shiny surfaces, diffuse color can be combined with specular color to create rough surface areas receiving diffuse lighting and smoother, more reflective surfaces receiving specular lighting.

➤ *Specular color*. Color that is illuminated by light as the light reflects from a surface. Specular color is the "highlight" or shiny portion of an object. Light striking a smooth surface will tend to produce bright reflections with parallel rays. Therefore, objects with smooth surfaces are said to have a high specular factor. Objects with matte surfaces are said to have a low specular factor.

➤ *Camera*. A user-defined object that establishes the location from which a scene will be viewed. Additionally, some shading software will allow you to identify a specific sized optical lens on the camera with which you will view the scene from that location.

➤ *Lens*. The viewing device of a camera. Like a real camera lens, a lower lens setting provides a wider but more distant view of the scene. A higher lens setting provides a zoom magnification effect with a correspondingly narrower range of vision.

➤ *Material*. A surface finish or pattern applied to the various features of a model for rendering purposes. Materials can be used to generate surface textures such as wood, plastic, marble, and various types of metal. See **Figure 10-20**.

➤ *Material map*. The orientation of a material surface pattern when it is applied to an object. Material maps simulate the composition and direction of the material in relation to the object when rendered. Coordinates and projection shapes are used to map or "tile" materials to object surfaces.

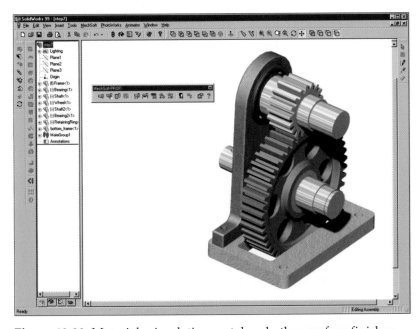

Figure 10-20. Materials simulating metal and other surface finishes are used to add realism to this rendered 3D model. (MechSoft.com)

Automatic Shading and Rendering Procedures

Many higher-end CAD systems with shading and rendering capability provide similar tools and procedures for creating shaded drawings. This chapter presents a generic overview of the automatic shading and rendering process. As previously discussed, the rendering process begins by developing a 3D model. The computer must have defined 3D coordinates for every surface feature of the object so that it can mathematically calculate the effects of camera position, lighting types and locations, lighting intensities, colors, and surface textures. Based upon these variables, the computer can develop the resulting surface-shaded view. Therefore, the basic rendering process involves drawing the 3D model, assigning surface colors or materials, locating lights and specifying lighting parameters, locating one or more cameras in relation to the object, and performing the rendering.

Most CAD systems have different functions for creating shaded drawings and actual renderings. Typically, a rendering is based on a scene with materials and lighting applied. In some cases, environmental effects and a background can also be added. A scene is then named or defined, and the rendering is created with the **Render** command. A more simple way to create a shaded drawing is to use the **Shade** command. This command creates a shaded view using the colors applied and default lighting. Materials are not rendered, and a default light source is used rather than user-defined lights. See **Figure 10-21**. With some options of the **Shade** command, smoothing is applied to object faces. However, the resulting shaded view is not as realistic as a rendering. Generating a shaded view also requires less time than performing a rendering.

Creating a view

When preparing a model for rendering, a three-dimensional perspective view of the model is first established. Numerous viewing tools, such as an orbit

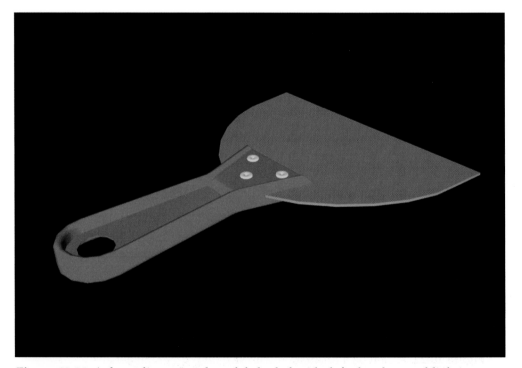

Figure 10-21. A three-dimensional model shaded with default colors and lighting. (Anthony J. Panozzo)

viewer, can be used in conjunction with zooming and panning functions. If a camera is used, a view matching the perspective viewing angle of the camera may be established. If the drawing includes a border and title block, you may want to place these objects on separate layers and turn off their display.

Viewports showing one or more orthographic views and a perspective view are generally displayed during the modeling process to help establish the best view of the object. See **Figure 10-22**. Often, viewports are created to show the top view, side view, front view, and perspective view. Using viewports with different views also simplifies the process of locating coordinates for lights and cameras. It is often useful to name and save each view that will be used later for rendering. This enables you to return to a previously defined view without having to use viewing tools to duplicate the view.

Creating materials

If materials are being applied to the model, they are typically assigned near the end of the modeling process. A *material* is a surface finish or texture applied to features of a model in order to define how object surfaces will appear when the scene is rendered. Materials provide a number of ways to add realism to a model. A material may be as simple as a basic display color applied to a surface. Settings can then be made to determine how the surface reflects light. Most CAD programs with rendering capability normally provide several predefined materials based on actual surface finishes, such as wood, metal, stone, or concrete. See **Figure 10-23**. If the existing types of materials do not suit your needs, custom materials can be created. In addition, bitmap files may be used as materials for advanced surface treatments. Some materials require you to apply them in orientations or scaled patterns based on the composition of the material. This is known as *mapping* or *tiling*. A *material map* uses coordinates or projection shapes, such as planar or spherical shapes, to define how the material will appear on a surface when rendered.

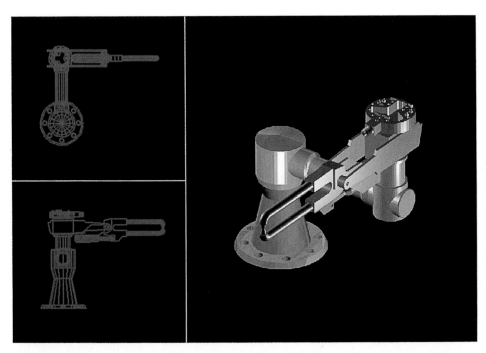

Figure 10-22. A viewport orientation with several orthographic views and a perspective view helps simplify the process of locating lights and establishing viewing angles for rendering. (Autodesk, Inc.)

Figure 10-23. Various surface finishes were applied to this model of a skyscraper to create a realistic scene. Bitmaps or predefined background scenes can typically be used for images such as the clouds in this rendering. Notice the textures of the various materials and the effects produced by the lighting. (Autodesk, Inc.)

Colors and other visual effects are normally applied to predefined and custom materials. For example, you may want to define a brick material with a sandy brown color or a polished brass material with the color gold. A number of common parameters are usually available to specify how a material appears when rendered. For example, you can control the amount of reflection, refraction, and transparency. Lighting is then added to generate the desired highlights and shadows.

Locating lights

When locating lights, they must be positioned to produce accurate illumination and shading effects in relation to the model. The same concepts used in manual shading and light projection are applied when locating lights in a scene to be rendered. If materials are applied, lights should be located to complement or enhance surface highlights. You will normally want to use a combination of point source lighting, direct lighting, and spotlighting. A *point source light* projects light in all directions and is similar to a lightbulb. A *direct light* projects parallel rays of illumination in a given direction. This is similar to the light provided by the sun. A *spotlight* projects light rays in a cone of illumination. Many rendered scenes require various sources of illumination. For example, a ceiling light may be used as a direct light to provide general illumination of an overall area, while a spotlight may be placed to focus high-intensity illumination on a specific area.

The parameter settings available for lights depend on the type of light created. Most CAD programs allow you to change the intensity, color, and types of shadows generated by each type of light. The amount of ambient light, or background light, can also be controlled. With spotlights or point source lights, the hotspot and falloff can be set. The *hotspot* of a light is the area of projection where the greatest amount of illumination is generated. *Falloff*, also known as *attenuation*, is the decrease in intensity of a light as the distance from the light source increases. The area of falloff is always greater than or equal to the area defining the hotspot. There are other parameter settings available for lights, such as the amount of diffuse and specular color affected. A rendering with a variety of lighting effects applied to colors, textures, and reflective surfaces is shown in **Figure 10-24**.

Lights must be named and located in the scene with exact XYZ coordinates. In most cases, an icon is used to indicate the location of a light when it is created. Each light should be given a meaningful name to identify the purpose of the light. For example, you might want to precede the name or number of a direct light with *DL* and the name or number of a point source light with *PL*.

Creating a camera

Once the lighting and materials have been defined, one or more cameras can be created by specifying a location with XYZ coordinates. A camera can be used to define how the scene appears when rendered. For example, you may want to define the scene so that it is viewed from the perspective of a camera. The camera can then be panned or rotated to change the view. You may also want to have the camera view a portion of the model by using clipping planes. *Clipping planes* are front and rear boundaries that "slice" through a model to create a partial view. Using clipping planes is similar to creating a section. Any features outside the clipping planes are not included in the resulting view.

As with a light, a camera must be named. Also, depending on the type of camera created, you may also have to specify the location of the target. Each camera located in a scene will produce a different view. Some camera functions allow you to make additional parameter settings. For example, the focal length of the camera may be adjusted by changing the lens size. This establishes the field of vision. Lens sizes are measured in millimeters. A lens size greater than

Figure 10-24. A variety of light sources were used to create the reflections and shadows in this rendering of a sunroom. Notice the background image and the transparent materials used to define the glass for the windows and tables. (Autodesk, Inc.)

50mm duplicates the effect of using a telephoto lens. This creates a smaller field of vision and allows you to zoom in to a closer view of the object. Using a smaller lens setting creates a wider field of vision. For example, if you want more of the surrounding area within the camera view, use a lens size such as 35mm.

Creating the rendering

Before performing a rendering, you must name and define the scene. Scenes are typically created using the **Scene** command. The process requires you to identify which view or camera the scene is to be viewed from. You must also specify which lights will be used to illuminate the scene. As with other named objects, scenes should be named to define what the resulting rendering will show. Once the scene is defined, the **Render** command is typically used to render the view. Some CAD programs allow you to make a preview rendering or a full rendering. Preview renderings are useful when you are uncertain how the final image will appear.

As previously discussed, 3D models are sometimes created in one CAD program and rendered in another. After drawing the model, it is exported to a different program. This is typically done when more advanced lighting functions or materials are required for presentation purposes. See **Figure 10-25**. A basic rendering of the welding machine in **Figure 10-22** is shown in **Figure 10-25A**. A high-quality rendering of the same model is shown in **Figure 10-25B**. A different viewpoint is used to display more detail, and the greater variety of materials creates a photorealistic appearance. Notice the enhanced shading details and reflections produced by the lighting in **Figure 10-25B**.

Renderings can also be created without assigning materials, lights, or cameras to the scene. If no materials are applied, and the scene does not include lights or a camera, default lighting and colors can be used to render the scene. The software assumes that the current view is what the camera will see and applies predefined lighting. This is similar to using the **Shade** command. The resulting rendering is generated relatively quickly.

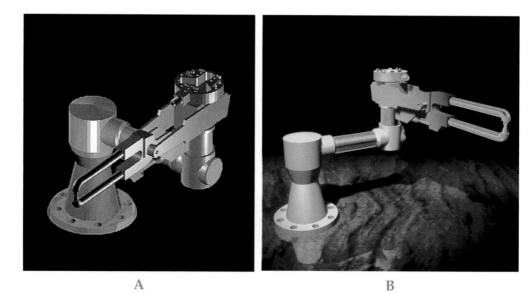

A   B

Figure 10-25. Models are imported into other programs when more detail is required for rendering. A—A simple rendering of a welding machine with basic materials and lighting applied. B—A different program was used to produce a photorealistic rendering of the same model with advanced material and lighting effects. (Autodesk, Inc.)

The major application of the **Shade** command is not to develop a shaded view for output to a printer. Shaded images do not appear as they do on screen when printed. Instead, shaded views are used to give the drafter/illustrator a quick impression of what the object will look like with basic surface shading applied. Renderings, by comparison, are designed to produce views with much higher quality. Because renderings require more sophisticated data calculation than simple surface-shaded views, the generation time takes considerably longer. If you are unsure of how the drawing will appear when rendered, you can often save time by applying the **Shade** command to produce a simple shaded view. However, this command does not render materials. If materials are already applied and you want to view the model with simple shading applied, it is best to make a preview rendering.

As with shaded drawings, renderings cannot be printed as they appear on screen. A rendered view must be saved in a usable format. Typically, renderings can be saved as bitmap (BMP), PostScript, TIFF, and JPEG (JPG) files. If you do not save a rendering to a file when it is created and then make any change to the drawing that results in a regeneration of the drawing database, you will lose the rendered view. Renderings are saved as files so that they can be viewed later or output to a printer. For future reference, you may want to record in a log book the name of each drawing that is rendered, a brief description of the contents, the name of the rendered file, and any names you have assigned to views, lights, and cameras.

Output Devices and Color-Shaded Reproductions

Simple shading and rendering only requires a CAD system and an appropriate output device. Laser and inkjet printers and different types of plotters are commonly used for production work. These types of output devices differ in the level of printing quality or *resolution*. Printing resolution is commonly measured in dots per inch (dpi). For example, output devices used for technical illustrations should have the ability to print a *minimum* of 300 dpi. Images printed at 300 dpi typically have good line quality. Higher-quality output devices are capable of printing images at 600 dpi, 1200 dpi, or 2400 dpi.

In more advanced commercial printing work, an imagesetter may be used to produce color illustrations. An *imagesetter* is an output device capable of converting images to various formats and printing them on paper, film, or printing plates. An imagesetter typically uses a *raster image processor (RIP)* to convert vector-based text and graphic elements to bitmap form. The resulting image is then output by the imagesetter to the appropriate printing media. Imagesetters are capable of producing graphics at resolutions exceeding 3000 dpi.

Output devices should be selected based on the requirements of the application. For example, most printers and plotters are suitable for producing color drawings with line contrast, line separation, and line surface shading. These types of illustrations can be output relatively quickly. However, higher-end devices may be necessary for color illustrations with stipple shading or solid fills. Color renderings also typically require high-resolution output. See **Figure 10-26**. Printing and plotting times normally increase for renderings and color-shaded line drawings.

When developing illustrations for offset printing, it is common to prepare an image master for platemaking. A *master* is a reproduction containing the image to be printed. A master may consist of a simple printout, a printing proof, or an image file. Output from a laser or inkjet printer can be used to

Figure 10-26. Photorealistic color renderings require printing devices capable of high-resolution output. (Autodesk, Inc.)

create a master. The image master is then used to make the printing plate. In color offset printing, a separate master is required for each printing color. The color masters are known as *separations*. With correct registration, the color masters are used to print the image. *Registration* is the correct alignment of different colors or images during the printing process. Creating image masters for multicolor printing is discussed in the next section.

An Example of Multicolor Printing

The following example illustrates a basic way to develop a technical illustration for reproduction. The two-color drawing shown in **Figure 10-27** is used in this example. This is a 2D-based axonometric illustration with CAD-drawn surface shading applied. The drawing is made using two different layers to define the masters. The two masters are needed for printing purposes. One will be used for the object lines and foreground shading printed in black, and one will be used for the surface shading printed in second color (blue).

To create the drawing, the two layers, named **Black** and **Color**, are first defined. The color black is assigned to the **Black** layer. The color blue is assigned to the **Color** layer. The lines defining the object and the shading patterns in the foreground are drawn on the **Black** layer, **Figure 10-27A**. Other shading patterns and solid fill areas are drawn on the **Color** layer, **Figure 10-27B**. Turning both layers on is similar to placing one transparent sheet on top of the other. This effect is simulated in **Figure 10-27C**.

Notice the small circles with intersecting lines in each example. These are called *registration marks.* These marks are used to properly align the images. When the masters are overlaid, the marks should be in perfect alignment. In the offset printing process, registration marks are used to align color separations and plates used in printing. The marks in **Figure 10-27** are created on a separate layer named **Register** with the color red. When each master is printed, the **Black** and **Register** layers are printed together, and the **Color** and **Register** layers are printed together. The masters are printed separately with the appropriate layers turned

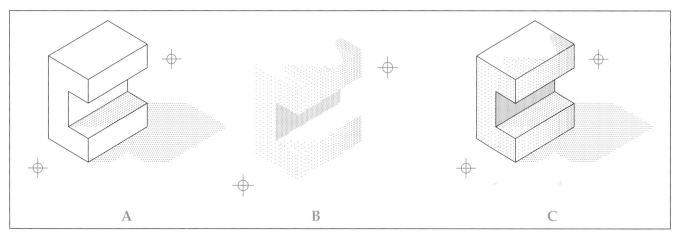

Figure 10-27. Creating masters for a two-color drawing. Registration marks are shown in each example for display purposes. Layers are turned off as needed for each master. A—Object lines and shading patterns in the foreground are drawn on a layer with the color black. B—Surface shading and solid fill areas are drawn on a layer with the color blue. C—The final object with colors correctly aligned.

on or off. After printing, the registration marks can be used to align the masters. The overlaid masters show how the final printed product should appear.

The same process can be used to create masters for drawings with additional colors. Each color is placed on a separate layer corresponding to a master. Then, each master is output for printing purposes. This is a simple way to prepare production work for multicolor illustrations.

Summary

A wide variety of illustration methods can be used when creating drawings with paint-based programs and CAD software. Computer software tools should be selected depending on the nature of the application. Paint programs often provide a simple way to create and edit color bitmap images. Shading and other effects can be applied using a number of drawing tools. Image editors provide advanced functions when special visual effects are used.

The types of drawings created with CAD programs range from simple 2D illustrations with user-drawn shading to full-color renderings with materials and lighting applied. Manual shading methods such as line contrast shading, line separation shading, stipple shading, and line surface shading can be replicated on 2D pictorials using CAD drawing tools. Methods for varying line thicknesses and other shading elements will depend on the CAD software you are using. However, the same concepts involved in manual line and surface shading are used in computer-drawn shading. While applying shading lines and patterns to a CAD drawing does not actually place shading on the drawing, the resulting effects give the pictorial a three-dimensional appearance.

More advanced types of illustrations can be created with solid modeling and surface modeling programs. CAD programs with rendering capability can be used to produce very realistic models. Basic shading functions can be used to initially view a model with color shading applied. As the modeling process progresses, lights, materials, background images, and other visual effects can be added to establish a scene for rendering. High-end modeling and rendering programs are more expensive to operate and require more advanced drawing and visualization skills. However, the use of modeling and rendering software in combination with core drawing skills gives the illustrator powerful tools to develop high-quality illustrations.

Review Questions

1. What is the main difference between illustrations developed in paint programs and drawings created in CAD programs?
2. What is the purpose of an image editor?
3. Explain two advantages to using layers when creating images in a paint-based program.
4. Briefly describe how colors are identified differently when using software with RGB color formation, CMYK color formation, and HSL color formation.
5. Explain why the color images displayed on your monitor may not have the actual color generated when an illustration is printed.
6. What is meant by the term *WYSIWYG*?
7. What is the purpose of a color management system?
8. Briefly discuss the major advantages and disadvantages associated with paint-based programs.
9. Why are 2D-based CAD drawings shaded with line shading and surface shading techniques using basic drawing commands rather than automatic computer-generated shading and rendering functions?
10. What is a *polyline* and how is it useful in creating line contrast shading in a CAD drawing?
11. How should surface shading lines be aligned within holes on pictorial drawings?
12. How should crosshatch patterns be normally oriented for line surface shading on objects in a pictorial drawing?
13. Briefly explain how elliptical arcs can be drawn using line surface shading techniques to represent rounded features in a CAD drawing.
14. How should surface shading lines be drawn to represent conical shapes on pictorial drawings?
15. List two ways to apply solid fill surface shading patterns using CAD drawing functions.
16. Define the following terms: *scene, ambient light, diffuse color, specular color,* and *material.*
17. List and describe the basic steps involved in preparing a 3D model for rendering.
18. What is the difference between the **Render** command and the **Shade** command?
19. What type of light created in a CAD program is most similar to a lightbulb?
20. Define *hotspot* and *falloff*. Which types of lights allow you to change parameter settings related to hotspot and falloff?
21. How can the focal length of a camera be adjusted when defining a scene?
22. Why are renderings saved as electronic files?
23. How is printing resolution commonly measured in relation to output devices?
24. What is *registration*?
25. Briefly explain the basic procedures involved in preparing image masters for multicolor printing.

For Problems 1–8, use the orthographic views shown and basic CAD drawing commands to develop axonometric drawings. For each problem, use the dimensions provided and orient the object to create the most realistic representation. Use the 2D-based shading techniques and CAD functions discussed in this chapter to apply line shading and surface shading. When necessary, combine different shading methods to create realistic highlights and shadows.

1.

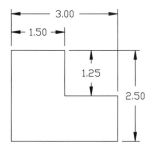
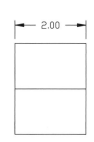

2.

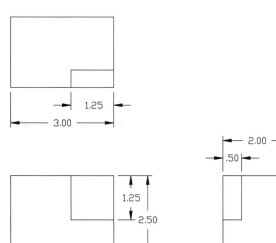

3.

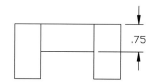

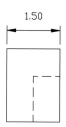

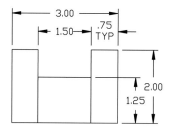

4.

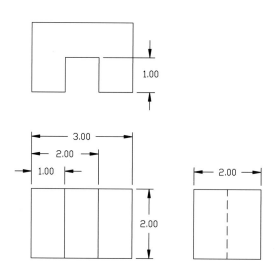

5.

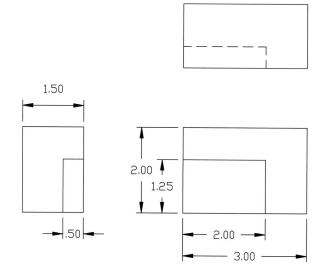

6.

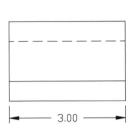

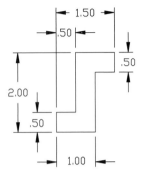

7.

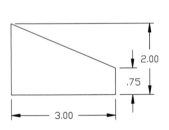

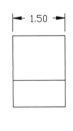

8.

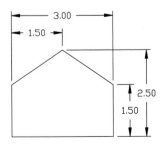

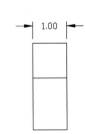

For Problems 9–16, use the orthographic views shown to develop three-dimensional models. Use surface modeling or solid modeling methods and primitive shapes when suitable. For each problem, use the dimensions provided and orient the object to create the most realistic view. Using the methods discussed in this chapter, prepare a scene for rendering by applying materials and lights. After completing each scene, perform a computer-generated shading or rendering as directed by your instructor.

9.

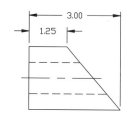
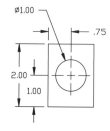

10.

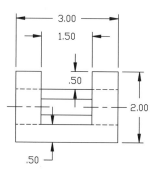
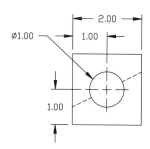

11.

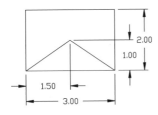

12.

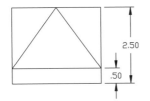

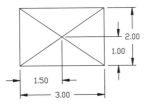

13.

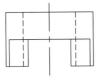

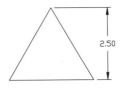

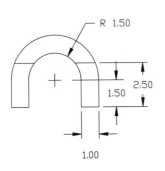

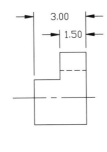

14.

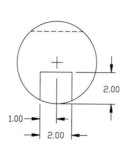

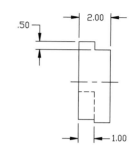

15.

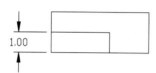

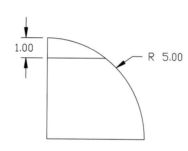

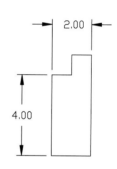

16.

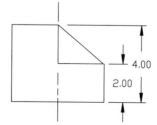

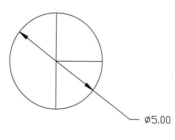

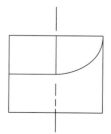

For Problems 17 and 18, use the major feature dimensions provided as a guide to develop three-dimensional models. Use your own dimensions where necessary. Use surface modeling or solid modeling methods and primitive shapes when suitable. Orient each drawing to create the most realistic view. For each problem, prepare a scene for rendering by applying materials and lights. You may also want to create ground and background objects, depending on the system you are using. After completing each scene, perform a computer-generated shading or rendering as directed by your instructor.

17.

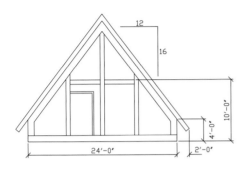
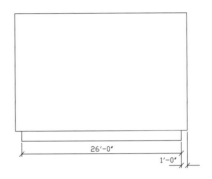

18.

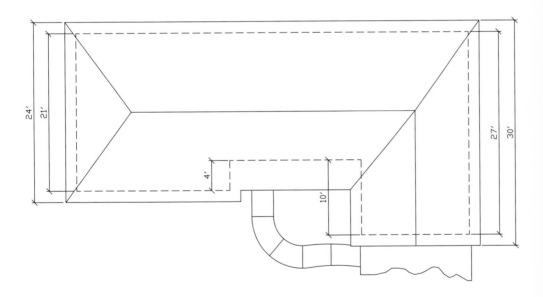

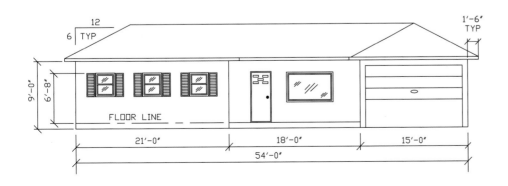

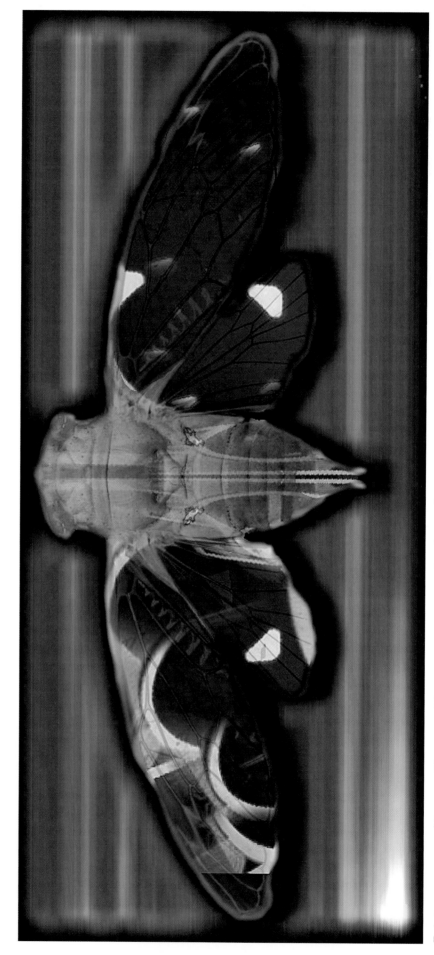

Paint programs and image editors provide numerous ways to simulate airbrushing and other painted effects. This cicada was created from an electronically painted image and a digital photo. After the two images were combined in an image editor, the background was added with an airbrush tool. (Jason Weiesnbach)

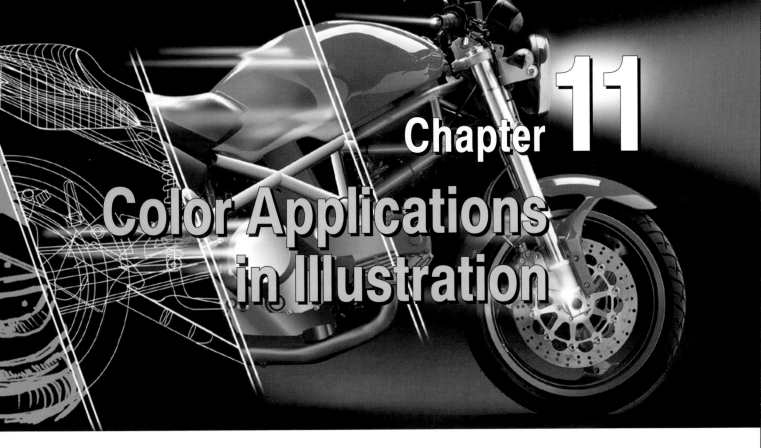

Chapter 11

Color Applications in Illustration

Learning Objectives

At the conclusion of this chapter, you will be able to:

☐ Describe the basic principles that define how color is perceived.

☐ Explain additive and subtractive color formation.

☐ List and describe the characteristics of color, including hue, saturation, and lightness.

☐ Explain the different devices and methods used for choosing colors in an illustration.

☐ Identify commonly used color models and explain their use.

☐ Differentiate between the primary ways in which images are created and describe the various formats and output devices used for printing color.

Introduction

Color illustrations are more prevalent than ever before, **Figure 11-1**. Therefore, color is a very important element to consider when deciding how to best communicate an idea graphically. Which colors will have the desired impact on the viewer, which colors look good together, and which tools are available to produce color are all things to consider when working with color illustrations. For example, the exterior and interior colors for a car will most likely be selected after a careful study of market trends in automobile colors. If the illustration is to target young car buyers, a "hot" or "flashy" color such as bright red may be used. However, middle-age buyers may respond better to a more conservative or subdued color such as beige. This chapter provides an overview of color theory and design considerations to apply when using color. Computer software color models and color printing applications are also discussed.

Light and Color

The perception of color is determined by how the brain interprets specific wavelengths of light that we see. If there is no light, there can be no color. Light can come directly from a light source, such as a blue gas flame. However, most color is perceived from light reflecting off an object, such as a red apple.

Like a radio signal, light is made up of electromagnetic waves. *Visible light* is the part of the electromagnetic spectrum we can see. Radio waves are part of the spectrum we cannot see, **Figure 11-2**. The *electromagnetic spectrum* consists of bands of wavelengths. One end contains long wavelengths, such as radio waves. In the middle are the medium wavelengths, known as visible light. These waves are visible to the human eye and are perceived as color. The corresponding colors range from red to violet and are primarily red, green, and blue. On the other side of the spectrum are very short wavelengths, such as X-rays.

Figure 11-1. Color helps to enhance the presentation of this illustration. (Honda)

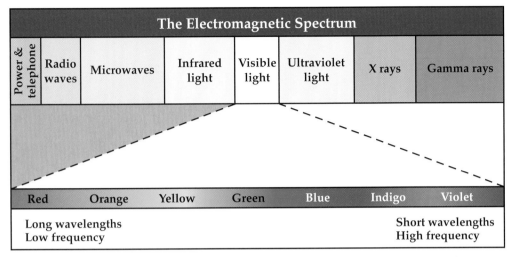

Figure 11-2. The electromagnetic spectrum is made up of bands of wavelengths. Visible light makes up a relatively small portion of the spectrum.

Most light sources, such as the sun, a flame, or a lightbulb, produce light in a combination of wavelengths. For example, Sir Isaac Newton discovered that a prism in sunlight divides the light into a rainbow of colors. This is usually called the *spectrum*. See **Figure 11-3**. All of the visible wavelengths are seen in approximately equal parts in what we perceive as white light.

There are two primary methods used to describe the formation of color—additive and subtractive. Additive colors and subtractive colors are often called "physicist's primaries" since they are based on the science of light and color. The additive and subtractive color formation methods are discussed in the next sections.

Additive Colors

In *additive color formation*, red, green, and blue wavelengths are combined to form other colors. When these colors are mixed in approximately equal amounts, they produce white light. See **Figure 11-4**. This type of color formation is often referred to as *RGB color* (RGB stands for red, green, and blue). The additive colors are called *additive primaries* or *primary colors* because they can be added together in varying amounts to create most other visible colors. For example, red and green light combine to produce yellow. The

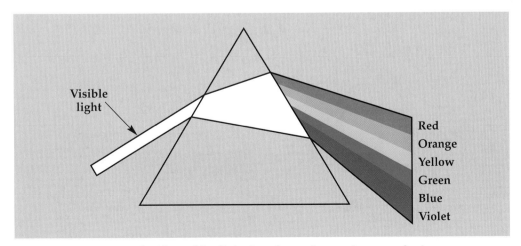

Figure 11-3. A prism divides white light into its various colors, producing a rainbow effect.

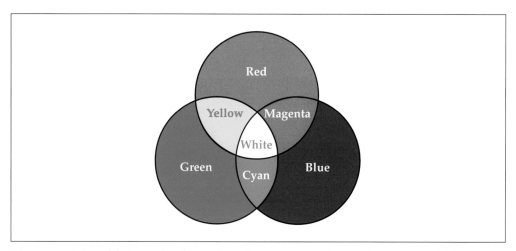

Figure 11-4. In additive color formation, the primary colors red, green, and blue produce white light when mixed in approximately equal amounts.

retinas of our eyes have special sensitivity to the additive primaries, allowing us to see full color.

Additive color formation is used by color television systems and computer monitors. Tiny red, green, and blue phosphor dots gradually brighten as they are struck by electrons. Various colors are formed based on which dots are "lit" and their level of brightness.

Subtractive Colors

Additive color formation is used for objects that transmit light directly to the eye, such as televisions. However, a color image on a surface *reflects* light. A different method of color formation is used to describe reflected color from a surface.

In *subtractive color formation*, some colors are reflected from a surface and other colors are absorbed or "subtracted." See **Figure 11-5**. For example,

Figure 11-5. When looking at a printed color, you see only the colors that are reflected as light even though all colors are present. In this example, the red and blue primaries from the white light are absorbed and only the color green is reflected.

cyan ink printed on white paper is seen as the color cyan because it absorbs other colors and reflects only cyan.

Cyan, magenta, and yellow are known as the *subtractive primaries*. Two of the three additive primaries make up each subtractive primary color. Only the additive primaries making up each subtractive primary color are reflected because each color absorbs other colors. For example, cyan ink absorbs red and reflects the blue and green wavelengths making up cyan. Magenta ink absorbs green and reflects the red and blue wavelengths making up magenta. Yellow ink absorbs blue and reflects the red and green wavelengths making up yellow. When two of the subtractive primaries are combined, they subtract all but one additive primary. When all three are combined, black is produced since all colors are absorbed, thus leaving no color (or black). See **Figure 11-6**.

Subtractive colors are used as ink pigments or dyes for printing purposes. For example, mixing cyan and magenta produces blue. Mixing magenta and yellow produces red. Mixing yellow and cyan produces green. Refer to **Figure 11-6**. Mixing all three produces black. However, since the combination of pigments is not perfect, mixing the three subtractive colors often creates a dark brown instead of black. Therefore, black is typically used as a fourth color in printing to ensure a good black and enhance the overall image. This type of printing, or color formation, is called *CMYK color* (CMYK stands for cyan, magenta, yellow, and black; black is designated by the letter K). CMYK color formation is used for printing color photos and the four-color images found in magazines, books, and other publications.

The colors blue, red, and yellow are often used to approximate the subtractive primaries cyan, magenta, and yellow. For example, artists use blue, red, and yellow as their primary colors when mixing inks because they can be mixed to form most other colors. These colors are called "artist's primaries."

Describing Color

How would you describe the color of each flower in **Figure 11-7**? The flowers on the right may appear orange to most people. To others, they may appear yellow or yellow-orange based on the lighting. Perception of color varies slightly from one person to the next. For this reason, color classification systems have been created to accurately describe color.

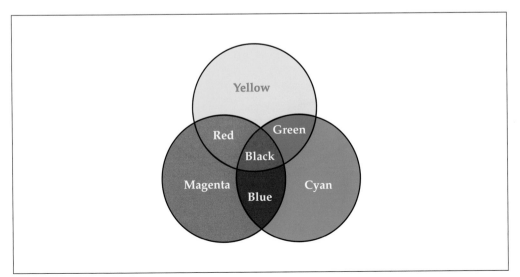

Figure 11-6. When combined, the subtractive primaries of cyan, magenta, and yellow produce black.

Figure 11-7. Individual perceptions of color are different. How would you describe the color of the flowers on the right? (Jack Klasey)

Some systems for describing color are based less on science and more on the way we actually perceive color. One such system is HSL, which stands for hue, saturation, and lightness (or brightness). See **Figure 11-8**. In this system, color has three dimensions (hue, saturation, and lightness) establishing the relationships among different colors.

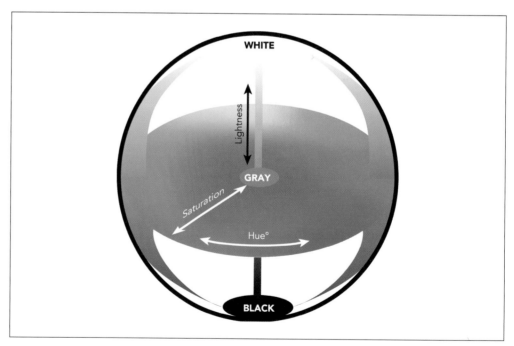

Figure 11-8. In the HSL system for describing color, color has three dimensions (hue, saturation, and lightness), as shown on this sphere. The different hues are represented by circular movement around the sphere. Saturation is defined by radial movement from the center of the sphere, and lightness is defined by upward or downward movement from the center. Hues become less saturated (more gray) toward the center of the sphere. Hues become lighter toward the top of the sphere, and darker toward the bottom. (X-Rite, Inc.)

Hue is the color of an object as defined by reflected light. It can simply be described as the name of the color. For example, a hue can be defined as red, green, yellow, or as a color with a system designation such as Pantone® Blue. Within a hue, such as green, a variety of slightly different colors can be created by changing the saturation and lightness. For example, a mint green and kelly green can be created within a green hue.

Saturation is the purity of a color, or absence of gray. Saturation is also called *chroma*. For example, both pink and scarlet have a red hue. However, scarlet is more saturated since it contains the most red. See **Figure 11-9**. Terms such as "vivid" or "dull" are sometimes used to give a rough estimate of the level of saturation. Pink is a "vivid" red while scarlet is a "dull" red.

Lightness describes how light or dark a color is. Lightness is also called *value*. An example of comparative lightness is a light blue and a dark blue. Lightness of a hue can be changed by adding white, black, or gray to it.

Color and Design

A variety of devices and designations are used to classify colors for design purposes. A *color wheel* is used to show the relationship among colors, **Figure 11-10**. It is often used to help determine colors that appear pleasing together. Color wheels are typically made up of primary colors and colors mixed from these primaries. Most colors can be produced from the primaries on the wheel.

The color wheel in **Figure 11-10** uses red, yellow, and blue as the primaries. When two primaries are mixed together in equal portions, a *secondary color* is formed. For example, mixing red and blue creates the secondary color violet. Yellow and blue are mixed to make green, and red and yellow are mixed to make orange. A color produced by combining a primary and secondary color is called an *intermediate color* or a *tertiary color*. Examples of tertiary colors are blue-green, blue-violet, red-violet, red-orange, yellow-orange, and yellow-green. When naming intermediate colors, the primary color is listed first.

Colors directly across from one another on the color wheel are called *complementary colors*. Orange and blue are complementary colors. Colors next to, or close to, one another are called *analogous colors.* Red and red-violet are analogous colors. A *triad* is made up of three equally spaced colors such as red, yellow, and blue.

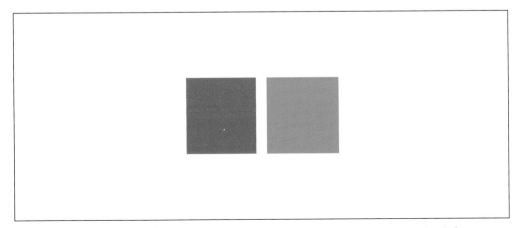

Figure 11-9. Even though these colors have the same hue, the color on the left is more highly saturated.

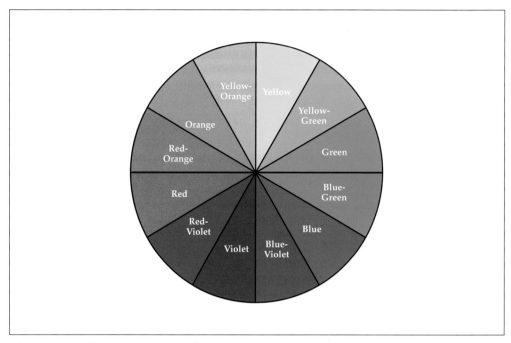

Figure 11-10. This color wheel is based on the artist's primaries—red, yellow, and blue. When two primaries are mixed, a secondary color is produced. When a secondary color and primary color are mixed, an intermediate color is produced.

Colors can be classified as warm or cool colors. Colors such as red, orange, and yellow are called *warm colors* because they tend to convey a feeling of warmth. Warm colors are often called *advancing colors* because they may appear to move forward and suggest activity. Colors such as blue, violet, and green are *cool colors* because they are often restful, relaxing, and calm. Cool colors are also called *receding colors* because they tend to move away from the viewer or stay in the background. Often, illustrations will have cool colors in the background to make the object in the foreground advance toward the viewer.

Colors tend to have different meanings in different cultures. For example, in some cultures, such as Western culture, red is often associated with danger or excitement. In Chinese culture, red typically signifies good luck. As the lightness or saturation of a hue changes, this also may change the meaning of the color. For example, darker colors tend to be associated with suspense, mystery, and tranquility. Lighter colors tend to be associated with youth, excitement, and happiness.

Choosing Colors

When choosing colors for a design, the general purpose and intended audience should be identified. If the purpose is to develop an illustration for an annual report with a seasonal theme, the theme should be a factor in choosing colors. In addition, if the illustration is to be of interest to a specific age group, certain colors might be most effective. In the case of an assembly drawing that shows how a mechanism is assembled, choosing colors that look good together and clearly show the different parts is more important than choosing colors for their meaning. See **Figure 11-11**. The following general guidelines should be used when choosing colors:

1. In many cases, one dominant color should be chosen for the illustration. Limit the number of colors in the illustration when possible. See **Figure 11-12**. For simple illustrations, try to use no more than three colors.

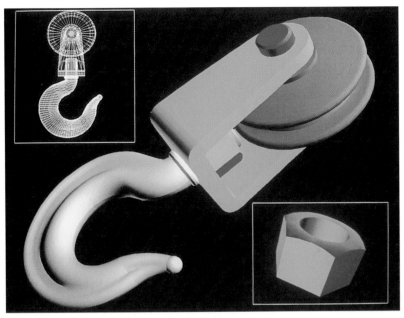

Figure 11-11. Each part of this assembly is shown clearly through the use of color. (Parametric Technology Corporation)

2. Color harmony can be achieved by using analogous hues or different lightness levels of the same hue. Achromatic colors can also be used to accomplish color harmony. These colors harmonize with all other colors, **Figure 11-13**.
3. When analogous colors are used, a complementary color can be applied sparingly as an accent to provide contrast.
4. Hue (color) is often less important than saturation or lightness. Decide what lightness is necessary to convey the intended mood. In addition, for impact, try varying lightness levels rather than hue. See **Figure 11-14**.

Figure 11-12. One color is dominant in this model of a wheel.

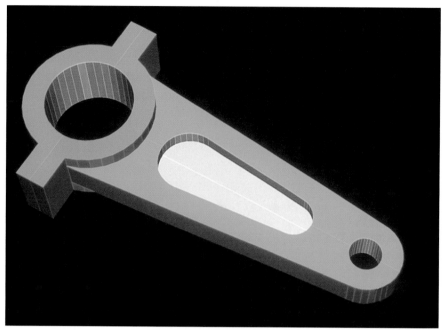

Figure 11-13. The achromatic colors black, gray, and white are used in this model for harmony. (Parametric Technology Corporation)

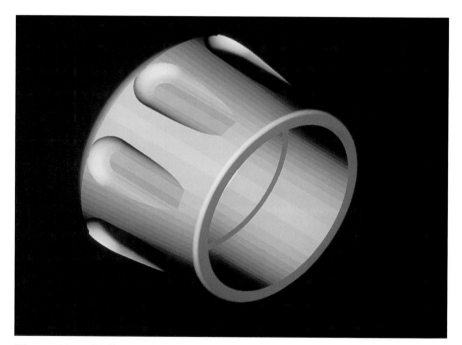

Figure 11-14. Different shades of the same hue, rather than different hues, are used in this illustration for impact.

5. Using a color triad often produces pleasing results. However, three equally saturated hues, such as a vivid blue, red, and yellow, tend to compete for dominance in an illustration and become distracting.
6. Cool colors work well for backgrounds since they tend to recede when viewed. Warm colors work well for the primary objects in a scene since they tend to move toward the viewer.
7. The intended audience for the illustration is very important. Color trends for the audience should be applied. Remember that colors have different meanings for different audiences.

8. If there are color standards used for a certain type of drawing, they should be applied. Color standards should always take precedence over other color considerations. For example, a drawing of the interior of an industrial plant with safety markings should use the specific color required, regardless of other design considerations.

Computer Software Color Models

As discussed in Chapter 10, a *color model* is a system used to describe colors. Many illustration software packages use color models as a way to define the desired colors for an illustration. Among the most common models are the RGB, HSL, and CMYK models. The model chosen often depends on the software, personal preference, and the precision of color needed for the final product.

The **RGB color model** defines colors as percentages of the three additive primaries of light (red, green, and blue). See **Figure 11-15**. A value of 100% for each of the primaries produces white. Color computer monitors use red, green, and blue phosphor dots to display images on screen. Color scanners and color film both use RGB color for capturing images. The RGB model is good to use if the color will be displayed by a monitor or a projection system.

The **HSL color model** defines colors based on hue, saturation, and lightness. These attributes were discussed earlier in this chapter. In some software programs, this model may be specified as HSV (hue, saturation, and value) or HLS (hue, lightness, and saturation). See **Figure 11-16**. In this system, colors are formed by adjusting the hue, saturation, and lightness individually. This model is used in many different types of software because it is a logical way to describe color.

The **CMYK color model** defines colors based on the subtractive primaries (cyan, magenta, and yellow) and black. A percentage is assigned to each color, **Figure 11-17**. CMYK color is used in process color printing. This is used for color printing in books, magazines, and similar publications.

As discussed in Chapter 10, a color display on a computer monitor frequently does not match the resulting color produced by an output device. Also, colors may change as images are converted from one format to another.

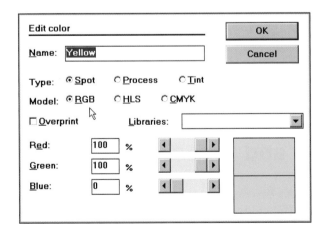

Figure 11-15. Colors are specified as percentages of the three additive primaries of light in the RGB color model. A value of 100% for both red and green produces yellow.

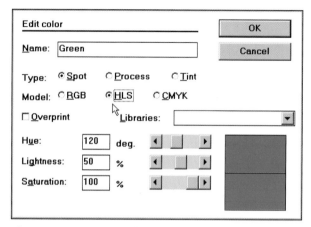

Figure 11-16. Many software programs allow you to define colors based on hue, saturation, and lightness.

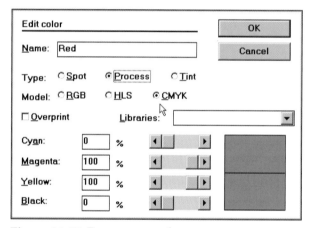

Figure 11-17. Percentages of cyan, magenta, yellow, and black can be adjusted using the CMYK color model.

In some cases, color management systems are used to control the consistency of color throughout the production process. When a color management system is not available, a simpler way to control color output is to use software with user specification options based on the Pantone® Matching System (PMS). This model allows you to specify names or numbers for colors corresponding to common ink colors used in the printing industry.

Color manuals and swatchbooks are used in the Pantone system and similar systems to select printing colors. A *swatchbook* is a color reference manual or guide with samples of all available colors in the system. Each color is specified with the corresponding identification name or number. See **Figure 11-18**.

Spot Color and Process Color

Printed color can be classified as *spot color* and *process color*. **Spot color** is a solid area of a specific color. **Process color** refers to four-color CMYK printing. Spot colors are printed with solid (opaque) inks. Process colors use transparent inks since the colors are designed to be overprinted, or layered over each other. A variety of software tools and color management systems can be used for matching spot and process colors during the production process.

Figure 11-18. Colors used for printing can be selected from a swatchbook by specifying the identification names or numbers.

Spot color and process color are often used separately in the same image. For example, an illustration may use process color to reproduce a color photo and spot color to print a company logo. As previously discussed, CMYK printing can be used to represent most colors. In this way, a CMYK color could be used as spot color on a printed product as well. However, if the spot color is highly important and must match a specific color, such as a logo color, then a separate ink color is used. The product will be more expensive to print because a fifth color will be printed. However, this ensures that the logo will have the exact color required for the client.

Line and Continuous Tone Images

There are two primary types of images that can be used for an illustration. These are line images and continuous tone images. A *line image*, also known as line art or a line illustration, has no gradation of tone; it has a uniform density of color, such as black. See **Figure 11-19**. The type on this page is an example of

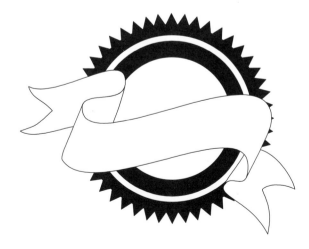

Figure 11-19. An example of a line image.

a line image. A single-color multiview drawing is another example of a line image.

A *continuous tone image* has gradations or variations of tone. For example, a black-and-white photo has black, white, and various shades of gray in the image. Black-and-white continuous tone images are also known as *grayscale images*. In most printing applications, a continuous tone image must be converted to an image consisting of a series of dots. The image can then be printed. You can see an example of this by looking at a newspaper photo with a magnifying glass. The printed photo is made up of a series of small dots. This is known as a *halftone*.

Continuous tone color images, such as color photos, are made up of a number of different colors. For commercial printing, color photos are converted to four color separations to prepare them for the CMYK process. Separations are then used to make individual printing plates, or printers. In this type of printing, a separate plate is used to print each CMYK color.

There are a variety of printing systems used for four-color printing. Some of the more common color output systems and printed formats are discussed in the next section.

Color Output

There are many different ways to print color illustrations. Color inkjet and pen plotters are commonly used to print line drawings created with CAD systems, **Figure 11-20**. Color inkjet and laser printers are used for photographs and other continuous tone images. When larger quantities are needed, color illustrations are printed with commercial printing equipment, **Figure 11-21**.

Depending on the application, color illustrations may also be output as proofs, film, or slides. If the required equipment is not readily available, the

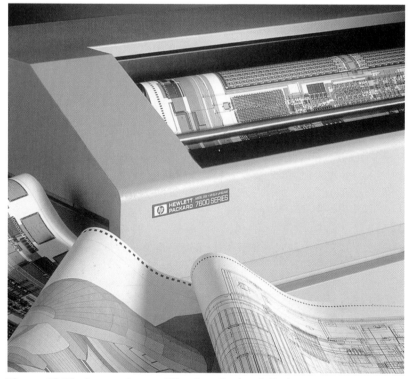

Figure 11-20. A pen plotter. (Hewlett-Packard Co.)

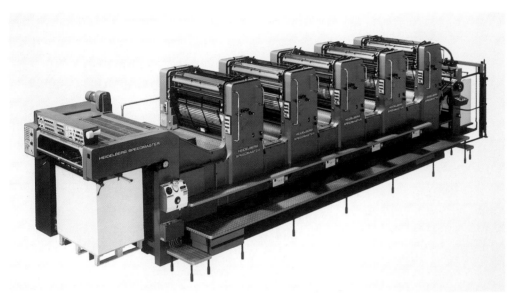

Figure 11-21. This printing press is used for printing in five colors—the four process colors, or CMYK colors, plus a fifth solid color if needed. (Heidelberg Eastern, Inc.)

electronic version of the illustration is taken to a service bureau, graphics company, or commercial printer that offers outputting services.

Some illustrations are created to be viewed electronically, such as on a Web page. The color model used for these illustrations will be RGB, and the file will normally be small in size so the page loads quickly.

Summary

A basic knowledge of color theory is important in understanding how we see color and how different design and printing technologies use this theory. Computer monitors use the additive primaries (red, green, and blue) to create the color images that we see on screen. In contrast, most output devices use the subtractive primaries (cyan, magenta, and yellow) to produce printed color images on a surface.

Color wheels assist in selecting colors that work well together. In many cases, specific guidelines are used in determining colors for a design. The intended audience plays a very important role in the choice of color. In addition, if there are certain color standards associated with a specific type of illustration, they should be a primary consideration.

Color models used in computer software are based on the various systems for describing color. The CMYK color model is typically used for four-color images that will be printed commercially. Single-color line art or spot color can be printed using a combination of the subtractive primaries or a separate color. In either case, color matching tools or color management systems should be used when it is important to produce accurate color.

In order to use color effectively in illustration, it helps to be aware of basic design guidelines. This includes being cognizant of the color preferences of the intended audience. You also need a basic knowledge of the color tools available. Be sure to experiment with different color in your designs. The end result will be an illustration that has greater visual appeal.

Review Questions

1. Explain in basic terms how color is perceived by the human eye.
2. The portion of the electromagnetic spectrum that we can see is known as what?
3. Define *additive color formation* and *subtractive color formation* and briefly explain the differences between the two.
4. In additive color formation, how is white light produced?
5. What colors are known as the *subtractive primaries*?
6. In theory, what color is produced when the subtractive primaries are combined in approximately equal amounts?
7. Define the terms *hue, saturation,* and *lightness.*
8. The colors pink and scarlet have the same red hue. Which color has more saturation?
9. In relation to a color wheel, what is a *secondary color*?
10. Colors that are directly across from one another on a color wheel are known as what?
11. List and describe three general guidelines that should be used when choosing colors for a design.
12. Which color model is used by computer monitors to display images on screen?
13. How are colors defined in the CMYK color model?
14. Explain the difference between *spot color* and *process color.*
15. What is the difference between a line image and a continuous tone image?
16. Briefly explain how continuous tone color images are prepared for offset printing.

Activities

1. Collect a number of discarded magazines from various sources and look through the color illustrations and advertisements. Pick several and describe how you think the colors were chosen using the color principles discussed in this chapter.
2. Many software programs, such as illustration programs and image editors, provide different color models that can be used for outputting color. Choose a program that you have access to, list the different models available, and use the help feature of the program to describe each one.
3. Using the color capabilities of a word processing program, design a simple announcement or advertisement using the color guidelines discussed in this chapter. Color features in a word processor include options related to borders, shading, font colors, and other elements. As an alternative, an illustration program or CAD program can be used.
4. Using four discarded color illustrations, locate examples of printed spot colors and process colors. Identify the colors by marking directly on the illustrations. Printed company logos are usually a good source for spot color.
5. Using a search engine on the Internet, locate at least three good sources of information about hue, lightness, and saturation. List your sources and describe how these terms differ.

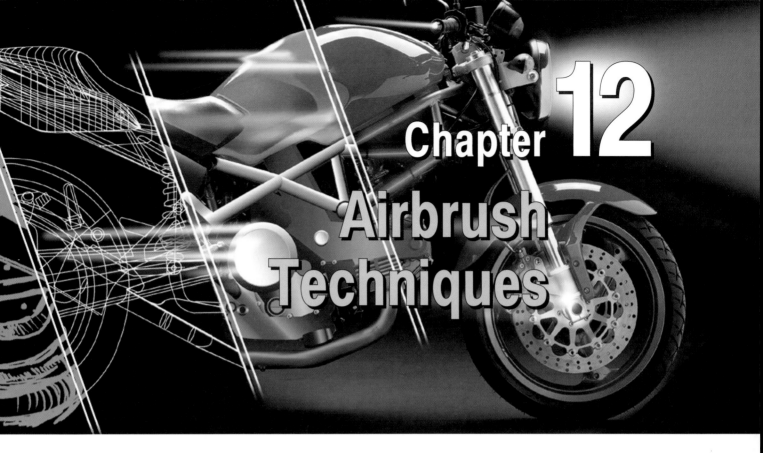

Chapter 12

Airbrush Techniques

Learning Objectives

At the conclusion of this chapter, you will be able to:

- [ ] Identify the principal applications of the airbrush process.
- [ ] List and explain the components of an airbrush system.
- [ ] Name and describe the use of common airbrush illustration materials and aids.
- [ ] Describe the procedures involved in preparing a drawing for airbrushing.
- [ ] Explain the principal techniques of airbrushing.
- [ ] Lay out, mask, and airbrush pictorial illustrations.
- [ ] List the health and safety guidelines related to airbrush illustration.

Introduction

Many people think of airbrushing in relation to artists spray painting T-shirts at a concert or fair. However, airbrushes have a much wider range of use in the illustration field. Airbrushes are used to prepare illustrations for catalogs, brochures, manuals, and magazines. Many photographs and advertisement illustrations are also modified with airbrushing prior to printing.

Effective use of an airbrush requires considerable practice and well-developed eye-hand coordination. Highly skilled airbrush illustrators are typically considered to be artistic specialists in the field of illustration. This chapter introduces the applications of airbrushing, the features of an airbrush system, the illustration materials needed for airbrushing, and the techniques involved in using an airbrush.

What Is an Airbrush?

An *airbrush* is a small, hand-held spray gun used to apply paints and inks with a high degree of control. See **Figure 12-1**. An extremely thin spray pattern can be created for details and object edges. A medium-width pattern can be created for shadow effects and confined surface features. A wide spray pattern can be created to paint surface texture effects on large surface areas. Airbrushes are primarily used in illustration to paint the surfaces of a drawn object.

An airbrushed illustration of an engine is shown in **Figure 12-2**. Various intensities of colored ink are used to develop lightly shaded surfaces, heavily shaded surfaces, and areas representing surface and feature edges. In addition, the colors that are used help the viewer understand parts of the engine.

Airbrushing is commonly used to touch up photographic images, **Figure 12-3**. This illustration shows a photo from a drafting supply catalog that has been retouched with an airbrush. Many photographs of objects in their natural environment lack enough contrast among the various surfaces and shapes. This makes

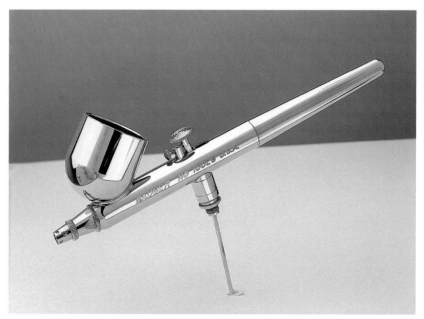

Figure 12-1. A double-action airbrush. Airbrushes are used to produce a variety of spray patterns and color densities. (Badger Air-Brush Co.)

Figure 12-2. High-quality shading and surface details provide a photorealistic effect in this airbrushed illustration. (David Kimble)

Figure 12-3. Photos of standalone objects are commonly retouched with an airbrush to enhance the appearance of surfaces and shapes. (Koh-I-Noor, Inc.)

it very difficult to visually separate the object from the background. In these situations, a fine spray pattern on the background features of the photograph can soften the image. This also brings the object toward the viewer. Although some illustrations are developed using only an airbrush, it is more common to use an airbrush in combination with technical pens or brushes.

Airbrush Applications in Illustration

There are a variety of illustration applications for airbrushing. Touching up photos is the most widespread application, **Figure 12-4**. Another application is

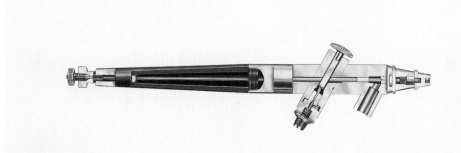

Figure 12-4. A touched-up photo of an airbrush used for a catalog illustration. The airbrush shading establishes fills and helps define the internal components. (Badger Air-Brush Co.)

spray shading of surfaces. This is similar to stipple shading. A very fine mist of ink or paint can be sprayed over an area to visually suggest that a solid surface exists. By making several passes of the airbrush to spray on more of the airbrush media, the appearance of a lightly shaded surface may be developed. Making multiple passes of the airbrush will increase the density of the media until a solid fill representing full shading has been accomplished. This shading technique is shown in **Figure 12-4** and is a very important application of airbrushing in illustration.

Objects developed with a technical pen or paintbrush appear artificial against a plain white or colored drawing surface. An airbrush can be used to spray colored media surrounding the exterior of the object to give the appearance that the object is sitting on a solid background surface. An airbrush is used for this type of background application because it is easier to create a uniform density of color with an airbrush than with a paintbrush. See **Figure 12-5**. This illustration shows a very complex line drawing of a transmission with airbrushing applied. In addition to background shading, internal parts have been airbrushed to provide shading and highlights. Without this shading, the individual components would be difficult to recognize.

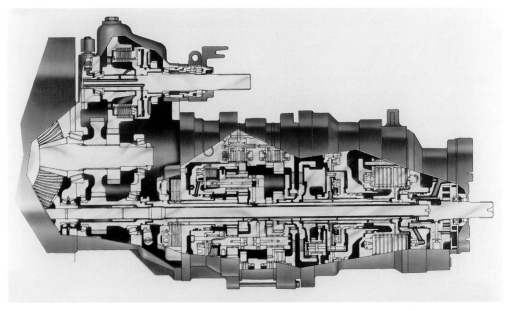

Figure 12-5. Airbrush shading is used for the background and internal components to enhance the details of this illustration. (John Deere & Co.)

Airbrush Equipment and Materials

An airbrush system consists of a source of compressed gas or air, a supply hose, and an airbrush. The system may also include additional components such as a pressure regulator and gauge, a dehumidifier, an air supply manifold, and other airbrushes. An airbrush system that is used on a regular basis should always be operated in a spray booth. A *spray booth* is a closed room or boxlike enclosure that allows control of production conditions. It is ventilated to remove paint fumes to maintain a safe and healthy work environment.

Air Supply Systems

An air supply system is selected to meet specific needs of the airbrush illustrator. The nature and volume of airbrush work to be done determines how much pressure and airflow is required. The different types of air supply systems available include pressure canisters, pressure tanks, industrial air compressors, and airbrush air compressors. The next sections explain the major features and applications of each of these systems.

Pressure canisters

A pressure canister is a small vessel or "can" containing compressed air. Pressure canisters are generally about the size of a spray paint can. Typical pressure canisters used for airbrushing are shown in **Figure 12-6**.

Pressure canisters are adequate for small airbrush projects or when airbrushing activities are only completed on an occasional basis. They are inexpensive and convenient. However, there are several disadvantages to using pressure canisters. First, they are normally used with low-cost systems operated without a pressure gauge or a pressure regulator. This limits the degree of precision usually necessary with an airbrush. Another disadvantage of pressure canisters is that their small size limits the amount of compressed gas

Figure 12-6. Pressure canisters used for airbrushing are inexpensive and used for smaller projects. (Badger Air-Brush Co.)

available. As the gas is sprayed through the airbrush, the gas pressure in the canister decreases. When the can empties, there is a very dramatic drop in pressure and rate of gas flow. This pressure drop can cause an unexpected loss of control of the airbrush spray pattern and can produce what some illustrators call *spubbles*. **Spubbles** occur when the spray from the airbrush strikes the illustration board surface as a series of large blobs, rather than as a finely atomized mist. The resulting spurts and bubbles can destroy several hours of work. Spubbles are caused by insufficient air pressure.

Several important safety guidelines should be followed when using pressure canisters. Always keep pressure canisters away from heat sources so internal pressures do not cause the container to explode. Also, *never* use a pressure canister designed for some purpose other than airbrushing. Although similar in appearance, pressure canisters for airbrush applications are *not* the same as canisters containing automotive air conditioning system refrigerant or similar canister products. Pressure canisters for non-airbrush applications can be very dangerous when connected to low-pressure airbrush attachments.

When using a pressure canister with an airbrush, wear safety glasses or goggles when attaching the valve to the top of the canister. Exhausted canisters should be disposed of properly.

Pressure tanks

Pressure tanks are containers designed to hold larger quantities of pressurized air or gas. Before high-quality air compressors and low-cost gas canisters were available, many airbrush illustrators used pressurized tanks. Some illustrators still prefer to use them. Although these tanks vary in size, the most common size used by illustrators is about the same as that of tanks used in welding.

There are four safe gases used for airbrushing. These are compressed air, helium, argon, and carbon dioxide. *Caution: Never use oxygen as a pressure source for airbrushing. Oxygen will greatly increase the potential for fire or explosion when in the presence of flammables such as oil-based airbrush paints. Never use a pressure tank containing oxygen or any other unsuitable gas for airbrushing.*

Compressed air is a very practical and inexpensive option for airbrushing. However, a compressed-air tank may have to be refilled often. Helium and argon are readily available, inert pressurized gases occasionally used for airbrushing. Inert gases are not flammable. In addition, inert gases do not create undesirable reactions with airbrush media, such as color changes or separation of pigments, binders, and thinners. Inert gases also do not increase drying time. However, they are not heavily used because they tend to be very expensive.

The most common gas used in a pressure tank for airbrushing is carbon dioxide (CO_2). CO_2 is readily available, relatively inexpensive, and comparatively safe to use. CO_2 tanks come in many different sizes. If you will be using CO_2 for a considerable length of time, you may want to purchase a tank. However, for only a limited time, a tank can be rented from a gas supply company. Small, easily transported, 20-pound or 40-pound CO_2 tanks are normally used for airbrush applications. The 40-pound tank holds 40 pounds of gas *by weight* and is approximately 9″ in diameter and 4′ tall.

Proper safety guidelines should always be followed when using pressure tanks. As with any pressurized gas, never directly breathe the gas vapors. Always wear proper eye protection and be careful that the pressurized gas does not blow objects toward yourself or others. The major safety concern in using a tank of pressurized gas is to make sure that the tank is secured in an upright position. The tank should have a sturdy restraining device to prevent it from falling over. If a pressure tank falls over and causes the control valve at the top of the tank to break off, it can literally become a rocket capable of flying through doors, walls, and people. Also, a tank laying on its side can allow liquid to enter the pressure valve. This can instantly destroy the pressure regulator diaphragm and allow a very dangerous, uncontrolled flow of gas through the regulator.

Air compressors

Air compressors designed for manufacturing and automotive repair shops are not usually used for airbrushing. These compressors are loud to operate, they produce high moisture levels in the air supply, and they "pulse" the airflow when operating without a supply reservoir. However, they do produce high pressure and a high volume of airflow, and they can be inexpensive.

Air compressors specifically designed for airbrush work are lightweight and operate at greatly reduced noise levels to avoid distraction of workers. They are portable and can be located close to the airbrush workstation. See **Figure 12-7**. A lightweight compressor such as the type shown in **Figure 12-7A** generally provides pressures up to 50 pounds per square inch (psi). Just as important as the pressure rating of a compressor is its rated airflow. *Airflow* is the amount or volume of air the compressor will supply at a rated pressure. Airflow is measured in cubic feet per minute (cfm). Most lightweight airbrush compressors provide airflow at a rate of .5 cfm to .75 cfm at a rating of 20 psi–35 psi.

If two airbrushes are to be used at the same time, the air compressor must be equipped with a storage tank to maintain an air supply at a steady pressure. Refer to **Figure 12-7B**. Airbrushes are not usually operated for a long period of time. Rather, they are in operation for a short period of time, followed by a period of inoperation, and then used again. This can cause pressure fluctuations and in turn may cause problems with another airbrush connected to the same air supply. A supply tank mounted to the air compressor provides a large volume of air to equalize air pressure fluctuations. If more than two airbrushes

A   B

Figure 12-7. Examples of airbrushing air compressors. A—A lightweight compressor with adjustable pressure up to approximately 50 psi. (Badger Air-Brush Co.)
B—A larger-capacity compressor equipped with a storage tank. (JUN-AIR USA, Inc.)

are to be used at the same time, a large-capacity industrial air compressor must be used.

Uncontrolled moisture or humidity can be very damaging to ink and paint. Normal moisture from the atmosphere is introduced by an air compressor. As pressure is released, the compressed air cools and this moisture condenses inside the system. A *moisture trap* can be installed in the supply line to trap and remove any moisture from the system. Airbrushing air compressors do not create as many problems with moisture buildup as industrial-type compressors because of their design.

Pressure Gauges and Regulators

A *pressure gauge* tells you how much pressure is in an airbrush system or pressure tank. A *pressure regulator* allows you to control the pressure of the air supplied to the airbrush. Systems using a pressure tank have two pressure gauges. One gauge tells you how much pressure is available in the tank. The other tells you how much pressure is supplied to the airbrush. See **Figure 12-8**. This illustration shows a typical compressor with a tank pressure gauge, a pressure regulator, and a supply line pressure gauge arrangement for airbrushing. Being able to adjust air pressure is important to good airbrushing.

The range of pressures required to use an airbrush varies according to the viscosity of the media being sprayed. *Viscosity* is a fluid's ability to resist flow. Fluids with a high viscosity are very thick, like honey. Thick-body pigments have high viscosity ratings and may require air pressure of 70 psi to draw the paint into the airbrush, atomize it, and spray it onto the painting surface. However, low-viscosity media, such as acrylics and inks, may only require air pressure of 10 psi to 25 psi. Without a pressure regulator and pressure gauge, setting this pressure is impossible. If two or more airbrushes are being used at the same time, a pressure regulator may be installed to allow a general system

Figure 12-8. Pressure controls on this airbrushing compressor permit precise adjustment of air pressure. (Badger Air-Brush Co.)

pressure of 40 psi to 80 psi in the supply lines. An additional pressure regulator at each workstation then allows individual adjustment of the pressure at the corresponding workstation. In this way, one illustrator can use an airbrush set to 15 psi while another illustrator with higher-viscosity media can use a variety of higher pressure settings.

Rigid Piping, Hoses, and Connectors

Rigid piping, hoses, and connectors can be used to transfer pressurized gas from the source to the pressure regulator and then to the airbrush. The use of nylon-reinforced rubber hosing or plastic hosing provides a great deal of flexibility. However, rigid piping does have several advantages. Airbrushing systems with an industrial-type air compressor normally have the compressor located some distance from the airbrush workstations. This type of compressor tends to be installed as a permanent fixture. Therefore, rigid piping is run to each workstation. Then, a quick disconnect coupling attaches a flexible hose to the rigid plumbing system. The rigid piping keeps the supply line out of the way.

Connectors link different sections of rigid piping together, attach hoses and rigid piping together, and link one hose to another. Piping connectors are attached with threads. Hose connectors may be threaded, or they may use quick disconnect mechanisms. *Quick disconnect couplings* are designed to snap together with a locking mechanism to provide a means of quickly connecting and disconnecting hoses. The smaller flexible hoses connecting the airbrush to the supply line or pressure regulator may also have either threaded connectors or quick disconnects. If a quick disconnect is unplugged from its matching connector, the pressure in the line or hose closes a check valve and the flow of gas is stopped. When the matching connector is inserted to reconnect the hose, the check valve is forced open to allow gas to flow again.

Both rigid piping and flexible hoses can be used to supply air to a workstation at pressures of 120 psi to 150 psi. Hoses from the workstation pressure regulator to the airbrush normally carry pressures of 15 psi to 60 psi. These low-pressure hoses tend to last much longer than higher-pressure supply hoses. Supply hoses should be checked regularly for wear. For example, flexible supply hoses tend to be used in areas where they can be rubbed or otherwise damaged. Inspect all hoses with normal operating pressure in them at least once a week for signs of wear, cuts, scuffed areas, or bulges that may indicate hose failure. A failed pressure hose can be dangerous because the flow of pressurized gas can cause a hose to violently whip around and injure people or cause property damage. If a hose fails unexpectedly, do not try to grab it as it whips around. Instead, warn everyone nearby to clear the area and then stop the flow of pressurized gas to the hose.

Leaks are also common problems with hoses. A small paintbrush and some soapy water will quickly locate leaks in a hose or coupling. Just paint the soapy solution on areas where a leak is suspected and watch for bubbles to form. Any bubbles that form will indicate the exact location of the leak so it can be repaired.

Airbrushes

Airbrushes can be classified in several different ways. The type chosen depends on the needs of the application. For example, airbrushes are classified as either *single-action* or *double-action* airbrushes. They are also classified as *internal-mix* or *external-mix* airbrushes. Airbrushes are further categorized as *gravity-feed* and *siphon-feed* airbrushes. These classifications are discussed in the following sections.

Single-action and double-action airbrushes

A *single-action airbrush* is a very simply designed airbrush that only requires the operator to press down on a trigger mechanism to obtain a spray. When the trigger is depressed, gas flows through the nozzle and gravity or siphon action mixes the air and media. Every time the trigger is depressed, a spray pattern is produced. See **Figure 12-9A**. The amount of spray that flows through the tip of the airbrush is controlled by either manually adjusting a needle on the rear of the airbrush or by screwing the nozzle tip in or out. Adjusting the needle into the airbrush reduces the amount of spray. Adjusting the needle out increases the amount of spray. The diameter of the spray pattern can be adjusted by changing tips. The simplicity of the single-action airbrush makes it good for beginners.

A *double-action airbrush* requires a little more coordination than the single-action airbrush because of its design. A double-action airbrush is shown in **Figure 12-9B**. Pressing the trigger down on the double-action airbrush releases gas through the nozzle, but no media. The further the trigger is depressed, the greater the flow of gas through the nozzle. To release media into the airstream, the trigger must be pulled back at the same time it is pressed down. The further back the trigger is pulled, the more media released. This allows you to finely control both airflow and media flow and create anything from a light-density line to an extremely dense and wide spray path.

Internal-mix and external-mix airbrushes

Most airbrushes have internal-mix operation. An *internal-mix airbrush* draws media inside the airbrush, mixes media with the pressurized gas, and forces spray through the tip. When the media flow is mixed with the pressurized gas, the attraction that holds the media particles together is disrupted. This produces a fine atomized spray as media and gas leave the tip. The finer you can make the media atomization, the finer the spray pattern will be. Fine airbrush spray patterns make it easier for the illustrator to obtain smooth gradations and softer blends of one color over another. The airbrush shown in **Figure 12-10A** is an internal-mix airbrush. Thorough cleaning is very important in keeping an internal-mix airbrush operating in a proper manner.

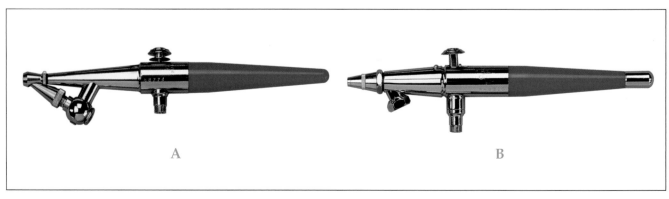

A   B

Figure 12-9. Airbrushes are commonly classified based on their design. A—A single-action airbrush. B—A double-action airbrush. (Paasche Airbrush Co.)

An *external-mix airbrush* draws media through a tube outside the body of the airbrush and then creates a mixture of media and gas at the front of the airbrush tip. The airbrush shown in **Figure 12-10B** is an external-mix airbrush. The external-mix method does not produce a mixture as finely atomized as the internal-mix method. Therefore, an external-mix airbrush produces a coarser spray pattern. Coarser patterns should not cause a problem on large-scale illustrations without intricate details. The external-mix design also has an advantage over the internal-mix design in creating stipple effects or illusions of surface texture. An external-mix airbrush can be adjusted to make the droplets of media large enough to see individual dots. This is sometimes difficult to do with an internal-mix airbrush.

Turbine airbrush

There are more advanced airbrush systems that may not be classified as internal-mix or external-mix airbrushes. For example, a turbine airbrush made by the Paasche Airbrush Company (pronounced *pash-ay*) is capable of producing highly detailed spray patterns. See **Figure 12-11**. This type of

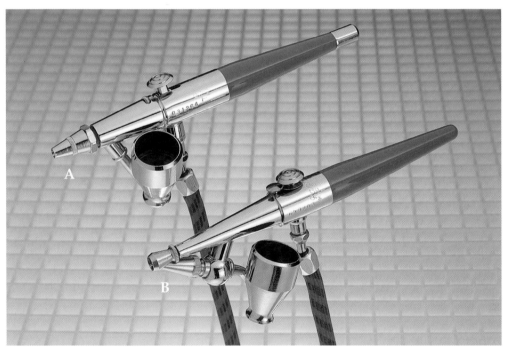

Figure 12-10. Airbrushes are classified based on how they draw and mix media. A—An internal-mix airbrush. B—An external-mix airbrush. (Paasche Airbrush Co.)

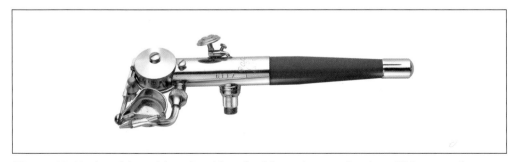

Figure 12-11. A turbine airbrush with a double-action mechanism. This type of airbrush permits extremely fine spray patterns. (Paasche Airbrush Co.)

airbrush mixes air and media by moving a needle in and out to feed small amounts of media into the airstream. As the airstream blows media off the needle, atomization occurs. This mixing procedure produces an extremely fine atomization of media. However, this type of airbrush is more complicated to operate than an internal-mix or external-mix airbrush.

Media supply methods

There are two methods used by airbrushes to draw media. Airbrushes draw media by gravity or siphon feed.

In a *gravity-feed airbrush*, gravity forces media to flow down from a reservoir into the airbrush, where a mixture of media and pressurized gas occurs. Gravity-feed airbrushes can be identified by the reservoir or "cup" on top of the airbrush or by a media chamber in the barrel of the airbrush. See **Figure 12-12**. Reservoirs for gravity-feed airbrushes are available in a variety of sizes.

A *siphon-feed airbrush* operates on the basis of siphon action or suction. Typically, a siphon-feed airbrush is mounted on top of a media jar. See **Figure 12-13**. A reservoir can also be used. The suction is created by a low-pressure effect called a *venturi effect*. The venturi effect is created by a narrowing of the air passage at the media entry port. This narrow passage is called the *venturi*. As gas travels through the venturi, the suction draws media from a feed tube.

Siphon-feed airbrushes have several advantages over gravity-feed airbrushes. One advantage is ease of cleaning. The removable reservoirs and media jars

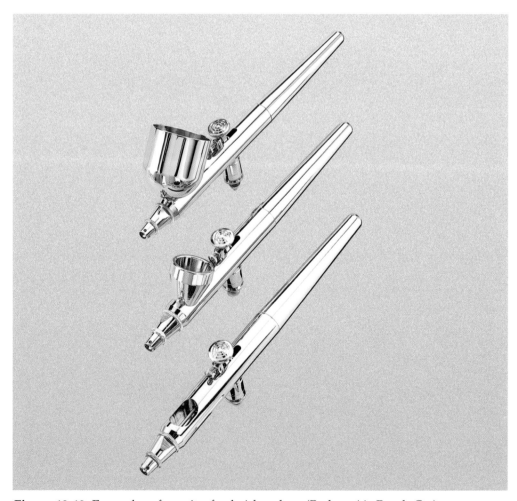

Figure 12-12. Examples of gravity-feed airbrushes. (Badger Air-Brush Co.)

Figure 12-13. A siphon-feed airbrush. Media is drawn from a mounted jar or reservoir. (Badger Air-Brush Co.)

allow easier access to the piping and containers that come into contact with media. Also, the removable reservoirs make it easy to remix media or change colors. This is done by having extra reservoirs with different colors of media. One reservoir is kept loaded with an appropriate cleaning fluid. To change colors, the current color media container is removed. The cleaning fluid container is installed to flush out any of the first color. Then, a reservoir with the new color is attached. Spray several trial passes of the new color on scrap paper. This will flush the cleaning liquid from the passages, load the mixing point with new color, and test the spray pattern and new media viscosity.

There are several different types of reservoirs used with gravity-feed and siphon-feed airbrushes. Barrel-type reservoirs usually hold about 1/32 oz. of media. These are not designed for large-scale projects. Top-mounted and side-mounted cup reservoirs hold about 1/16 oz. to 1/2 oz. of media. Media jars for siphon-feed airbrushes normally hold 1 oz. to 3 oz. of media. Larger containers allow uniform color and viscosity to be maintained. Differences in mix ratios can result in visible color variations when work dries. These differences are seldom noticeable on fine detail. However, large expanses of background tend to visually magnify any color and density variations. Slight changes in the mix ratios of thinning agents may occur when adding media to small reservoirs. Small reservoirs also allow the illustrator to thin, tint, or shade media in the cup to subtly change the hues of adjacent spray areas. This produces gradual color transitions without wasting large quantities of media, which may occur if a large reservoir is used.

An airbrush with a top-mounted, gravity-feed reservoir cup or a siphon-feed media jar can normally be used in the right or left hand. Side-mounted reservoirs are available with right-handed or left-handed airbrushes. The reservoir should be on the thumb side of the airbrush when holding it in the proper position.

Pressure adjustments

Gas pressure is primarily adjusted at the pressure regulator. Pressure should not be confused with flow rate. For example, suppose the pressure regulator is set to 15 psi. Pressing the trigger a very slight amount allows just a little gas or air to pass through the tip at 15 psi. Full depression of the trigger allows maximum flow. However, the pressure regulator is still set to maintain 15 psi. In other words, no matter what the flow rate, pressure stays the same.

Note: The trigger is a restriction on the system. When the trigger is pressed, this decreases the restriction. The more the trigger is depressed, the less restriction there is. Therefore, with the restriction reduced or removed from the system, the pressure actually drops. However, the pressure regulator always attempts to maintain the pressure it is set at. This compensates for the slight change in pressure when the trigger is depressed.

Airbrush adjustments

Although there are a variety of manufacturers and types of airbrushes, the principal adjustment controls used with airbrushes are generally similar. See **Figure 12-14**. This illustration shows a cutaway view of a single-action, internal-mix, siphon-feed airbrush with a reservoir bottle attachment. Single-action airbrushes usually have an adjustable media flow needle control located either at the front or back of the airbrush barrel. The closer the needle is to the seat, the less media flow through the opening. Some single-action airbrushes have an adjustable head and a fixed needle. The maximum size of the circular spray pattern is determined by the size of the tip and the distance from the illustration surface.

Double-action airbrushes do not have a needle adjustment. Instead, pulling back on the trigger changes the position of the needle. A cutaway view of a double-action airbrush is shown in **Figure 12-15**. Pressing straight down on the trigger opens the metering pen in the air valve at the bottom of the airbrush. Pulling back on the trigger (to the right in **Figure 12-15**) pushes the entire needle assembly back. This moves the tapered end of the needle off its seat.

External-mix airbrushes generally have a tapered media adjustment and a spherical mixing chamber immediately adjacent to the tip of the airbrush. Refer to **Figure 12-10B**. Media and gas are fed into the mixing chamber. The flow rate is adjusted with the nozzle located between the mixing chamber and the tip of the airbrush.

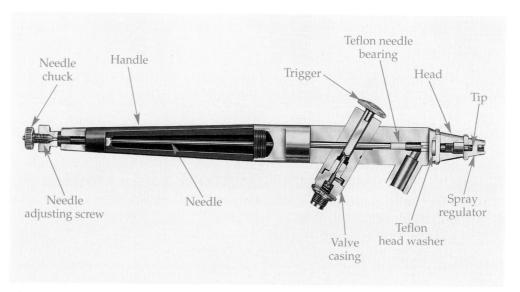

Figure 12-14. A cutaway view of a single-action, internal-mix airbrush with typical adjustment controls. (Badger Air-Brush Co.)

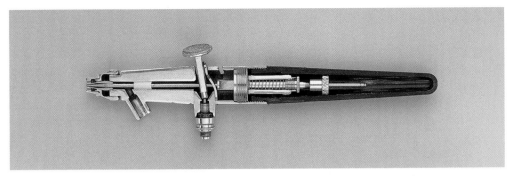

Figure 12-15. A cutaway view of a double-action, internal-mix airbrush. (Badger Air-Brush Co.)

Airbrushing Materials and Tools

There are a variety of materials and tools available that make airbrushing a practical technique for the illustrator. The following sections provide a brief overview of some of the principal airbrushing materials and their uses. Materials and tools such as commercially prepared frisket sheets, cutting and burning tools, paper, board, and media are available at art supply stores.

Frisket sheets

In many airbrushing applications, specific areas and details of an illustration are airbrushed while masking other areas. *Masking* is covering areas you do not want to spray. There are a number of ways to mask areas of an illustration. Frisket sheets are frequently used for this purpose. A *frisket* is a commercially prepared sheet of plastic acetate with an adhesive backing. See **Figure 12-16**. Frisket sheets need to be transparent so that light construction lines on the illustration are visible.

Many illustrators make their own frisket sheets. They can be made from sheets of clear acetate 3 mm to 5 mm thick. A light coat of spray-mount adhesive or rubber cement is applied to the back. Thin plastic drafting film can also

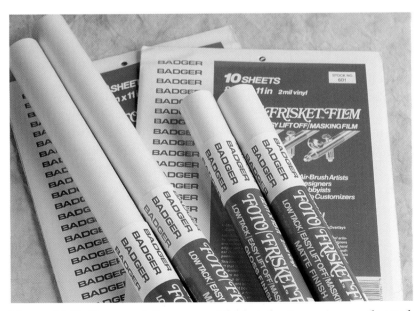

Figure 12-16. Commercially prepared frisket sheets are commonly used to mask illustrations for airbrushing. (Badger Air-Brush Co.)

be used instead of clear acetate. However, drafting film is designed to resist tearing. Therefore, use care when cutting shapes out of the film.

Avoid using thin paper to make a frisket. Lightweight paper does not work well because it tends to lift off the illustration board. This allows media to be sprayed under the mask. Also, some types of heavier paper act as a wick. This means that fine "trails" of airbrush media are drawn under the edges of the paper mask, creating a fuzzy edge.

Cutting and burning tools

Frisket sheets and masks are cut with an art knife or with a stencil burner. An art knife is lightweight and easy to hold, and it can be rotated to cut arcs and circles. Dull blades can be quickly replaced. If there are many curved shapes, a special art knife with a blade mounted in a swivel is used. As the tip of the blade is moved along a curved outline, the blade rotates in the swivel so it is always cutting parallel to the direction you are pulling it. When using an art knife to trim frisket material, it can also be used to lift the material so that it can be easily peeled from the illustration board.

A special tool called a *stencil burner,* or *hot knife*, is used in some applications to burn away the acetate material in order to create a sculpted hole in the mask or stencil. A stencil burner is used in place of a regular art knife. This tool requires a little practice to develop accuracy and proficiency. However, it produces a good edge for freehand lines. An art knife is better for straight lines. The best application for a stencil burner is large mask openings.

Airbrush paper and illustration board

Most illustrators develop a preference for certain types of paper and illustration board. However, some applications determine specific combinations of materials that must be used. For example, fabric requires special types of spray media.

Illustration board is available with a smooth finish or a slightly textured surface. Hot-pressed board has a smooth finish. Cold-pressed board has a textured finish. For either type, use double-thick board. This prevents warping when spraying wet airbrush media on the surface.

There are a few paper-based materials used with airbrushed media. These paper materials are normally used for small illustrations to be photographed, rather than exhibited as the original airbrushed illustration. Paper costs much less than illustration board. However, only a light amount of spray media can be applied without warping or wrinkling the paper.

Some materials, such as canvas, must be primed before applying media. Priming smoothes the rough texture and provides a better bond for the airbrushed material. Large pieces of illustration board must also be primed. This makes the board more stable and less likely to warp. Illustration board should also be primed when using watercolor or acrylic paints.

Spray media

There are many different types of paints and inks used in airbrushing. Many times, what is used is based solely on an illustrator's preference. Common types of spray media include watercolor, gouache, acrylic paint, water-soluble paint, dyes and inks, enamel, lacquer, and textile paints.

Watercolor is a tinted, dry powder that is mixed with water. Watercolors can be mixed to create any color needed. Watercolors are most often used where smooth transitions, soft effects, and muted tones are more important than color intensity. Watercolors are transparent and can be airbrushed over other colors to obtain blended effects.

Gouache (pronounced *gwash*) is created by mixing finely ground color chalk in water and a gum binder. Although gouache is a type of watercolor, the ground chalk in the solution makes it opaque.

Acrylic is a polymer-base paint with exceptional opacity and color clarity. See **Figure 12-17**. Acrylics can produce vivid colors and accept fine-line detail once dried. They also provide exceptional depth of color that cannot be created with most inks or watercolors. However, acrylics tend to be thick even when a thinning agent is used. This can make them difficult to spray. This also means that an airbrush with a large tip, large media passageways, and large orifices must be used. *Water-soluble paint* has similar properties to acrylic. Many illustrators prefer to use water-soluble paints because they are easy to thin, easy to clean from the airbrush, and generally less expensive.

Inks and *dyes* are historically among the most frequently used types of airbrush media. Inks and dyes are easy to airbrush. They can be sprayed in an extremely fine mist to give a slight haze effect, or they can be applied completely opaque. Most inks can be used directly out of the bottle without diluting or thinning. Inks and dyes are also available in transparent hues. In addition, inks and some dyes can be mixed to create custom hues and intensities.

Enamel is a paint used for metals, plastics, and porous materials, such as wood. Although enamels require considerable thinning to apply with an airbrush, they provide good range of color and opacity. They also have better durability than most other media for exterior applications.

Lacquer is a paint used primarily for architectural model building or applications involving metal. Lacquer is quick drying and can produce an extremely high gloss. High gloss can give the illusion of depth. Lacquer is available in a variety of transparent and opaque hues.

Textile paints are heavy-bodied media used for buildup and bonding with fabric. The primary technical illustration application of textile media is canvas

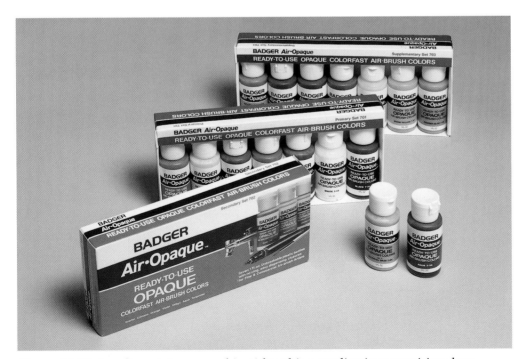

Figure 12-17. Acrylic paints are used in airbrushing applications requiring deep, opaque colors. (Badger Air-Brush Co.)

airbrushing. Textile paints must be airbrushed using larger tips, passageways, and orifices. The thickness of textile paints can sometimes make them quite difficult to clean from an airbrush. Therefore, never let textile media start to dry in an airbrush.

Techniques of Airbrushing

Many airbrushing applications require a high level of skill. See **Figure 12-18**. This chapter presents an overview of common techniques and processes used in airbrushing. This will help you get started using an airbrush and will allow you to determine your level of interest in this specialized field of illustration. If you want to develop high-level airbrushing skills, you may want to research specialized instructional texts that deal exclusively with the many techniques used in airbrushing.

The following sections present basic layout and masking procedures and the fundamental techniques involved in using and adjusting an airbrush. When practicing these techniques on your first projects, focus on developing skills by spraying very basic shapes. Once you gain practice, you will be able to produce more complex shapes and contours. As with other manual illustrating skills, you will find it necessary to develop fundamental airbrushing skills before progressing to more complex levels. Practice is the only way these skills can be developed.

Layout and Masking Procedures

The first step in creating an airbrush illustration is to develop a sketch and pencil-shaded drawing. This allows you to see what type of media you will

Figure 12-18. Airbrushed illustrations provide an excellent way to display internal components in cutaway drawings. This type of visual detail requires a high level of skill and technique. (David Kimble)

want to use and determine the techniques required. It also allows you to determine colors, the order in which media will be applied, and any special masking or stencil designs that are needed.

Using an illustration board, draw light construction lines to lay out the illustration. Many illustrators use slides or photographs and a projector to project the object onto the surface of the illustration board. The projector makes it easy to increase or decrease the scale of the illustration because you can simply move it closer to or farther from the board. A photocopied image can also be used with a projector to enlarge or reduce the preliminary artwork. The object lines are then traced onto the illustration board. Once the object lines are drawn, a thin coat of transparent primer can be used to "lock" the pencil image and establish a better surface for adhering the airbrush media.

Next, airbrush the background of the illustration and allow it to dry. A paper mask with a "window" defining the area of the illustration can be applied. This protects the perimeter areas from accidental spraying and keeps normal soiling off the background areas. Tape down the edges of the mask with drafting tape. This prevents the mask from lifting and allowing overspray. A frisket is then used in the opening to outline the object. An example of using masking materials to prevent overspray on an illustration board is shown in **Figure 12-19**.

Some illustrators who work on highly artistic projects prefer to use little or no frisket masking. This requires well-developed airbrush skills and very good eye-hand coordination. Other illustrators use an area mask held in place with one hand while airbrushing with the other hand. This allows them to hold the mask at varying heights above the painting surface and create soft, subtle edges.

When first operating an airbrush, use full masking and frisket material. Also, label the frisket areas with the color, density, and order in which you will lay them out. For example, the first area you plan to airbrush may be labeled

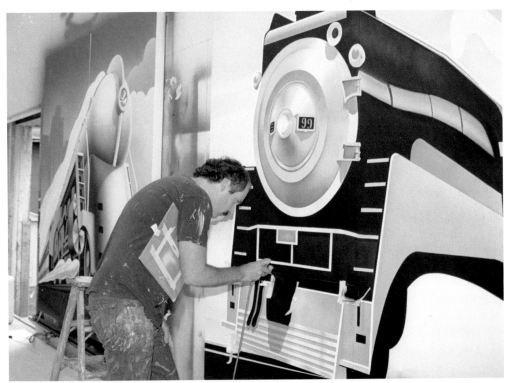

Figure 12-19. This illustrator is using cardboard masks to fill in details on the locomotive. (W.E.T. Studios)

1GHV. This indicates the first area, green color, and heavy density. Use a labeling system that makes sense to you. If the airbrush illustration has complex details or tiny areas, make a large-scale sketch and place the labels on it.

Next, cut the frisket material from the areas that are to be airbrushed first. Try to keep the removed section in one piece. This allows you to replace it and prevents you from cutting the illustration board where it will not be hidden by airbrushed boundaries. Frisket material can normally be reused at least once or twice if the buildup of media is not too great. Only light pressure is needed to penetrate the plastic. Too much pressure on the knife can cut the illustration board. Always lightly trace around a cutting edge with a smooth object, such as the end of a knife handle. This ensures that the frisket is sealed to the surface.

Media Preparation and Airbrush Adjustments

After board preparation and illustration layout, the airbrush must be prepared for use. First, prepare the media. This may only consist of shaking a container jar of vendor-prepared paint or ink for several minutes. Media should be shaken thoroughly to mix the pigmentation and thinning agent. In other instances, you may have to mix a specific type of media with the appropriate thinning agent. Some types of media use water, while others use special chemical thinning agents. Be sure to use the correct thinning agent. An incorrect thinning agent may react and create a "glob." Generally, the media viscosity should be somewhere between that of motor oil and water. After adding the thinning agent and mixing, pour the media mixture through a fine paint filter. This removes any large particles that might clog the airbrush. You can make your own filters with clean, discarded nylons. Cut three patches and place them on top of each other. Rotate the weave pattern of each layer slightly. Then, pour the media mixture through the nylons.

Next, adjust the spray pattern on the airbrush with a small amount of water or thinner. Use the same thinning agent that the media type uses. Adjust the air supply pressure and depress the trigger to test the spray pattern on scrap stock. If a minor change is needed, adjust the spray regulator nozzle at the tip of the airbrush. If a larger change is needed, adjust the position of the needle. You will then also have to fine-tune the spray pattern with the spray regulator on the tip.

When the pattern is satisfactory, load a small amount of media into the airbrush reservoir. Do another test on a scrap piece of the material you will be airbrushing on. This allows you to make sure the media mixture is properly thinned. You may need to make a minor adjustment to the spray regulator nozzle. You are now ready to begin airbrushing.

Applying Airbrush Techniques

Technique is a large factor in the quality of airbrushing. Developing good technique requires practice. See **Figure 12-20**. It is always important to keep the airbrush at the proper distance from the surface. It is also important to move the airbrush in the correct direction and equally important to control the speed of the movement.

Hold the airbrush in your dominant hand in a loose, comfortable grip. See **Figure 12-21**. For a wide, thinly dispersed spray pattern, hold the airbrush away from the surface. Hold the airbrush very close to the surface to spray a very thin, dense line. By varying the distance from the surface, you can control the density of the spray pattern. You can create many different effects with this

Figure 12-20. Examples of commercial airbrush work. Much practice and skill is required to create the unique colors and shading patterns in these types of illustrations. (W.E.T. Studios)

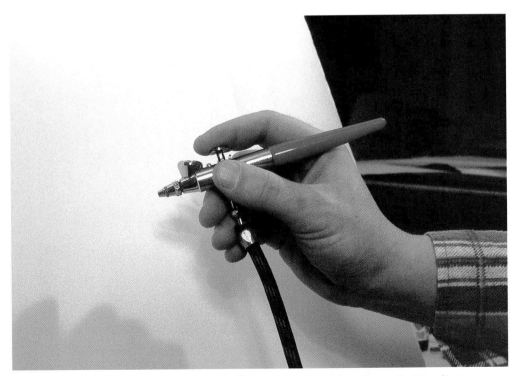

Figure 12-21. An airbrush should be held in a comfortable grip so that its distance from the illustration can be easily adjusted.

technique. However, if you hold the airbrush too close to the board without moving it, a "puddle" will form. This puddle will have cobweb-shaped spokes all around the pattern. These puddles are very difficult to touch up. In most cases, an airbrush eraser must be used to remove the puddle. Airbrush erasers are discussed later in this chapter.

Hand motion is important in developing spray pattern contours appropriate to the shapes being airbrushed. Make a "practice run" with the airbrush first without spraying media. This helps you get a feel for the motion. Then, make another pass, this time spraying. Always begin a spray pass well before the true starting point. Continue spraying well past the end of the area to be airbrushed. The frisket prevents overspray. Starting before the beginning of the area allows you to adjust the trigger position. Ending beyond the area ensures that you do not release the trigger too quickly and leave a "dry" spot. Also, try to avoid varying the direction your arm has to move. Instead, rotate the surface if possible.

The speed that you use to move the airbrush determines the density of the spray pattern. The faster you move the airbrush across an area, the less dense the spray pattern. The slower you move the airbrush, the denser the pattern. However, it is better to create a more dense pattern with several quick passes than with one slow pass. A transition from light to dark can be created by slowing your hand motion gradually through the pass. See **Figure 12-22**.

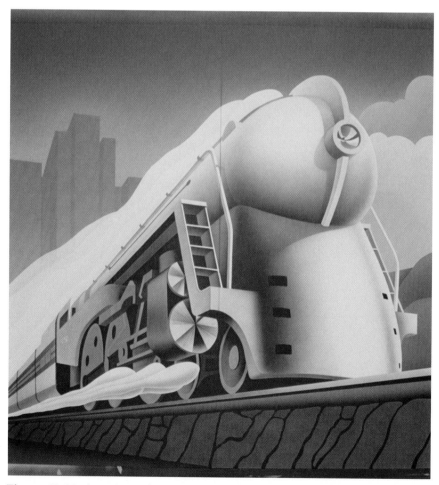

Figure 12-22. A variety of color transitions and shading effects can be achieved in airbrushing by varying spray patterns. Spray patterns are controlled by manipulating the painting direction, the speed of movement, and the distance of the airbrush from the surface. (W.E.T. Studios)

Often, joining edges of airbrushed surface areas appear too abrupt. It is difficult to spray adjacent surfaces with the same density when using a frisket. To correct this, make a very fine "last pass blend" along the edge. The pattern should be dense at the edge, and it should diminish very quickly to no spray. Use a very light touch on the trigger or the edge will disappear. You only want a hint of shading to soften the abruptness. See **Figure 12-23**. This transition can be created by using a file folder mask to keep the darker surface from receiving any overspray.

Figure 12-23. Custom shading details and highlights created with specific airbrushing effects are apparent in this illustration. (David Kimble)

Removing Media

It is seldom possible to correct an airbrush mistake by simply scraping off media with the tip of a razor knife. Using a cotton swab with a thinner liquid can produce a blob or smear. If you cannot correct the problem by repainting the area with an airbrush, you will probably need to use an airbrush eraser.

An *airbrush eraser* or *media blaster* is used to remove portions of media that have been airbrushed onto a surface. See **Figure 12-24**. An airbrush eraser is a miniaturized sandblaster. Its operation and performance is almost identical to that of an airbrush. However, extremely fine beads of abrasive are sprayed instead of paint or ink. These beads etch or "blast" airbrush media from the illustration surface. However, an airbrush eraser can also etch the illustration board. Therefore, always keep the airbrush eraser moving. Also, use proper safety with an airbrush eraser. *Always wear safety glasses or goggles and a filter mask when using an airbrush eraser or similar tool.*

Airbrush Cleaning Procedures

Airbrush instruction manuals usually include cleaning instructions. Most of the procedures for cleaning airbrushes are simply precautions exercising common sense. The following cleaning procedures should be followed for all types of airbrushes.

➤ Never let media start to dry in the airbrush. Always empty media from the reservoir immediately after airbrushing. Flush or wipe out as much of the excess media as possible. Then, load a straight thinning agent in the reservoir and spray it through the airbrush. This flushes media from the internal passages. Finally, disassemble the airbrush and clean each part using a clean thinning agent.

➤ Use a small, moderately soft paintbrush to clean crevices, corners, and passageways in the airbrush. Never use a wire or sharp instrument. This will scratch the airbrush and give media a place to adhere the next time the airbrush is used.

➤ If thinner does not remove dried media, use an ultrasonic cleaner with thinner as the cleaning fluid.

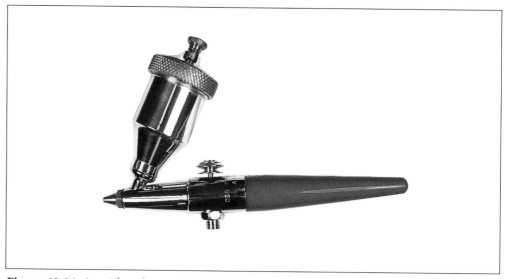

Figure 12-24. An airbrush eraser appears similar to an airbrush, but it is used to remove media. (Paasche Airbrush Co.)

Airbrush Workstation Environment

Much of the success in using an airbrush can be attributed to maintaining the proper environment in the area of operation. Controlling the airbrushing environment can add significantly to the quality of your work. More importantly, it can help ensure your health and safety. Using a spray booth, installing the proper filtration and air handling systems, and maintaining the appropriate lighting levels will all help you improve production in airbrushing applications.

Spray Booths

As previously discussed, spray booths are ventilated work areas used to control conditions for airbrushing. See **Figure 12-25**. There are a number of excellent reasons that you should perform all airbrushing in a spray booth. Most importantly, it is easier to control potential safety and health hazards within a spray booth rather than outside the confines of a booth. Airbrushing in a confined space permits easier air ventilation, reduces disruptive drafts generated by forced air heating and cooling systems, and helps eliminate glare from direct light entering windows. Spray booths also keep overspray from drifting to surrounding work areas. They can usually be purchased from an art supply store or a manufacturer of airbrushing equipment.

If you are on a limited budget, you can make your own spray booth with a large, heavy-duty cardboard box. Boxes used to pack large appliances, such as washing machines and clothes dryers, are most suitable. For ventilation, you can attach a flexible clothes dryer venting hose. These types of hoses are generally inexpensive. Attach one end to the top or back of the box and the other end to an existing venting system or to a separate vent that carries fumes to a safe exhaust area. On the sides of the box, make holes large enough to insert a long dowel rod approximately 3/8" to 1/2" thick. This will allow you to prop the illustration board upright inside the spray booth so that you can spray at a vertical angle. For example, you may want to position the board at a 45° angle.

Figure 12-25. An airbrushing spray booth with a ventilation unit. (Paasche Airbrush Co.)

If too much shadow is created by the top of the box, you can cut away part of the top to allow more illumination inside.

Spray booths for commercial sale often have their own lighting systems and ventilation fans. Airbrushing spray booths are similar to arc welding booths, except they typically have a sheet metal base for ease of cleaning. Although they require greater expense, airbrushing spray booths provide optimal production conditions and more control over environmental hazards. If not already equipped, most spray booths can be modified by installing venting and filtration systems, air supply plumbing, and small shelves or hanging devices for media bottles, airbrushes, safety glasses, masks, and other production supplies.

Multiple Airbrush Workstations

Workstations with individual pressure supply lines are commonly used in applications requiring multiple airbrushes. These types of workstations typically have a manifold layout, **Figure 12-26**. A pressure regulator controls the supply of air from the compressor to the workstation. Each airbrush receives the same working pressure. However, individual adjustments can be made with the controls on each airbrush. A manifold system provides great flexibility to the illustrator because different pattern widths corresponding to the different airbrush spray nozzles are immediately available. Even if variable size nozzles

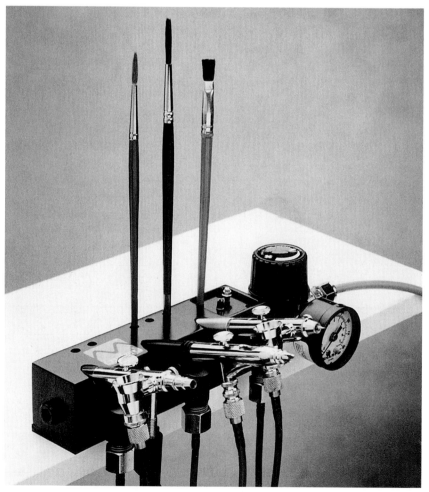

Figure 12-26. A multiple airbrush workstation increases the number of options available to the illustrator and helps improve productivity. (Air Nouveau)

are not needed, different painting or ink colors can be used. Multiple airbrush workstations are used by illustrators who require flexibility and the equipment to produce a high rate of work.

Air Handling and Filtration Systems

An air handling system draws and removes tiny droplets of sprayed media from the air. This reduces the buildup of paint on surfaces around the spray area. It also lessens the chance of inhaling fumes. The most common air handling system consists of an exhaust fan that draws air into a hood over the airbrush workstation. The air is then vented through a duct system and outside the building. Filters are placed in the ductwork to trap media and prevent spray from being released into the environment.

Self-contained air handling units are also available. These units combine a fan and filtration system. After the air is filtered, it is then vented back into the work area.

Masks and Eye Protection

A respiratory mask and safety glasses or goggles should *always* be used when airbrushing. A wide variety of respiratory masks are available to choose from. Masks filter out any sprayed media that the ventilation system does not remove. Inhaling sprayed media is dangerous to your health. Safety glasses and goggles prevent sprayed media from entering your eyes.

There are three basic types of respiratory masks. These are paper-based dust masks, canister-style filter respirators, and vapor respirators. Paper-based dust masks can trap dust and most mists. However, they cannot protect you from extremely fine mists. Canister-style filter respirators draw air through a filter located in a canister. These types of respirators can trap dust and most mists, but are not suited for very fine mists. Vapor respirators are very different from dust and mist filtering masks. Vapor respirators have a special high-density filtration system to remove very fine mists. They also have exhalation valves that open to vent heat and moisture when you exhale. As you inhale, the valves close and force air entering the mask to pass through disposable filters. See **Figure 12-27**.

Figure 12-27. Vapor respirators provide special protection from fine airbrush mists. (Paasche Airbrush Co.)

Safety glasses or goggles should be worn at all times when using an airbrush. A variety of types are available. Safety glasses and goggles with protective shielding on the sides should be used in airbrushing applications. These protect the illustrator from eye contact with direct sprays of media and mists. Soft, flexible safety glasses and goggles are relatively comfortable to wear. Goggles can be worn over prescription eyeglasses and are easy to clean. To provide maximum protection from airbrush mists, full visors can be worn.

Airbrush Workstation Lighting

Working with an airbrush requires lighting that is both easy on the eyes and bright enough to illuminate the object or surface. However, lighting sources that do not change your perception of color and density are also very important. Airbrush illustrations created under high-intensity fluorescent lighting may have a different appearance when viewed in natural light. A way to approximate sunlight is to use a combination of four fluorescent lights located above the painting area and a 250 watt to 300 watt incandescent light located above and to one side of the area.

Summary

An airbrush is a small, hand-held spray gun capable of producing a wide range of spray patterns with a high degree of precision. The principal applications of airbrushing include photograph touch-up and spray shading. In most airbrushing applications, airbrushes are used in combination with other illustration tools to create and shade designs.

Various types of airbrushing systems are available. Airbrushes are classified as single-action or double-action, internal-mix or external-mix, and gravity-feed or siphon-feed. An airbrush system requires a number of other components. These include a source of compressed air or gas, air supply hoses or piping, a pressure gauge, and a pressure regulator.

A number of illustration materials can help make airbrushing easier and more effective. These include masks and frisket sheets, art knives, and stencil burners. Many different types of media can be used with an airbrush, including watercolor, gouache, acrylic, inks and dyes, enamel, lacquer, and textile media. An airbrush eraser can be used to remove mistakes from an illustration.

The basic procedure for airbrushing is to lay out the drawing, airbrush the background, apply a mask and frisket material, label each area to be airbrushed, prepare the airbrush, and airbrush the illustration. Be sure to remove unused media from the airbrush when finished. Also, clean the airbrush and its components thoroughly before placing it in storage.

Always maintain a clean, safe workstation environment. Whenever possible, use a spray booth with proper ventilation and an air handling system. *Always* wear safety glasses or goggles and respiratory protection. In addition, use appropriate lighting when airbrushing so that the illustration will appear correctly in the display environment.

1. Identify two primary applications of airbrushing in technical illustration.
2. List at least four components of an airbrush system.
3. What is a *spray booth*?
4. Name four common air supply systems used in airbrushing applications.
5. What are *spubbles* and how are they caused?
6. Name three gases that can be safely used for airbrushing.
7. Briefly discuss the advantages and disadvantages of using an industrial air compressor for airbrushing.
8. Define *airflow*.
9. How does moisture occur in airbrush plumbing and why is it a concern?
10. What is the difference between a pressure gauge and a pressure regulator in regard to an airbrush system?
11. How are leaks located in the plumbing of an airbrush system?
12. What is the main difference between a single-action airbrush and a double-action airbrush?
13. What is the main difference between an internal-mix airbrush and an external-mix airbrush? Which type generally produces a more fine spray pattern?
14. What two basic methods are used by airbrushes to draw media into the airbrush body?
15. What airbrush adjustments and controls are used to vary the media flow and spray pattern with a single-action, internal-mix airbrush?
16. What is a *frisket* and how is it used in the airbrush process?
17. What tool can be used besides an art knife to cut frisket material?
18. List four common types of media used with airbrushes.
19. Briefly describe the process used to lay out and mask an illustration for airbrushing.
20. Explain why you should begin a spray pass with an airbrush well before the true starting point and why the spray should continue past the end point.
21. How does an airbrush eraser work and what safety precaution should you always observe when using one?
22. What is the purpose of a multiple airbrush workstation and what are two advantages to using a multiple airbrush system?
23. What types of respiratory masks can be used in airbrushing applications?

Drawing Problems

Each of the following problems can be completed by referring to the object shape and shading shown. First, draw the shape on illustration board using very light construction lines. Then shade the shape with an airbrush using a light-colored paint or ink. Use each example to guide your application of media as you color and shade the object. Pay particular attention to highlights created by the reflection of the light source and the darker shading created by shadows.

Many objects have surface features that are made up of one or more of these shapes in combination. These basic shapes make good projects for shading. Use them to develop your airbrushing skills. For more practice, repeat the airbrushing process for the shapes after you are done, adding holes going into or through each shape.

1. Box

2. Sphere

3. Cylinder

4. Cone

5. Wedge

6. Torus

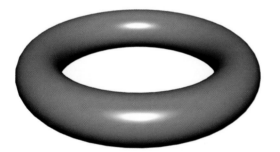

Illustration Applications and Production

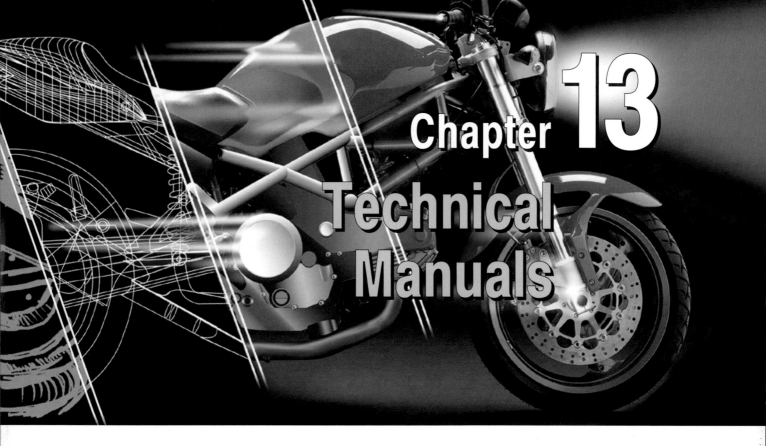

Chapter 13

Technical Manuals

Learning Objectives

At the conclusion of this chapter, you will be able to:

☐ Identify the primary types of technical manuals.

☐ Describe the factors to consider when determining the format and content for a technical manual.

☐ Explain the different types of technical illustration drawings commonly used in technical manual production.

☐ Apply special drawing techniques to create technical manual illustrations.

☐ Compare the traditional and contemporary methods of technical manual production.

☐ Explain copyright and trademark principles and the procedures used to protect created work.

Introduction

A *technical manual* is a publication produced by a company to provide operating, assembly, installation, or service instructions for a product, **Figure 13-1**. Technical manuals are written and illustrated in a manner that allows users to fully understand the content. Text alone is not always sufficient to provide this information. Technically accurate illustrations usually accompany text material in a technical manual. Therefore, technical manuals require considerable technical illustration work.

This chapter introduces the different types of technical manuals and the common illustration methods used in technical manual production. Traditional and electronic publishing methods are discussed, as well as copyright and trademark guidelines.

Types of Technical Manuals

Many products require a technical manual. For example, a stove has installation and operating instructions, and usually a parts list. Other products, such as bicycles and computer systems, have assembly instructions included as well. Very complex items may have several separate technical manuals. For example, a car has an operating manual for the owner. Service and maintenance manuals are also needed by the auto dealer so that the car can be serviced. In addition, a parts manual is needed to identify all the various parts for replacement purposes.

Catalogs have some of the same characteristics as technical manuals. Some catalogs contain technical information for parts and products. In fact, a parts manual is sometimes referred to as a parts catalog. However, catalogs usually contain both technical and nontechnical material. Catalogs often show items that can be purchased, while technical manuals are used as reference documents. Although catalogs differ from technical manuals, their illustrations are prepared in similar ways.

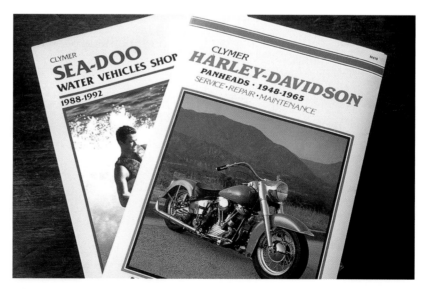

Figure 13-1. Technical manuals are designed to provide instructional material and product support through the use of text and graphics. (Intertec Publishing Corp.)

Technical manuals can be categorized as assembly/installation manuals, owner's/operation manuals, parts identification lists, and maintenance manuals. In all cases, these can be complete publications or parts of other publications.

Assembly/Installation Manuals

An *assembly manual* provides step-by-step instructions on how a product is put together, **Figure 13-2**. Assembly instructions are sometimes included as a part of other manuals. Illustrations are critical for this type of manual because they help clarify written information. In some cases, illustrations can replace written instructions completely. Poorly created assembly instructions can be a source of frustration for the consumer and can result in returned items or loss of sales.

An *installation manual* provides instructions on how a product is installed, placed, or mounted, **Figure 13-3**. For example, home stereo and aftermarket car

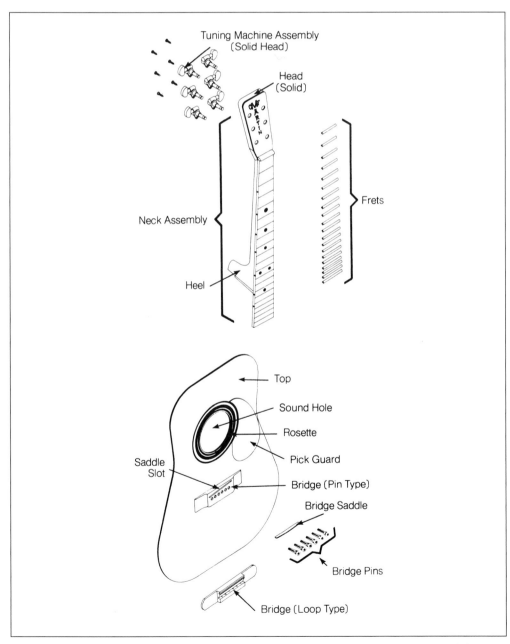

Figure 13-2. Along with step-by-step instructions, this illustration can be used for assembling a guitar. (The Martin Guitar Company)

Figure 13-3. Detailed technical illustrations typically accompany instructions in an installation manual. (Cerwin-Vega, Inc.)

speakers are supplied with a manual that tells the user how to locate and wire the speakers. A new disk drive for a computer is provided with similar instructions.

Owner's/Operation Manuals

An *operation manual* provides instructions on how a product is to be used, **Figure 13-4**. The product might be a very simple item, such as a radio, or something as complex as a car. It is critical that the information in this manual is correct so that the item is used correctly. Incorrect information may result in an accident, injury, or even death.

An *owner's manual* usually contains operation instructions, but other technical information as well. See **Figure 13-5**. An owner's manual for a car

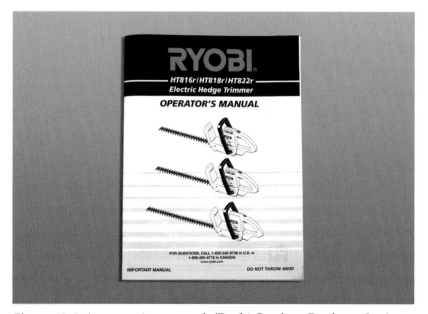

Figure 13-4. An operation manual. (Ryobi Outdoor Products, Inc.)

Figure 13-5. An owner's manual. (Fender Musical Instruments Corp.)

might contain a parts identification chart and information on simple maintenance that can be completed by the owner, such as adding fluids. Specifications such as fuel capacities, engine displacement, and vehicle weight may also be included.

Parts Identification Lists

A *parts identification list* usually consists of a number of illustrations with part numbers keyed to a list of the parts organized by number, name, and manufacturer's part number. See **Figure 13-6**. Parts lists are included in owner's manuals and service manuals. They are also used by companies that carry replacement parts for the product, such as auto parts stores. Often, these parts lists are placed on microfilm or stored electronically.

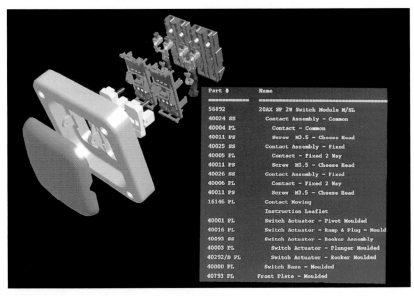

Figure 13-6. An illustration and related list used for parts identification. (I-DEAS Artisan Series)

Maintenance Manuals

A *maintenance manual* provides instructions for the care or repair of a product. Preventive maintenance, such as a schedule for changing a car's oil, is one example. Routine maintenance information is normally a part of most owner's manuals. Repair instructions, on the other hand, are typically included in separate manuals. See **Figure 13-7**. Several maintenance manuals may be needed for one product, depending on how complex it is. Specifications are also usually part of the maintenance document. For example, tire size and pressure are specified on maintenance charts for cars, trucks, and motorcycles.

Determining Format and Content

When designing technical manuals and similar documents, both the purpose and audience must be identified. For example, the purpose might be to convey assembly information for an entertainment center. In addition, the audience might require a low reading level, such as an 8th-grade level. If the assembly instructions have a 12th-grade reading level and no illustrations, chances are likely that many individuals will have problems with the assembly and many of the products will be returned to stores.

When analyzing the audience, a variety of factors must be taken into consideration. What is the age, sex, level of education, nationality, and native language of the audience? What are the differences in culture that might require a different approach in presenting the information? Something as simple as color can cause a problem with interpretation since colors have different meanings in different cultures.

The purpose of the document and needs of the audience are combined to determine the format. For example, suppose an owner's manual is needed for a newly designed car so the owner has information about the various functions of the car. This is the purpose. A manual that fits in the glove compartment of the car is used. This is the format, or method of delivery. This format is decided

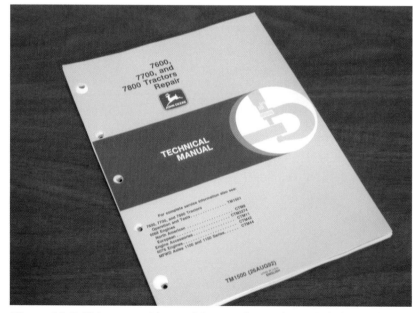

Figure 13-7. This manual is used for service and repair information. (John Deere & Co.)

on because it will hold the quantity of information needed and is small enough to be stored in the glove compartment. Also, the audience (made up of car buyers) normally expects this format for owner's manuals.

Many technical documents, such as an owner's manual, are written in the simplest language possible. A simple writing style makes it easy to translate the content into other languages. Since many technical terms are used, it is important to use consistency in identifying each item, even if several names are normally used. This will reduce confusion and make translation easier. Another important guideline is to avoid slang terms and idioms such as "change of heart." These tend to have meaning in only one culture.

The text and graphics in a publication should be logically arranged and located close together. This makes it easy for the reader to refer to both elements together, **Figure 13-8**. Figures should also be clearly marked both on the illustration and in the text to avoid confusion among illustrations.

Page Size and Orientation

Technical documents can be oriented either horizontally or vertically. A horizontal orientation is called a *landscape orientation.* A vertical orientation is called a *portrait orientation*. For example, this book uses a portrait orientation. Most publications are oriented vertically. The orientation is determined after the purpose of the publication is set and the audience studied.

Many illustrations are created in a horizontal format, even for vertical-format publications. The illustration is then reduced and laid out to match the vertical format. On occasion, it may be necessary to turn the illustration horizontally on the page, **Figure 13-9**. The reader then has to turn the publication 90° in order to read the illustration. Rotated illustrations usually cover the entire page.

A wide variety of sheet sizes are used for technical documents. Determining the sheet size is in part dictated by tradition, since readers expect certain sizes of documents for certain applications. For example, business communications are normally in an 8 1/2" × 11" format and tabloids are

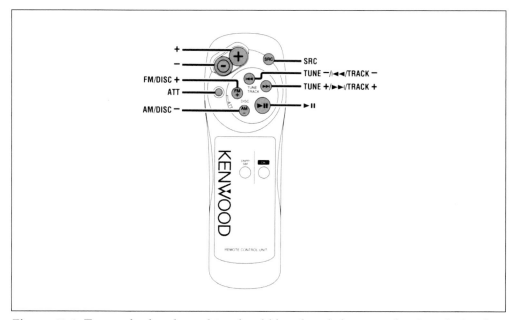

Figure 13-8. Text and related graphics should be placed close together in a design for ease of reference. (Kenwood USA Corporation)

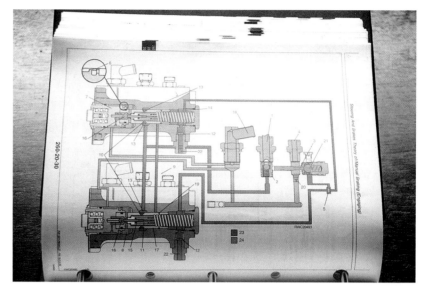

Figure 13-9. This horizontal page layout was needed in an otherwise vertically oriented manual. (John Deere & Co.)

normally in an 11″ × 17″ format. Publication page size does not necessarily limit illustration size. Illustrations can be reduced and enlarged as necessary to fit the space available. In fact, it is usually a good idea to create illustrations over-size. This is because reduction helps hide minor flaws in the illustration.

Types of Technical Illustration Drawings

There are a variety of ways to draw the same product, depending on the purpose of the drawing. The most common types of drawings used in technical illustration are visible-object drawings, section drawings, exploded drawings, phantom-view drawings, transparent-view drawings, action-view drawings, and schematic drawings. These are discussed in the following sections.

Visible-Object Drawings

A *visible-object drawing* is a pictorial drawing of a product that uses object lines to show the parts assembled, or each part separately, **Figure 13-10.** The

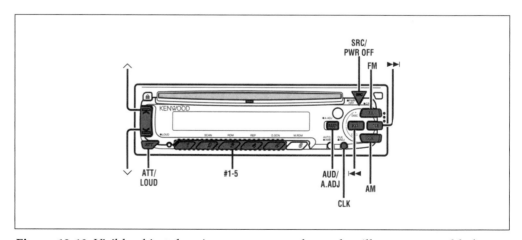

Figure 13-10. Visible-object drawings are commonly used to illustrate assembled parts or products. (Kenwood USA Corporation)

purpose of this type of drawing is to give the reader a visual understanding of the object. Drawings of this type are often used at the beginning of manuals or major sections to visually introduce the reader to the subject matter.

Section Drawings

A *section drawing* shows internal details that would normally be hidden from view, **Figure 13-11**. Section drawings are often called *cutaways* because parts of the object appear to be sliced away. Section lines are used to indicate the cut surfaces. Section drawings are frequently used to show detailed assemblies and internal working parts.

Exploded Drawings

An *exploded drawing* is a pictorial drawing that shows an object disassembled, but with the parts aligned so that it is easy to determine how they go together. See **Figure 13-12**. Each part in the drawing is normally assigned a number, and the part number and name are provided in an accompanying parts list.

Phantom-View Drawings

A *phantom-view drawing* uses phantom lines to show an alternate position for a mechanism. For example, a phantom drawing might be used to illustrate the handle of a faucet in an open and closed position. An example of a phantom-view drawing is shown in **Figure 13-13**.

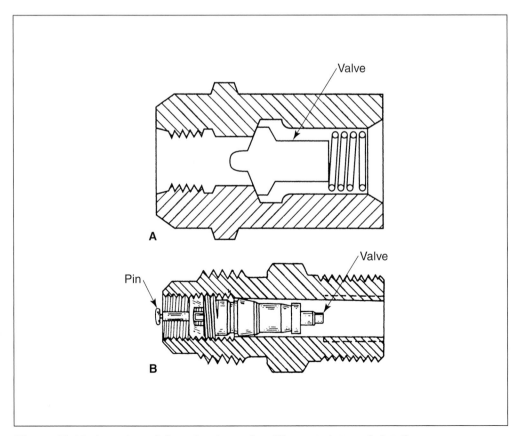

Figure 13-11. A sectioned drawing is used to illustrate internal details.

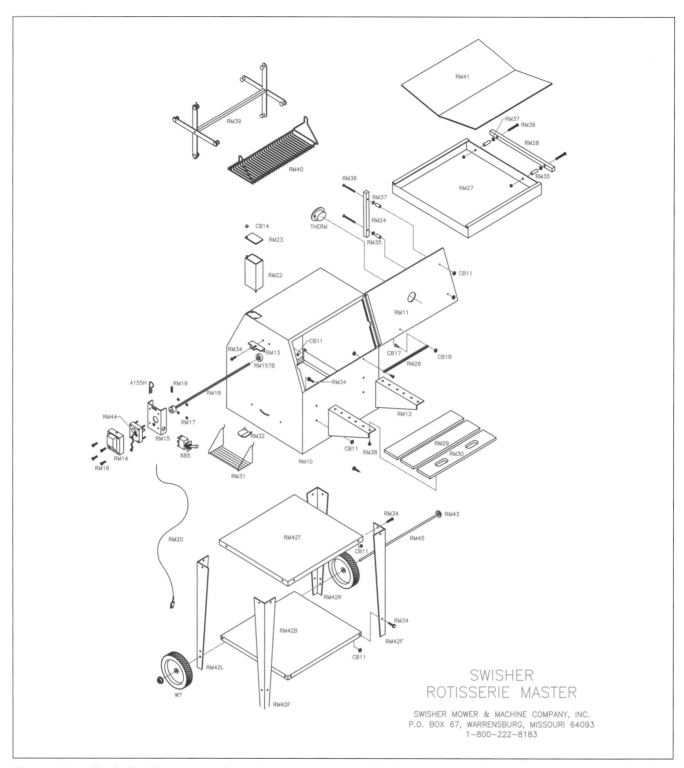

Figure 13-12. Exploded drawings indicate how parts are assembled.

Transparent-View Drawings

A ***transparent-view drawing*** is similar to a section drawing in that internal features are shown. However, the complete outer surface of the drawing is transparent, **Figure 13-14**. This drawing method is frequently employed for advertisements of products, since it is important for the customer to be able to identify the product by its overall shape while at the same time view the features within. Products such as cars, trucks, and cameras are often drawn in this manner.

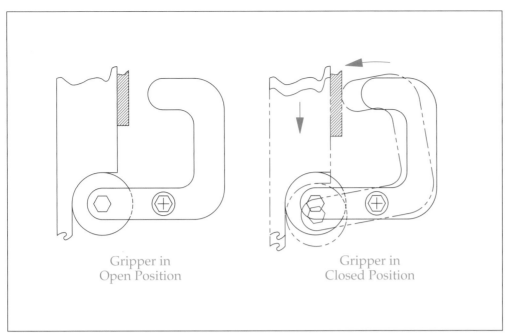

Gripper in
Open Position

Gripper in
Closed Position

Figure 13-13. A phantom-view drawing shows alternate positions for moving parts.

Figure 13-14. A transparent-view drawing uses a transparent outer surface to show the internal features of a product. (Ford)

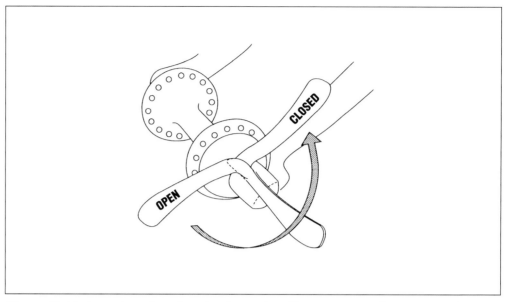

Figure 13-15. An action-view drawing is used to illustrate a specific action performed by equipment or personnel. (Trek Bicycle Corp.)

Action-View Drawings

An *action-view drawing* is used to depict action of the equipment or action of people using the equipment. The equipment action might consist of the flipping of a switch or the turning of a knob, **Figure 13-15**. An example of human action might be an illustration of a mechanic performing an engine test or turning a wrench.

Schematic Drawings

A *schematic drawing* shows a diagram of a functional system, such as an electronic circuit or a hydraulic system, **Figure 13-16**. Schematic drawings can be simple or complex. A wiring diagram for a television is an example of a complex system.

Specialized Drawing Techniques for Technical Manuals

A number of illustration methods are used to develop the common types of drawings designed for technical manuals. The two most common types of illustrations used in technical manuals are visible-object pictorial assembly drawings and exploded drawings. Visible-object drawings used to develop assemblies and subassemblies can be created using axonometric, oblique, and perspective drawing techniques. These techniques are discussed in Chapters 5 through 8 of this text. Transparent-view drawings and action-view drawings also normally use standard pictorial drawing techniques. Special techniques used to develop these types of drawings and exploded drawings for technical manual illustrations are discussed in the following sections.

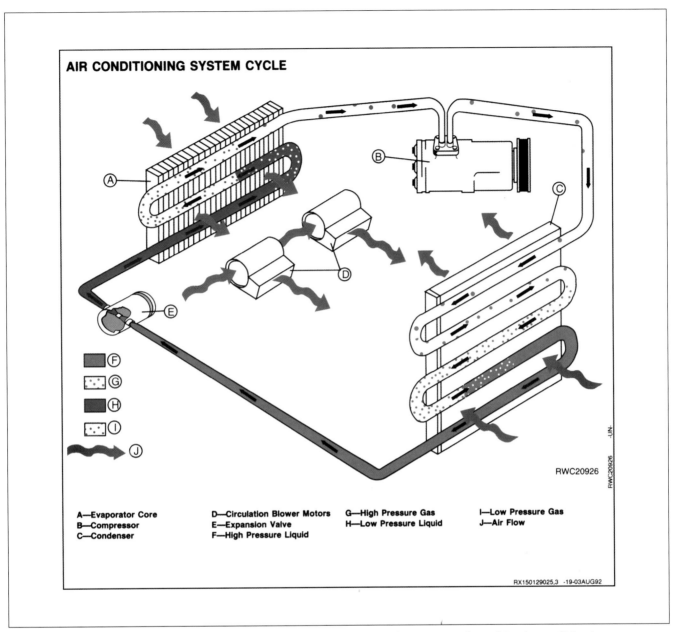

AIR CONDITIONING SYSTEM CYCLE

RWC20926

RX150129025,3 -19-03AUG92

A—Evaporator Core
B—Compressor
C—Condenser

D—Circulation Blower Motors
E—Expansion Valve
F—High Pressure Liquid

G—High Pressure Gas
H—Low Pressure Liquid

I—Low Pressure Gas
J—Air Flow

Figure 13-16. A schematic drawing shows a diagram of a functional system, such as this air conditioning system. (John Deere & Co.)

Developing Exploded Drawings

Exploded drawings are primarily used in technical manuals, parts catalogs, and other printed documents to show how parts go together. They are often used with a chart to allow users to locate a specific part name, part number, and vendor. An exploded drawing of a portable hand drill is shown in **Figure 13-17**. Observe that the drawing is an isometric drawing. This is typical of exploded drawings. In some cases, surface shading may also be applied to the drawing to distinguish one part from another. These types of drawings must be proportionally accurate. However, some measurements may be adjusted to make the parts more recognizable. For example, two parts may be almost identical in appearance except for a very small hole in the side of one

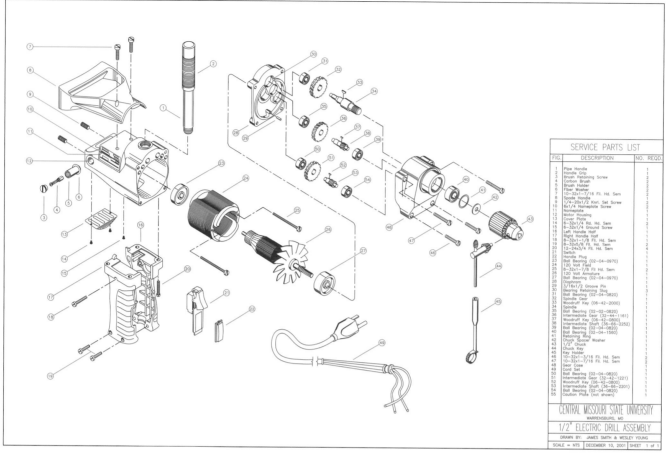

Figure 13-17. Exploded drawings are developed in isometric form. A list identifying the name of each numbered part is provided.

part. The part with the hole may be slightly rotated to better display the hole. Also, the size of the hole may be increased to make it more noticeable.

When planning an exploded drawing, make sure to leave room to fit all of the individual pieces in their exploded positions. An isometric drawing of the object may fit very comfortably on an A-size or B-size sheet. However, an exploded view of the same object may require a D-size sheet. As a rule of thumb, an exploded drawing requires four to five times the space of an isometric drawing of the assembled object.

Enlarged details of the exploded drawing of the drill are shown in **Figure 13-18**. Study the various objects and shapes used to draw each part. All features are developed using isometric drawing techniques. The following steps are used to manually create an exploded drawing:

1. Using individual sheets of paper, create an isometric drawing of the visible portion of each major part in the assembly.
2. Create an isometric drawing of each minor part using only as much detail and shape description as necessary to recognize the part. Do not draw standard nuts and bolts. They can be drawn later using isometric templates.
3. Using a large drafting table, lay each of the individual part drawings on the table in their approximate locations. Then, move the part drawings close together to consolidate them into the smallest space practical. Choose a drawing sheet size.
4. Use a drafting machine to precisely align each part drawing and tape the drawings down to the drafting table.

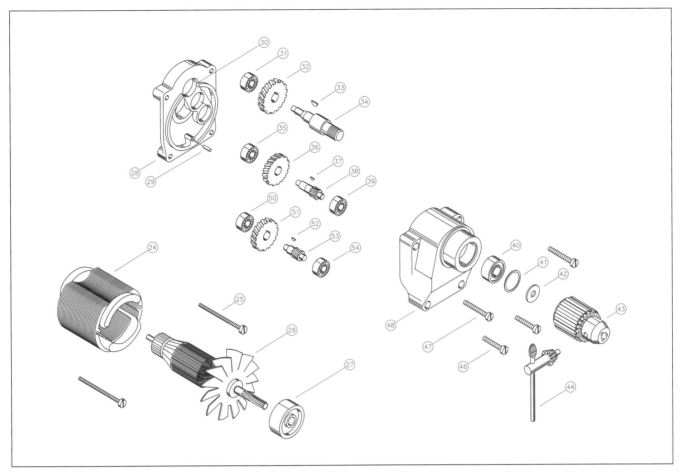

Figure 13-18. Enlarged views showing details of several part assemblies from Figure 13-17.

5. Place a drawing sheet centered over the parts, align the bottom edge with the drafting machine, and tape the sheet down. Trace the underlying part drawings on the top sheet.

6. Add callout bubbles. These are usually 5/16″ to 3/8″ diameter circles. Develop a symmetrical alignment of the callout bubbles. Add leaders to the bubbles. All leaders should be placed at approximately equal angles. Curved leaders are actually preferred over straight-line leaders because variations in the leader angle are not as distracting.

7. Starting with the part that all other parts are attached to, number the callout bubbles in either a clockwise or counterclockwise direction.

8. Develop the parts list in a sequential order beginning with the first numbered part.

Developing Transparent-View Drawings

Most manually developed transparent-view drawings are created as isometric drawings. However, true 3D transparent-view drawings can be created using the modeling tools of a CAD system. To manually create a transparent-view drawing, start by creating preliminary views of the objects. You can draw all interior features first or all exterior features first. This is simply personal preference, so use the method that works best for you. Next, place the preliminary drawings under the final drawing sheet and trace them. By using preliminary drawings, you can make slight modifications to various shapes, details, and positions.

Developing Action-View Drawings

Action-view drawings typically depict people or objects in some type of motion or activity. Drawing human forms is normally considered to be one of the most difficult tasks in technical illustration. There are several important principles to apply when drawing human forms. These include maintaining proportion and making the figure look "solid" (three-dimensional). There are templates available to help draw human forms. These templates have a wide variety of human forms in different positions as well as standard parts of the body, such as hands and feet.

Some CAD systems have 2D and 3D human shapes programmed into the software, **Figure 13-19**. You can modify these figures to fit your needs and then export the image or add other objects directly in the program. The image can be manipulated as needed and rendered to add skin, clothes, surfaces, or color. See **Figure 13-20**.

Another way to create computer-generated human forms is to use scanned photographs. Simply photograph a human subject in the posture, position, action, or activity required for the illustration. Then, scan the image as a bitmap file and import it into an image editor. Finally, alter the image and add messages, logos, or other graphic features as needed.

Using Clip Art

Graphic objects in technical manual illustrations can sometimes be replicated using clip art. **Clip art** is a sampling of commercially prepared images that can be "cut" and "pasted" into your own illustrations. Using clip art can save time because the object does not have to be manually drawn. Clip art is usually royalty free. When you purchase the clip art, you are purchasing the right to use clip art images in your illustrations. Clip art is available in printed books for manual layout. It is also available in a variety of electronic file formats for 2D or 3D illustration work, **Figure 13-21**.

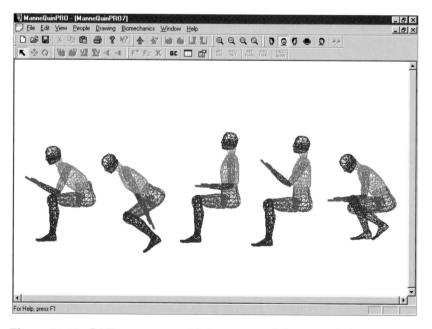

Figure 13-19. CAD programs with human modeling capability allow you to create figures in a variety of orientations. (Courtesy of NexGen Ergonomics, Inc., www.nexgenergo.com)

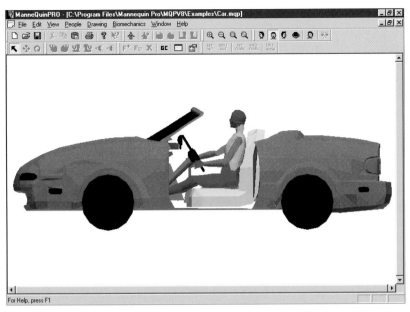

Figure 13-20. A sample rendering of a scene created in a CAD program with human modeling tools. (Courtesy of NexGen Ergonomics Inc., www.nexgenergo.com)

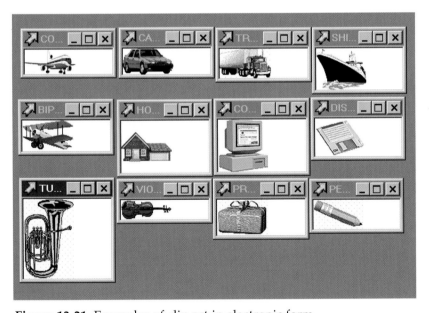

Figure 13-21. Examples of clip art in electronic form.

Computer-Based Technical Manual Production

Traditional production work for technical manuals involves manual drawing and layout methods. In this type of illustration work, text is set by hand and illustrations are obtained as manual drawings, clip art, or photographs. The text is output on galleys. *Galleys* are high-quality, camera-ready sheets of copy that are placed in a layout. Manual line illustrations are scaled photographically if needed. Any black-and-white or color photos used for production are converted to halftones for printing purposes. A *halftone* is an image converted to a series of dots. The dots making up the image simulate the shades of gray or colors in the original image. Once the text and illustrations are ready, a layout artist creates a paste-up mechanical for each page.

Manual layout is rarely used today. Traditional production work has been almost completely replaced by electronic page composition programs. See **Figure 13-22**. In electronic page composition, text is entered into the computer and imported into the layout program. The text may be typed or scanned from hard copy. Line art is created electronically or scanned, and it is imported into the layout program as well. Photographs can be taken with a digital camera and converted into image files. If printed photos are used, they can be scanned. The resulting photographic images are also imported into the layout program. The graphic designer then performs the same functions involved in manual layout and arranges the elements into an effective design. However, since all of the elements are in electronic form, layout tasks are much easier and less time-consuming.

Once all of the elements are arranged in the design, a hard copy is output on a high-resolution laser or inkjet printer. The hard copy is proofread, and any changes are then made in the layout program. Once the layout is finalized, film is created from the electronic files using an imagesetter, which produces higher resolution than a printer. Printing plates are then made from the film. In some cases, printing plates are made directly from the electronic files, thus eliminating the need for film output.

Steps in Document Production

The steps in creating a technical document using electronic methods vary from company to company. However, there are a few general steps that are applied. These steps are as follows:

1. Determine the purpose and format of the document.
2. Determine the design specifications. These include typefaces, document orientation, and page size.

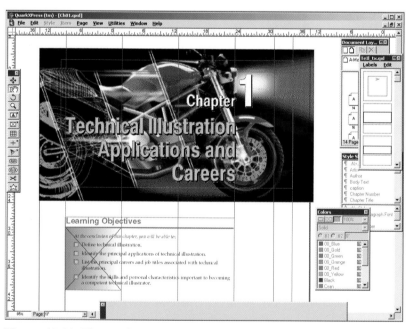

Figure 13-22. Electronic page composition programs are used to create complete design layouts with text and graphics.

3. Generate the text and illustrations to be used in the design. Convert any photographic images to an electronic format.
4. Import all text and graphics into the page composition program. Arrange the elements to create the layout.
5. Print a hard copy of the document. Proof the content and make corrections.
6. Send the finalized document to a printing firm so that it can be printed and bound.
7. Distribute the printed document.

Copyright and Trademark Principles

Copyrights and trademarks protect original works that are created. These works may be text-based works, illustrations, or photographs. When working with a technical illustration, you need to be aware of how your work is protected. You also need to be aware of how materials that you may use in your illustration, such as photographs, are the protected property of the individual or company that created them. The following sections discuss copyrights and trademarks.

Copyright Protection

A *copyright* is a form of protection for an original work, such as a written work, illustrations, artwork, and movies. In the case of a written work, it may be published or unpublished. Protection is automatic and begins as soon as the work is in tangible form (for example, on paper or electronic media). In other words, an idea in your head cannot be copyrighted. Illustrations and text are copyright protected. Other original works, such as music and sound recordings, are also copyright protected.

Although copyright protection is automatic, it is a good idea to register a copyright for commercial work. Registering a copyright is discussed in this section. Whether or not you register a copyright, you should include a copyright notice with your work when you publish it. This ensures that no one can mistake your work for copyright-free work. Always place the copyright notice in a visible location on your work.

A copyright notice is made up of several elements. First, the word Copyright, the abbreviation Copr., or the copyright symbol © appears in the notice. Sometimes both the word and symbol are used together. In addition, the year of first publication is shown. Finally, the owner of the copyright is listed. An example of a copyright notice might be:

© 2002 by ABC Company

A copyright can be registered with the U.S. Copyright Office. Registering a copyright provides protection in case legal action is needed for copyright infringement. A special application form corresponding to the type of work is used. Copyright forms are available from the U.S. Copyright Office (located in Washington, D.C.) and from other sources. Different forms pertaining to different types of work can be downloaded by visiting the Web site www.copyright.gov. Some public libraries also provide copyright forms. Complete the appropriate form and return it. There is a small fee required to complete the process.

Using Copyrighted Material

When using materials from another source, you must determine whether the work is in the public domain or copyrighted. Works in the *public domain* have no copyright, or the copyright has expired. Public domain works can be freely copied. If a work does not have a copyright notice, it may still be copyrighted. If it is copyrighted, you must get permission from the copyright owner to use the material.

When requesting permission to use a work, many companies have a standard *permission form* to fill out, **Figure 13-23**. A permission form usually asks you to state what you are going to use the material for, where it will be distributed, and whether or not it will be used for commercial purposes. In some cases, companies also ask that you pay to use their material. Almost always, a credit line such as "Material reprinted with permission of…" is included with the copyrighted material.

Goodheart-Willcox Publisher

18604 West Creek Drive • Tinley Park, IL 60477-6243

Photographic Release Form

The undersigned release(s) to the Goodheart-Willcox Company, Inc. of Tinley Park, Illinois the right to use the following photograph(s) in the text ___**Technical Illustration**___ by **John A. Dennison and Charles D. Johnson**, which is expected to be published ___**2003**___.

(code numbers or descriptions of photos)

The credit line for said photograph(s) should be:

Signed:_____ Date:_____
Signed:_____ Date:_____

If applicable, indicate need for return:

Figure 13-23. A sample permission form.

Trademark Protection

A *trademark* is a unique symbol, name, or slogan used to identify the source of a product or service. Examples are Microsoft and Pepsi. A trademark used to identify the source of a service is called a *service mark*.

Trademark rights begin when a trademark symbol or name is used to identify goods or services. Rights also begin if there is an intention to use the trademark in the future and an application is made to register the mark. Like a copyright, a trademark is protected automatically. However, registration is important if legal action is needed to prevent infringement. Registration also ensures that the trademark is not already being used by someone else. The U.S. Patent and Trademark Office registers trademarks. Application forms can be obtained by visiting the Web site www.uspto.gov. A sample drawing must be sent along with an application to register a trademark. See **Figure 13-24**.

When a trademark has been created, the symbol ^{TM} or ^{SM} can be used to show that it is a trademark. If the trademark is registered, then the symbol $^{®}$ or the phrase "Registered in U.S. Patent and Trademark Office" is used. Using this identification ensures that everybody knows it is a trademark, even though there is automatic protection without a trademark symbol or phrase.

In a publication, trademark symbols are normally left off. Instead, a statement is placed in the front pages of the publication indicating the trademark. An example is:

123 is registered in the U.S. Patent and Trademark Office by The 123 Corporation.

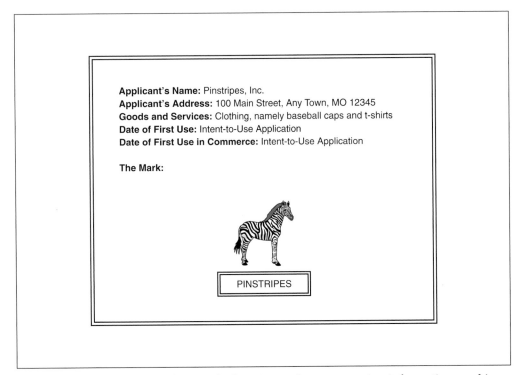

Applicant's Name: Pinstripes, Inc.
Applicant's Address: 100 Main Street, Any Town, MO 12345
Goods and Services: Clothing, namely baseball caps and t-shirts
Date of First Use: Intent-to-Use Application
Date of First Use in Commerce: Intent-to-Use Application

The Mark:

PINSTRIPES

Figure 13-24. An example of a sample drawing and accompanying information used in a trademark application. (U.S. Department of Commerce, Patent and Trademark Office)

Summary

Technical illustrations are widely used in technical manuals. Therefore, it is important for a technical illustrator to have an understanding of the various types of these publications. In addition, an illustrator needs to understand the purpose and audience in order to determine what kind of document to develop, along with the format and content. For example, a plumbing manual prepared for a professional plumber requires different text and illustrations in comparison to those required for one written for a homeowner.

It is important for a technical illustrator to understand how the final illustration will be incorporated into a publication. This may help the illustrator make decisions when creating the illustration, such as choosing a portrait or landscape orientation.

As is the case with many other types of publications, technical manuals are commonly produced using electronic methods. The same basic concepts that apply to manual production work apply to electronic production work in the way text and illustrations are placed on a page. However, layouts are easier to create and revise with electronic page composition programs.

Copyrights and trademarks provide legal protection for original works. Although protection in both cases is automatic, it is a good idea to register those works created by a company. Always be sure that any material you use is free from copyright or trademark protection, or that you have permission from the owner to use the material.

1. Of the four common types of technical manuals, which type would be most commonly used for a new bike that is to be shipped unassembled?
2. How does an installation manual differ from an operation manual?
3. What is a *parts identification list*? Give two examples of where parts lists are typically found.
4. Briefly discuss the factors considered when determining the purpose and audience for a technical document.
5. What is the difference between a visible-object drawing and a transparent-view drawing?
6. What is the primary purpose for using a section drawing?
7. Give three examples of applications that would require a phantom-view drawing.
8. What type of drawing is used to depict the action of equipment or the action of people using equipment?
9. A complex diagram showing the functions of an electronic circuit is an example of what type of drawing?
10. Briefly discuss the process used for laying out and developing an exploded drawing by hand.
11. What is a *halftone*?
12. How are photographic and line art images prepared for layout in an electronic page composition program?
13. Define the terms *copyright* and *trademark*.
14. Identify the elements making up a copyright notice.
15. How can a copyright be registered?
16. Why are works in the public domain freely available for use?
17. Give two reasons why it might be necessary to register a trademark.

1. Locate examples of three different types of technical manuals as described in this chapter. List the name of the manual, the publisher, and the date of publication.
2. Using one of the technical manuals found in the previous activity, describe its purpose and audience in a brief report. Explain why it is designed in a certain manner and identify the types of content that serve the purpose of the publication. Describe how you think these elements affect the presentation format. Identify who you think the typical reader is and describe how this affects the content and reading level.
3. Design and create one of the types of technical illustrations discussed in this chapter. This can be done by hand or with a CAD program.
4. Download a copyright registration form from the U.S. Copyright Office on the Internet and complete it for a proposed technical manual.
5. Locate four examples each of copyrights and trademarks. For each one, list the product it protects and describe the way in which the protection is identified.

Logo designs should represent a business in a positive manner and appeal to the target audience. This logo was created for a skateshop. It was produced electronically in a paint program. (Jason Weiesnbach)

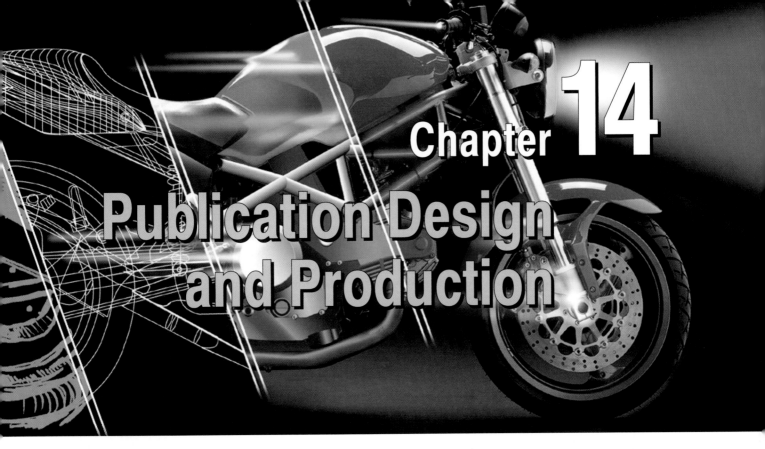

Chapter 14

Publication Design and Production

Learning Objectives

At the conclusion of this chapter, you will be able to:

☐ Outline the production steps involved in creating a printed product.

☐ Use design principles and elements to produce effective page designs.

☐ Explain manual and electronic layout techniques.

☐ Identify common hard copy output devices and explain how their uses vary.

Introduction

Technical illustrators should have some background in publication design and production for several reasons, **Figure 14-1**. First, the technical illustrator may also be performing electronic publishing tasks. Second, technical illustrators work and communicate with individuals involved in printing and publishing. Knowledge of the publication production process allows illustrators to communicate intelligently. Third, the production process may place limitations or restrictions on illustrations. A familiarity with the process will help illustrators avoid mistakes.

This chapter introduces the various production tasks involved in the publishing process. Design principles and image types are discussed, as well as outputting procedures typically used for producing illustrated publications.

Production Steps

There are several general steps to follow when developing a printed product. The product might be a catalog, manual, newspaper, brochure, book, or newsletter. Specific steps vary among companies. However, the following general steps always apply:

1. Image design.
2. Image generation and assembly.
3. Image carrier preparation.
4. Image transfer (printing).
5. Product finishing.

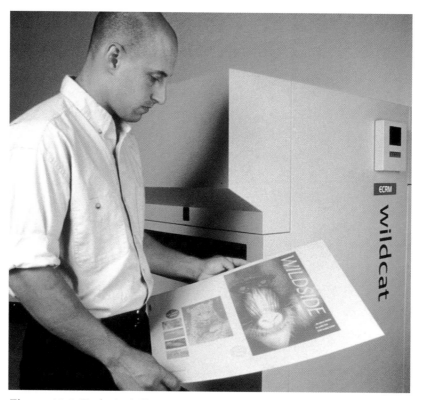

Figure 14-1. Technical illustrators need some understanding of publication design and production. (ECRM Imaging Systems)

Image Design

Image design is a critical step in the creation of any printed product, and should be done carefully. If the design is poorly done, then the final product will be inadequate.

Image design actually begins by determining the purpose of the document and the audience. Once the purpose and audience are determined, sketching can begin. Thumbnail sketches are made first. *Thumbnail sketches* are small, proportionally correct sketches of design ideas. They are usually done on a grid sheet or on a computer to aid the sketching process, **Figure 14-2**. When a computer is used, design options (not actually sketches) are made full size. A simple way to lay out space for several hand-drawn thumbnail sketches is to fold the final-size paper in half for both width and height. This creates four smaller, proportional areas for thumbnail sketches. Typically, at least four design options are created.

Thumbnails are used to create a rough layout. A *rough layout* is a full-size representation, but it still lacks some detail. In many cases, the rough will be sufficient to determine if the design is appropriate. The next step is the comprehensive layout. A *comprehensive layout* includes color, illustrations, and text, **Figure 14-3**. This allows the customer to see exactly how the final product will appear.

Image Generation and Assembly

After the publication is designed, image generation and assembly takes place. In this step, the images and text are generated. These elements are then pulled together to create the publication. This step is also called the *layout stage* because all the elements are placed on each page either manually or electronically with a page composition program. See **Figure 14-4**. At the end of this step, an original of the publication is complete.

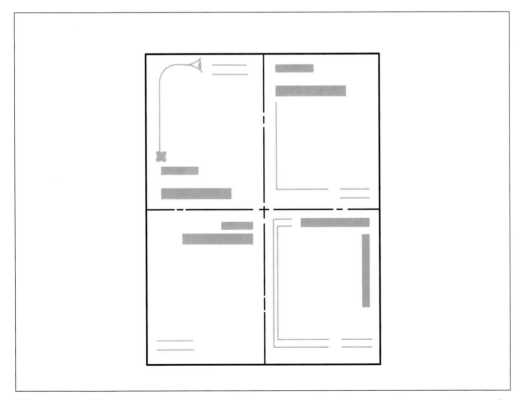

Figure 14-2. Thumbnail sketches provide a proportionally correct representation of a design.

Figure 14-3. Color is being manually added to this comprehensive layout. (Pioneer Graphics)

Figure 14-4. An example of a design created with an electronic page composition program. (Xyvision, Inc.)

Image Carrier Preparation

Multiple printed copies are necessary for most publications. This requires an image carrier that can be used to print the product. See **Figure 14-5**. The *image carrier* is placed on a printing device and is used to actually place the image on the page. For example, in offset lithographic printing, the image carrier is the plate.

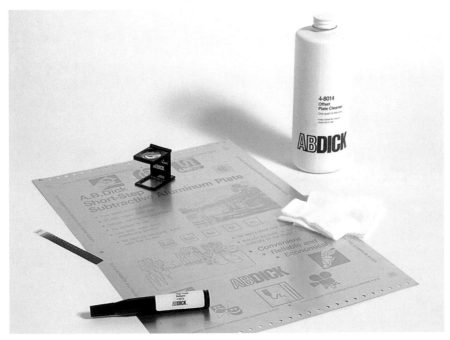

Figure 14-5. A lithographic plate is used as the image carrier in offset printing. (A.B. Dick)

Image Transfer

The image transfer is the actual printing process. For example, in offset lithography, the plate (image carrier) is placed on a printing press. Ink adheres to the image area on the plate and is transferred to paper. This completes the image transfer step, **Figure 14-6**.

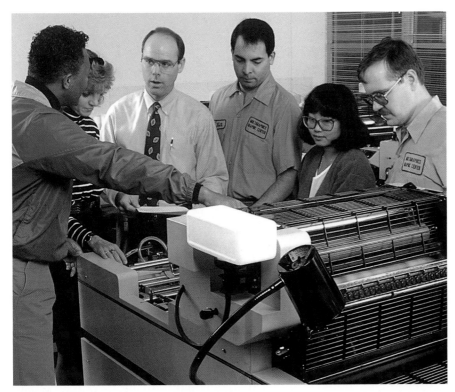

Figure 14-6. Image transfer (printing) is done on a press, such as this offset lithographic press. (AM International, Inc.)

Product Finishing

The final step in preparing a printed product is product finishing. This refers to all of the steps necessary to place the printed pages in publication form. Product finishing might involve paper cutting and trimming, folding, and binding the pages together. Other operations, such as adding a cover and packaging inserts inside, are also finishing steps.

Principles and Elements of Design

When designing a publication, it is important to incorporate elements that attract reader attention while following conventional guidelines for the type of publication being produced. Design strategies vary by the type of publication. For example, design considerations used with newsletters differ from those used with catalogs and books. Books require a specific design for the first pages of chapters. Chapter opener graphics serve the purpose of identifying the chapter, but they also entice the reader to read further.

If you look at several newspapers, each has unique characteristics. However, they all have a similar basic appearance. The front page of a normal broadsheet newspaper is approximately $14'' \times 22\ 1/2''$ and set in multiple columns. An identifying nameplate is at the top. Other design elements include headlines, subheads, text, and illustrations. For emphasis, kickers and pull quotes might be used.

The following sections discuss the common design principles and elements that are used in the development of printed publications. Good design principles allow you to create effective visual materials. A number of common design elements can be used in accordance with these principles to generate reader interest and communicate the message of the publication.

Design Principles

There are several standard design principles that provide some general guidance in the development of any publication. These principles are *proportion*, *balance*, *emphasis*, *contrast*, *rhythm*, and *unity*. They are used in combination to create good designs. These principles are not rules set in stone. Rather, they are general guidelines to follow in order to create a pleasing design.

Proportion

Proportion is the size relationship of one part of a design to the size of the whole design. When you make thumbnail sketches, you should have some idea of the proportional relationship defined by the size of the finished product. In other words, what is the width-to-height ratio of the finished product? For a square-shaped product, this proportion will be 1:1 and your sketch will need to be made in a 1:1 ratio. If $8\ 1/2'' \times 11''$ paper is to be used, the ratio is 8.5:11. A traditional ratio used to set proportion, known as the *golden section* ratio, is commonly used to create visually pleasing designs. This ratio, discussed in Chapter 3 of this text, is 1:1.618.

The elements on the page also should be in proportion with one another. For example, headlines should be larger than body copy. In addition, if an illustration is placed on the page, elements in the illustration, such as trees and flowers, should be in proportion to one another.

Balance

Balance is the sense of visual equilibrium in a design. There are two types of balance—formal and informal. *Formal balance* occurs when all elements on both sides of an illustration are equal and the design is symmetrical. See **Figure 14-7**. *Informal balance* occurs when elements on one side of an illustration are "heavier" and the design is asymmetrical. See **Figure 14-8**.

Formal balance is a more conservative design strategy. Informal balance tends to be used for a more modern appearance. For example, the cover of a new computer catalog might be designed with informal balance. An invitation to a bank reception would most likely use formal balance.

Formal and informal balance refer to horizontal balance. However, vertical balance is also important, **Figure 14-9**. Designs are normally more pleasing to the eye if they are slightly above the mathematical center. This visually pleasing center point is known as the *optical center*.

Figure 14-7. An example of formal balance. Elements surrounding the optical center create a symmetrical appearance.

Figure 14-8. An example of informal balance. Elements on either side of the optical center are arranged differently, creating an asymmetrical appearance.

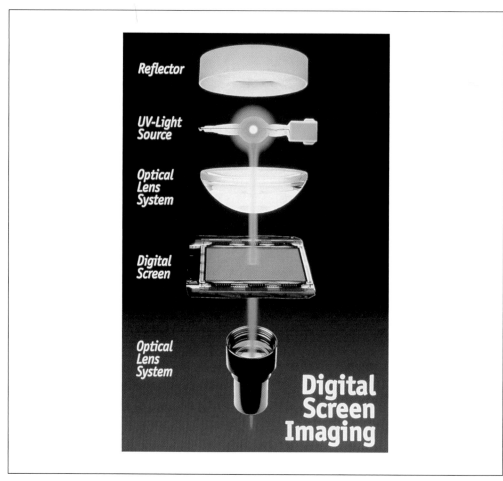

Figure 14-9. An example of vertical balance. The elements in the design are arranged in a way that generates visual interest.

Emphasis and contrast

Emphasis is used to make one part of a design stand out, or draw the greatest interest, **Figure 14-10**. The portion of a design receiving the greatest emphasis is intended to be the first element you see when looking at the design. For instance, a large, bold headline in a newspaper is a point of emphasis. Emphasis is also used to a varying degree in other areas of the design so that we look at other parts after viewing the center of interest. Size, color, and shape can all be used to emphasize a part of the design.

Contrast is closely related to emphasis. *Contrast* is the variation of elements in a design to draw attention or provide meaning. For example, you are attracted to headlines in a newspaper because they have a larger type size than the text copy. When you see a photo of a candle in a darkened room, your eyes are attracted to the light of the candle because it is in contrast to the dimly lit room.

Rhythm

People with a good sense of musical rhythm can keep time by tapping their feet in time with the music. This same idea of repeated action can be used in design. *Rhythm* is the sense of movement or repetition. Simply repeating certain elements in the design can provide a rhythmic effect, **Figure 14-11**. A photo of ripples on water is a good example of this principle. The concentric circles of the ripple are the repeating elements.

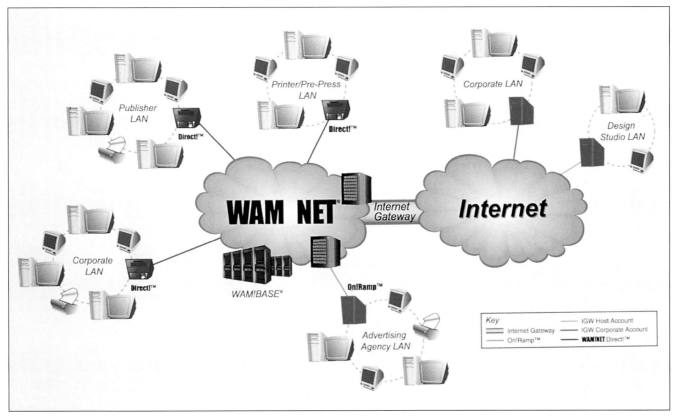

Figure 14-10. What is the center of interest (emphasis) in this design? (WAM!NET, Inc.)

Figure 14-11. Rhythm is achieved in this illustration by simulating the flight pattern of the airplane. (Macromedia FreeHand)

Unity

Unity is perhaps the most important principle of design. **Unity** is achieved when the entire design is tied together as a whole. For example, different type styles on the same page should be harmonious unless contrast is being used for emphasis. Simply stated, all the elements in the completed design should look pleasing together.

Design Elements

A design is developed using the elements of color, line, shape, and texture, **Figure 14-12**. Text on a page creates lines. These lines create columns. Columns are rectangular in shape. Color or texture may also be added to the page. *Texture* is an optical simulation of a physical surface texture. An example of texture is the shading of drawn solids with lines and dots. There are also certain page elements that have become standard in publications over time. These are discussed in the following sections.

Headers and footers

Headers and footers are identifying elements that are repeated on every page at the top and bottom, respectively. A *header* appears at the top of a page. A *footer* appears at the bottom of a page. For example, a header or footer might include the page number. Often, a header or footer includes a running head-line. This is known as a *running head.* A running head is used to inform the reader of the chapter or section currently being read.

Headlines and subheads

Headlines are headings on a page. They are commonly referred to as *heads*. Headlines provide a general idea of what is being presented until the next major head. *Subheads* divide the material under a headline into smaller sections. Subheads are often divided again. The different types of heads are often called Level 1 heads, Level 2 heads, Level 3 heads, and so on. A typeface and point size are selected for each head and then used consistently throughout a publication. For example, a first-level head may appear in all caps at the center of the column. A second-level head may also be all caps, but it may be placed on the left margin. A third-level head may be placed in upper and lowercase letters on the left margin.

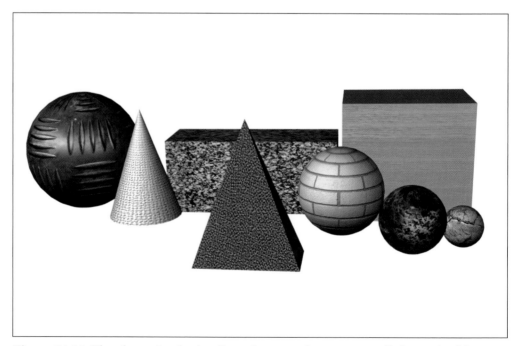

Figure 14-12. The elements of color, line, shape, and texture are all shown in this design.

Body copy

Body copy is the text of a publication. It is usually placed on the page in columns. The number of columns used can be varied, but a general rule of thumb is to have column widths that will hold at least four words. Columns can be unequal in width if desired.

As with a headline, text needs to have a specific typeface and size. A common specification for text is to use a serif typeface 12 points in size. In addition, specifications need to be made for the line spacing (leading), paragraph formatting (indents), and type alignment. Type alignment classifications include left (ragged right) and justified. Justified type has evenly aligned left and right margins. This establishes a more formal appearance. In order to justify the lines, space is added between words and words are hyphenated as needed at the end of each line. Justification is often used because it accommodates more words in a line than other alignment methods. However, flush left (ragged right) is thought to be easier to read since word spacing remains constant and the eye can more easily follow the lines on the page because of their unequal length. Typically, tradition should be your guide when determining the type alignment.

When a story is continued on another page in a newspaper or magazine, a jump line is used. A *jump line* tells the reader where the story continues.

Kickers, pull quotes, and sidebars

A number of common elements are used in publications to enhance the design and generate reader interest. These include kickers, pull quotes, and sidebars. A *kicker* is a short phrase normally placed near a headline and used to summarize the text or provide clarification of the headline. A *pull quote* is a quote from the text that is set apart to promote interest. See **Figure 14-13**. A *sidebar* is a small article placed on the side of a page that relates to the main article.

Here's our one word summary of why you want suspension: **CONTROL**. Read more if you want to, but that's the story in a nutshell.

You want suspension so you can see the trail. You want it so you can hold a line when a mountain switchback transforms your bike into a jack hammer. You want it to keep your wheels from locking and planting your face in the sand.

...that delicate balance of speed, braking and handling on punishing terrain...

You want it to keep fatigue from turning your arms and legs into linguini as you tough out that last climb.

Whether original equipment or "after shock," adding a suspension isn't something you do solely to smooth the ride. If comfort were your only concern, you wouldn't have a mountain bike, just a chaise lounge. Maintaining control...that delicate balance of speed, braking and handling on punishing terrain...that's why you want suspension!

Trek's solution: three air sprung/oil damped shocks designed for specific performance levels. Within those broad ranges, forks are rider-adjustable to deliver the requisite stiffness and impact absorption.

Figure 14-13. A pull quote is used to enhance a design and generate reader interest. (Trek Bicycle Corp.)

Illustrations

Illustrations should be placed close to where they are referenced in the text. This helps prevent confusion and makes it easy for readers to look at the illustration as they read the text. The width of an illustration can be sized to align with a column, or it can extend over several columns. Boxed borders and drop shadows can be used with illustrations for emphasis. A *border* is simply a ruled line around the illustration. A *drop shadow* is a second outline placed slightly off to the side of an illustration to create a shadow appearance. A caption may be placed next to, or below, the illustration for clarification. *Captions* contain a summary of the illustration. They also credit the source if the illustration is copyrighted by another individual or company.

Runarounds are sometimes used with illustrations. A *runaround* is an arrangement for text so that it follows the outline of an irregularly shaped illustration. See **Figure 14-14**. This enhances the design and reduces the white space on the page.

White space

It is important to take into consideration the blank areas of a page. These blank areas constitute *white space.* This may seem like a strange concept. However, effective use of white space can help create a very effective publication. A common problem in design is trying to place too much information on a page. This can eliminate almost all white space except for that in the margins. A page with very little white space can be hard to read. Carefully used, white space makes a page much more inviting to the reader.

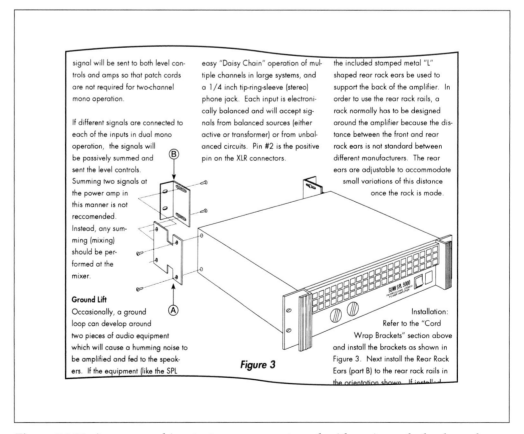

Figure 14-14. A runaround is a text arrangement used with an irregularly shaped illustration. (Fender Musical Instruments Corp.)

Graphic elements

Ruled lines and borders are frequently used graphic elements. *Ruled lines* are used to separate parts of a page. They can be thick or thin. Borders are usually placed around parts of a page to separate them from the rest of the page. In some cases, borders are placed around the entire page.

Other graphic elements can be used in combination with bordered areas. For example, borders are commonly used with sidebars. The background of the boxed area can then be filled with a tint to help set it off. In this instance, text within the bordered box can be set as reverse type. *Reverse type* is white type set against a colored background.

Cover page elements

The *cover page* of a publication is designed to communicate what is inside. It should also attract reader attention to the publication. This is done by carefully blending large type and illustrations for maximum impact, while keeping practical considerations in mind. If the publication is to be displayed on a magazine rack, for instance, the name should be positioned at the top so it can be seen.

In the production of periodicals such as newsletters, newspapers, and magazines, a nameplate is used on the front page. The *nameplate*, also known as the *flag*, contains information such as the name of the publication, the date, and the cost. The nameplate is designed to be used over and over again with minor changes.

Another important element often found on the cover of a corporate publication is the logo, **Figure 14-15**. A *logo* is an identifying symbol for a company or business. It is placed on all publications by the company.

Preparing Images

As discussed in Chapter 11, there are two types of images used in publications. These are line and continuous tone images. Different methods are used to prepare the different types of images for printing. These methods are discussed in the following sections.

Line Images

Line images have a uniformly dark image. Most technical drawings are classified as line images. Line drawings should be done in ink or with another method that will produce sharp, crisp lines. For example, a pencil does not produce lines that are dense enough, or dark enough, for most publication purposes.

Figure 14-15. A logo is designed to represent the name of a company. In this case, the symbol is made up of company colors and the abbreviated name.

Continuous Tone Images

Continuous tone images include photographs, paintings, and airbrushed illustrations. These illustrations are made up of a series of tones. For example, a black-and-white photo has portions of black and white in the image, but the majority of the image is made up of different shades of gray. Each tone requires a separate tone pattern when printed on a press. Therefore, continuous tone images must be modified to use in printing.

To use a continuous tone image in printing, it is first electronically scanned and then converted into a halftone. A *halftone* is a print made up of a series of dots that simulate the tones in the original image. The halftone dot structure is used to reproduce the continuous tone original. When the final printed image is viewed, our eyes cannot actually see the dots, but the dots and the paper color are blended so that you see the original tone. See **Figure 14-16**. Dots are more closely spaced, or larger, in darker areas of the image. In lighter areas of the image, dots are further apart, or smaller.

Scaling and Cropping

Scaling is reducing or enlarging an illustration proportionately. Page composition, image editing, and CAD programs all have scaling capabilities. In manual layout, scaling is done photographically. In this instance, a *proportion wheel* can be used to determine enlargements and reductions of an illustration. See **Figure 14-17**. The original height or width is located on the inner disk and aligned with the desired printing size on the outer disk. While keeping the wheel in this position, the height, width, and percentage of enlargement or reduction can be read from the wheel.

Cropping is the removal of material from one or more edges of an illustration to delete unneeded material. For example, a photo might contain two pieces of equipment side by side, but only one needs to be shown. The photo can be cropped

Figure 14-16. An enlarged area of this four-color photograph shows the dot structure used to simulate tones in the image.

Figure 14-17. A proportion wheel can be used to size images in manual layout.

to remove the unneeded equipment. Another use for cropping is to make an illustration fit a predetermined space. This may be necessary when a publication has illustrations of uniform size. Cropping will normally change the height-to-width ratio of the image. Therefore, cropping is usually done before scaling.

Cropping is typically done with a page composition program. A box or window is drawn around the image area to be kept using the software's cropping function, **Figure 14-18**.

Figure 14-18. Cropping tools in page composition programs are typically used to remove unwanted material from images. (Ford)

Image Conversion

For page layouts done with a page composition program, images must be in electronic form. An image can be created electronically (for example, with a CAD program) and imported directly into the program. However, if the image is created manually or is in the form of a photograph, it needs to be converted to electronic form. Scanners are used for this. *Scanners* pass light over an illustration and measure the amount of light reflected. See **Figure 14-19**. In this operation, colors and shades of gray are converted to digital form to assemble an electronic image. There are hand-held, flatbed, drum, and film scanners. Drum scanners produce the highest-quality reproductions.

Welcome to the Print Shop

In the beginning of this chapter, the general steps used in producing a printed product were discussed. Let's observe how these steps might be carried out in a print shop.

Assume you are working with a small company that produces catalogs, and you have just completed the page layout. All line and continuous tone images have been placed on the pages. You have printed out the pages on a laser printer in order to proofread them and make corrections. However, you plan on giving the printer the completed publication in electronic form so that printing plate materials can be output with an imagesetter. This provides a much higher-quality image than what you are capable of producing.

If you needed only 100 copies of the catalog, as with a previous job, you might be able to use a simpler process to produce the publication. For example,

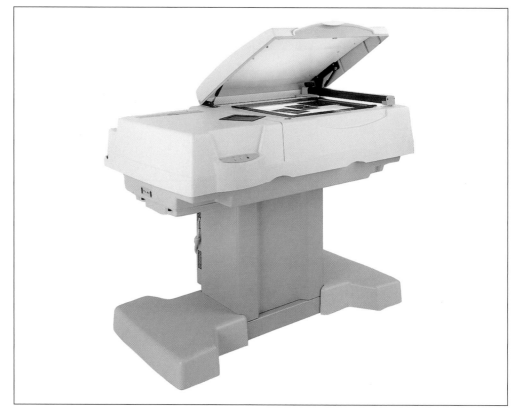

Figure 14-19. A flatbed scanner. (Fuji Photo Film USA, Inc.)

a quick printing service is usually suitable for smaller print jobs. However, 10,000 copies are needed for this job, so you will need to have it printed by a commercial printer. In this case, the catalog will be printed on an offset lithography printing press, sometimes simply referred to as an offset press. This is one of the most common forms of printing.

After you give the materials to the printer, the related electronic files are checked and changes are made. If more than one color is to be printed, a plate is made for each color. In the case of a color photo, for example, four plates would be needed in order to print in exact register cyan, magenta, yellow, and black. These four colors will produce a full-color illustration on the printed page.

In offset lithography printing, a press is used to print from plates mounted on cylinders. This is also called *planographic printing* because printing is done from the smooth surface of a plate. This is accomplished by applying a dampening solution to the nonimage area of the plate and then applying ink to the image area. The water repels the ink in the nonimage area, thus ensuring a clean image on the page. When printing begins, the right-reading image on the plate is reversed on another cylinder called a blanket cylinder, and the blanket cylinder makes contact with the paper for actual printing. This is referred to as the offset process because the image is transferred or *offset* to the paper through the contact of two cylinders, rather than directly from the plate.

In most cases, signatures will be printed. A *signature* is one entire section of the catalog, or any other publication with a large number of pages. It is created by printing multiple pages on a large sheet and then folding and cutting the sheet to make the smaller pages.

After printing, finishing operations occur. This involves folding and cutting, and binding the signatures into the finished product. There are different types of binding used depending on the needs of the product. Common binding methods include adhesive, mechanical, and sewn binding. In mechanical binding, sheets are held together by fasteners such as wire, staples, and metal or plastic rings. Sewn binding is the most durable form of binding.

There are many variations of the printing methods previously discussed. However, this information should give you a general understanding of the steps involved.

Hard Copy Output Devices

There are a variety of devices that can be used for outputting illustrations and other material for a publication. Output devices can be categorized as vector or raster devices. Each type has advantages and disadvantages. The two types of output devices are discussed in the following sections.

Vector Output Devices

Vector output devices produce drawing and text images as a series of straight lines. Entities such as arcs and circles are produced by combining very short lines. The pen plotter is the most common vector output device, **Figure 14-20**. Its output closely simulates the appearance of a hand drawing. Special pens are placed in a pen carriage and used to draw the illustration. Depending on the application, the pens might be fiber-tipped, liquid ink, or ballpoint pens.

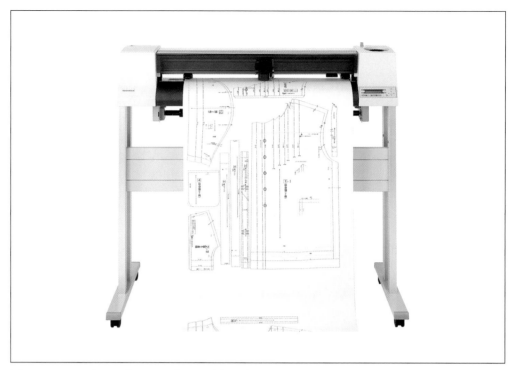

Figure 14-20. An eight-pen plotter. (Graphtec)

Pen plotters are classified as either *single-pen* or *multiple-pen* plotters. For single-pen plotters, the pen is inserted manually into a gripper. In order to plot with a different color, the pen must be changed. Multiple-pen plotters hold pens in a rack or rotating carriage. The pen holder moves to the rack and the gripper clutches the appropriate pen. When a pen is no longer needed, the plotter returns the pen to the rack and retrieves another. Pen positions are numbered. Before plotting, you define which pen number corresponds to a certain linetype or color on screen, **Figure 14-21**. Pen plotters provide very high resolution, but they are slower than other output devices.

Pen plotters are also classified as flatbed, drum, and microgrip plotters. With a flatbed plotter, the paper remains stationary while the print head moves horizontally and vertically over the surface. With a drum plotter, paper is placed on a cylinder. The drum rotates over the surface as the pen moves left to right. A microgrip plotter also moves the paper horizontally as the pen moves vertically. Small rollers grip the paper at the edges to move it.

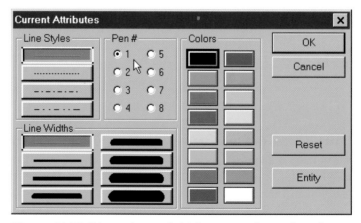

Figure 14-21. Line color and thickness can be assigned to specific plotter pens using pen options in a CAD program.

Controlling vector output

There are a number of guidelines to follow when using a vector output device. These relate to line quality, color, and lettering. Common methods for controlling vector output are discussed below. Vector output devices produce high-quality images, but they are slower than raster output devices. This must be taken into account as well.

➤ **Line quality.** To produce proper line quality, make sure the correct type of pen is being used with the media you select. For example, high-quality lines require liquid ink. Drafting film is a good choice for media with this type of pen. Another consideration in relation to line quality is plot speed. Some colors tend to be light if the pen speed is too fast and the pen size is small. A slower speed may produce a darker color.

➤ **Color.** When plotting in multiple colors, it may be necessary to stop the plotter between colors. Some colors may bleed when wet. If a color is allowed to dry before the next color is plotted, bleeding will not be a problem.

➤ **Lettering.** Some typefaces may close up when plotted using wide-point pens. A fine-point pen (0.3 mm) usually takes care of this problem and produces higher-quality letters.

Raster Output Devices

Raster output devices produce drawing and text images as a series of dots, **Figure 14-22**. The dots, known as pixels, are arranged in rows. Printing resolution is determined by the number of dots displayed. In many cases, resolution is specified in dots per inch (dpi).

In general, raster output devices are faster than vector output devices. Typical raster output devices include electrostatic, thermal, laser, and inkjet devices. Although raster output is typically associated with printers, raster-based technology is also used by different types of plotters. Plotters are used for large-format printing, whereas printers are generally used for output with smaller sheet sizes. Common raster output devices are discussed in the following sections.

Electrostatic plotters

An electrostatic plotter works much like a plain paper copier. The paper is first given a negative electrical charge where the image will be printed. Positively charged toner is then applied. The toner is attracted to the image area. Finally, heat is used to dry the toner. Color electrostatic plotters are available, as well as single-color units.

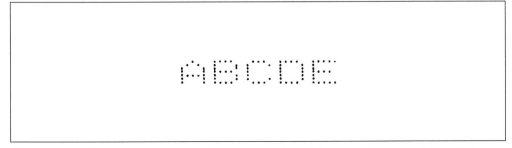

Figure 14-22. Raster output is made up of a series of dots, as shown in this inkjet printing example. (Domino Amjet, Inc.)

Thermal plotters

Thermal plotters create plots through the use of heat. In direct thermal plotting, heat is applied to the image area on chemically coated paper to create the image. In thermal transfer plotting, a printhead is used to heat an inked ribbon, which transfers the image to paper. Thermal transfer plotters produce better plotting quality than direct thermal plotters.

Laser printers

A laser printer is an electrostatic printer. Its operation is similar to that of an electrostatic plotter. See **Figure 14-23**. In this case, a light-sensitive belt or drum is given a positive electrical charge. The image is then scanned on the drum using a laser light beam that creates a negatively charged image area. Then, a positively charged toner powder strikes the drum and adheres to the image. The drum is rolled against a substrate, which transfers the toner. Heat is then used for fusing the toner to the paper.

Laser printers normally output images ranging from 300 dpi to 1200 dpi. This is relatively good output for technical illustrations. Laser printers are also faster than other output devices, such as plotters.

Inkjet printers

An inkjet printer forms an image by spraying ink onto paper through small nozzles or jets, **Figure 14-24**. The printed image is made up of dots. Inkjet printers are quick and provide an inexpensive way to create color illustrations. Inkjet plotters are similar to inkjet printers, with the exception that they are typically used for much larger sheet sizes. See **Figure 14-25**.

Figure 14-23. A laser printer. (Hewlett-Packard Co.)

Figure 14-24. An inkjet printer. (Epson America, Inc.)

Figure 14-25. An inkjet plotter. (Roland DGA)

Controlling raster output

As is the case with vector output, there are a number of ways to ensure high-quality production when using a raster output device. Several methods can be used to control line quality and lettering:

➤ **Line quality**. Lines printed by raster output devices are made up of a series of dots. This means that the lines do not have actual segments. Low-resolution devices often produce jagged lines. The dots on the jagged line represent the "best fit" of the line with the dots available. As output resolution increases, the line "jaggies" are reduced. A similar problem, though not as obvious, is slight line width variations in the printed image. Again, this is usually a result of printing lines with the best fit possible at the available dot resolution. One solution is to change line thicknesses, which may make this problem less noticeable.

When printing bitmapped images, it is helpful to try to print at the same resolution of the original bitmap. In this way, there is a close match between the original bitmap resolution and the printer resolution, and the dot pattern will be approximately the same. This should result in a closer match between the original bitmap and the final illustration.

➤ **Lettering**. Raster output devices are capable of producing a wide variety of typefaces and styles with accuracy. However, at times, there may be a problem with on-screen fonts matching the printed fonts. This problem can be solved by using font utility software or a page description language such as PostScript. PostScript fonts are built into PostScript-compatible printers. This means that the on-screen font should print exactly as it appears.

Summary

A knowledge of the production processes used in publishing is important for technical illustrators. Being familiar with page composition techniques and printing technologies helps illustrators make good decisions when developing materials for publication. In some cases, illustrators may also be responsible for layout tasks and other duties making up the production process.

Whether a publication is created manually or using computer software, a series of common production steps are followed. These include image design, image generation and assembly, image carrier preparation, printing, and product finishing. Image design is a critical first step, since a poor design cannot be improved in any of the later production phases.

Good design principles are key to developing a sound page design. Elements making up the design should be in correct proportion, and proper emphasis should be given to the elements so that the design has unity. A number of standard elements, such as borders, white space, and reverse type, can be incorporated to achieve a pleasing design.

Graphic images can be classified as line and continuous tone images. The methods used to prepare images for printing depend on the type of image being reproduced. In page composition programs, a scanner is frequently used to convert images to electronic form.

Common output devices used to produce technical illustrations and other materials for publication include pen plotters, laser printers, and inkjet printers and plotters. Output devices are classified as vector and raster devices. Depending on the type of device used, various methods can be used to ensure high-quality output.

1. Give two reasons why technical illustrators should have a knowledge of the publication production process.
2. List five common production steps involved in developing a printed product.
3. Why is image design such an important step in creating a publication?
4. Explain the difference between thumbnail sketches and a rough layout.
5. In offset lithographic printing, what is the image carrier used to print the product?
6. What are the two types of balance used in a design and how do they differ?
7. A large, bold headline in a newspaper is an example of which design principle?
8. Define *unity*.
9. What is a *kicker*?
10. Why is white space an important element in page design?
11. White type set against a colored background is known as what?
12. List two ways to prepare a continuous tone image for printing.
13. Define *cropping*.
14. What occurs when an image is scanned?
15. Briefly explain the difference between electronic and manual page composition.
16. What is the difference between vector and raster output?
17. What type of output device closely simulates a hand drawing?
18. When printing bitmapped images, why is it helpful to print at the same resolution of the original image?

1. Using a search engine on the Internet, locate and research information about the services provided by a large commercial printing firm. Using the production steps described in this chapter as a starting point, describe the services that are offered in a brief report. Identify the printing and finishing methods that are available and describe the types of originals accepted from customers. As an option, you can interview personnel from a printing company to assemble this information.
2. A graphics company that designs technical illustrations wants to advertise its services with an 8 1/2″ × 11″ flyer that can be distributed around town. You agree to do the work. Make thumbnail sketches and design a rough layout for a page with at least two illustrations. The flyer should include limited information about the company, such as contact information. Use proper text fonts and sizes so that the flyer attracts attention from a distance.
3. Using discarded publications, find graphic examples that apply the design principles discussed in this chapter. These include proportion, balance (informal and formal), emphasis, contrast, rhythm, and unity. Identify how the principles are applied directly on each image or on a separate page.
4. Using a discarded publication, identify the common page elements used in the publication that are discussed in this chapter. Record this information by marking directly on the pages of the publication.

5. Using a search engine on the Internet, locate companies that manufacture or sell printers and plotters. In a brief report, discuss the different types. Identify at least two that you believe would be suitable for a small company that creates technical illustrations.

Glossary

A

Absolute coordinates. Coordinates specifying a location in relation to the origin (X=0,Y=0,Z=0).

Acrylic. A type of paint with a polymer base and exceptional opacity.

Action-view drawing. A drawing that depicts equipment in action or the action of someone using the equipment.

Additive color formation. A method of color formation in which red, green, and blue wavelengths are combined to form other colors. Also called *RGB color formation*.

Adjustable ruling pen. A pen with two adjustable nibs used to draw a variety of line widths.

Aerial perspective. A view of an object or scene from above the normal elevation.

Agate. A unit of measurement for type, approximately equal to 5 1/2 points. See *point*.

Airbrush. A small, handheld spray gun used to apply paints and inks with a high degree of control.

Airbrush eraser. An illustration tool used to remove portions of media that have been airbrushed onto a surface, similar to a miniaturized sand blaster. Also known as a *media blaster*.

Airflow. In reference to airbrushing, the amount or volume of air a compressor will supply at a rated pressure.

Alphabet of lines. An industry standard, developed by the American Society of Mechanical Engineers (ASME), that classifies the different types of lines used in drawings.

Ambient light. The evenly distributed background light that is cast throughout a scene.

Analogous colors. Colors positioned next to, or close to, one another on a color wheel.

Angular perspective. A type of pictorial drawing in which the principal planes are inclined to the picture plane and receding axis lines converge at two vanishing points. Also called *two-point perspective*.

Applications software. Computer programs that perform specific tasks.

Appliqué shading. Adhering images from a printed transfer sheet to a surface in a drawing for shading purposes. Also called *transfer shading*.

Architect's scale. An open-divided scale marked in feet, most often used for drawing buildings.

Assembly manual. A technical manual that provides step-by-step instructions on how a product is put together.

Atmospheric effect. An environmental effect produced by light reflecting in different directions, causing details to blur and colors to fade as viewing distance increases.

B

Balance. The sense of visual equilibrium in appearance. *Formal balance* occurs in symmetrical illustrations (one side of the illustration is a mirror image of the other side). *Informal balance* exists when the opposing sides are not identical but contain approximately the same number of lines or mass, or a balance of colors or textures.

Binary code. A code in which all characters are represented as a string of zeros and ones, used in computer processing.

Bitmap graphics. Images made up of patterns of dots, or pixels. Also called *raster images.*

Blending stump. A double-ended soft felt pad used to blend or smudge drawing materials.

Break lines. Lines that indicate part of an object has been removed.

Burnishing tool. A pen-like tool with a smooth steel or plastic ball on the tip used to apply dry transfer material.

C

Cabinet oblique. An oblique drawing in which the receding axis is drawn at half scale.

Cartesian coordinate system. A coordinate system used by most CAD programs in which point locations are specified using three perpendicular axes (the X axis, Y axis, and Z axis).

Cavalier oblique. An oblique drawing in which the receding axis is drawn at full scale.

CD-R. An optical storage device. CD-R is an abbreviation for *compact disc-recordable.* Data can be read, written, and erased, but previously recorded sectors of a CD-R cannot be written to again.

CD-ROM. An optical storage device that can store large amounts of data. CD-ROM is an abbreviation for *compact disc read-only memory.* Data can be read from a CD-ROM, but new data cannot be written to it.

CD-RW. An optical storage device. CD-RW is an abbreviation for *compact disc-rewritable.* Data can be read, written, and erased, and previously recorded sectors can be written to again.

Centerlines. Thin lines used to show the center of circles and arcs. They are drawn as a series of lines with a long dash followed by a short dash.

Central processing unit (CPU). The core of a computer system, where processing occurs.

Chain lines. Thick lines with alternating long and short dashes placed next to surfaces that are to have some treatment, such as hardening.

Chamois. A piece of very soft leather used to blend or smudge drawing materials.

Civil engineer's scale. A decimal scale that divides inches into parts that are multiples of 10, often used for drawing highways and maps.

Clarity. A design principle, usually pertaining to text content, that measures how understandable something is.

Clip art. A sampling of commercially prepared images designed for use in illustration work.

Clipping planes. In a CAD drawing program, front and rear boundaries that "slice" through a model to create a partial view.

CMYK color model. A color production system that defines colors based on the subtractive primaries (cyan, magenta, and yellow) and black.

Color model. In regard to computer software, a system used to describe colors.

Color wheel. A device used to show the relationship among the primary, secondary, and intermediate colors.

Combination scale. A scale that contains many of the most commonly used scales, such as a mechanical engineer's scale, an architect's scale, and a civil engineer's scale, and may include both US Customary and SI Metric measurements.

Commands. In regard to computer-aided drafting, instructions to the computer.

Compass. A drafting instrument used to draw circles.

Complementary colors. Colors positioned directly across from one another on a color wheel.

Composite solid. In a CAD drawing program, a combination of solids created with solid editing operations.

Composition. The process of using design principles to arrange individual elements into a completed product.

Comprehensive layout. A full-scale illustration with all colors, textures, text, and other features needed to develop an illustration as a production-ready piece. A comprehensive normally evolves from an illustration rough.

Construction lines. Very light lines used to initially lay out a drawing. The lines are drawn lightly so they do not show on copies.

Continuous tone image. A type of image that has gradations or variations of tone, such as a black-and-white photo.

Contrast. A design principle that is achieved when elements are varied to draw attention or provide meaning.

Coordinate method. In relation to isometric drawing, a method of defining curves on foreshortened or angled surfaces in which points on a grid are transferred from an orthographic view to the pictorial drawing.

Copyright. A form of protection for an original document.

Cover page. The front page of a publication. The cover page is designed to communicate what is inside and should attract reader attention.

Cropping. The removal of material from one or more edges of an illustration to delete unneeded material.

Cutting-plane lines. Lines that indicate where an imaginary cut is made on the object. Arrows on the end of the line indicate the direction in which the viewer is looking at the section.

D

Diffuse color. Color that is illuminated by the distribution of diffuse light as light strikes an object's surfaces.

Digitizer. A computer input device similar to a graphics tablet, used to convert drawings into digital form.

Dimension lines. Thin lines typically placed between extension lines, with arrowheads at each end. The actual dimension is placed in a break along the line.

Dimetric drawing. An axonometric pictorial drawing in which two of the three angles created by the intersections of the axes are equal, but the third angle is different. Measurements along two of the axes use the same scale, but the third axis is drawn at a different scale.

Direct light. In a CAD drawing program, a light that projects parallel rays of illumination in a given direction.

Display controls. In a CAD drawing program, tools used to change the way a drawing appears on screen. Also called *view controls.*

Dividers. Drafting instrument with two adjustable, pointed legs used to measure and transfer distances.

Double-action airbrush. An airbrush that operates by depressing and pulling back a trigger mechanism in the same motion.

Drafting brush. A soft brush used to sweep away eraser shavings from paper.

Drafting film. Moisture-proof plastic sheet with one side roughened to accept ink and pencil lines.

Drafting machine. A combination drafting instrument used for drawing horizontal, vertical, and angled lines.

Draw programs. Graphics programs that create images using vector graphics.

Drawing media. The paper, film, or other surface to which lead or ink is applied to create an illustration.

Drop shadow. A second outline placed slightly off to the side of an illustration to create a shadow appearance.

Drum plotter. A plotter that uses a cylinder to move the paper forward and back while pens move over the surface to produce the image.

E

Edge defined surface. In a CAD drawing program, a mesh surface that fills an area between connected boundaries.

Electric eraser. An eraser with an electric motor that automatically moves the eraser from side to side, used for erasing large areas.

Electromagnetic spectrum. The entire range of wavelengths of electromagnetic radiation. The spectrum ranges from short wavelengths (such as gamma rays and X rays) to long wavelengths (such as radio waves) and includes visible light.

Elevation line. In perspective drawing, a line used in place of an elevation view to locate vertical features.

Emphasis. A design principle that is used to make one part of a design stand out, or draw the greatest interest.

Entities. In a CAD drawing program, the elements that make up a drawing, such as points, lines, shapes, dimensions, and text. Also called *objects.*

Erasing shield. A small, thin metal plate with various cutouts, used to isolate an area of a drawing to be erased. The erasing shield protects the material that is not being erased.

Exploded drawing. A pictorial drawing that shows an object disassembled, with the parts aligned in a manner that indicates how they fit together.

Extension lines. Thin lines used to extend dimensions from the objects in the drawing. A short space is placed between the object and extension line, and the line extends beyond the dimension line.

External-mix airbrush. An airbrush that operates by drawing media through an outside tube and creating a mixture of media and pressurized gas at the front of the airbrush tip to produce a spray pattern.

Extrusion. In a CAD drawing program, a solid model created from basic 2D geometric shapes.

F

Falloff. The decrease in intensity of a light as the distance from the light source increases. Also known as *attenuation.*

Fineline pencil. A pencil similar to a mechanical pencil but designed to accommodate a single lead width.

First-angle projection. An orthographic projection method in which the sides of the object are projected to the sides of an imaginary glass box and away from the viewer. See *third-angle projection.*

Fixed tip. A nonmoving fineline pencil tip, often used when working with rules and templates.

Flatbed plotter. A plotter that holds the paper stationary while pens move over the surface to produce the image.

Focal point. Any important aspect or feature on an illustration upon which focus is intended.

Font. A typeface classification containing all the characters of a single typeface style.

Footer. A design element appearing at the bottom of a page, used to identify the page number or information about the material being read.

Foreshortening. A term used to describe a drawing in which a feature appears smaller than true size in an orthographic view. Features that are not parallel to the projection plane are foreshortened.

Four-center method. In relation to isometric drawing, a method used to locate center points for arcs defining an isometric circle.

Frisket. A commercially prepared sheet of plastic acetate with an adhesive backing, used for masking illustrations for airbrushing.

Frontal view. The front orthographic view of a three-dimensional object.

Full-sliding sleeve. A fineline pencil tip that retracts completely into the pencil as the lead is used. Most often used for writing or freehand lettering rather than drafting purposes.

G

Galleys. High-quality, camera-ready sheets of copy that are placed in a layout in traditional production work.

General oblique. An oblique drawing in which the receding axis is drawn at a scale other than one-half or full size, typically a three-quarter scale.

Ghost. Graphite embedded in paper. Can be caused by applying too much pressure while erasing.

Golden section. A traditional design concept that is used to set a proportion of 1:1.618.

Gouache. A type of watercolor produced by mixing finely ground color chalk in water and a gum binder.

Graphics software. Application programs used to create images.

Graphics tablet. A computer input device that consists of a flat surface with a wire grid underneath. Used with a stylus or puck.

Gravity-feed airbrush. An airbrush that draws media by gravity feed from a reservoir or chamber in the barrel.

Grid. In a CAD drawing program, a configuration of dots or lines on the screen arranged in rows and columns, used much like graph paper.

Grid sheets. Sheets of paper printed with a grid pattern that serves as a guide for sketching and drawing. The grid may have a rectangular or isometric pattern, or a different pattern.

Ground line. In perspective drawing, a line representing the elevation where the object sits in relation to wherever it touches the picture plane in the plan view.

Guidelines. Very light lines used to make sure that lettering is uniform in size. The lines are drawn lightly so they do not show on copies.

Gum eraser. A drafting instrument useful for removing light lines and smudges, less abrasive than a rubber eraser.

Gutters. The vertical spaces between columns of text.

H

Halftone. An image converted to a series of dots for printing purposes.

Hard drive. The main data storage device of a computer. It consists of metal platters in a sealed case.

Header. A design element appearing at the top of a page, used to identify the page number or information about the material being read.

Headline. A heading on a page used to provide a general idea of what is being presented until the next major heading. Commonly referred to as a *head*.

Hidden lines. Thin, dark lines used to represent edges hidden from view.

Horizon line. In perspective drawing, a line indicating the location of the horizon with respect to the elevation of the ground line. The horizon line is where receding lines converge and represents "eye level."

Horizontal views. The top and bottom orthographic views of a three-dimensional object.

Hotspot. The area of projection from a light source where the greatest amount of illumination is generated.

HSL color model. A color production system that defines colors based on hue, saturation, and lightness.

Hue. The color of an object, or name of a color, as defined by reflected light.

I

Illustration board. Heavy paper specially designed to accept wet materials without buckling. *Hot-pressed illustration board* has a smooth finish and is not very absorbent. *Cold-pressed illustration board* has a textured surface and is more absorbent.

Image carrier. A printing plate or intermediate used to print multiple copies of an image on a press or other printing device.

Imagesetter. An output device used to print high-resolution images in various formats on paper, film, or printing plates.

Inclined surfaces. In orthographic projection, surfaces perpendicular to two projection planes, but inclined to all others. They appear as lines where they are perpendicular to the projection plane and as foreshortened surfaces in other views. See *normal surfaces* and *skewed surfaces.*

Ink risers. Plastic disks attached to the bottom of a template so that the entire template does not rest on the drawing.

Inkjet printer. A hard copy computer output device that forms an image by shooting droplets of ink onto the page.

Input. With regard to computers, data entered into the system, such as through a keyboard or mouse.

Input device. A computer hardware device, such as a keyboard or mouse, used to get information into the computer.

Installation manual. A technical manual that provides instructions on how a product is installed, placed, or mounted.

Intermediate color. A color produced by combining a primary and secondary color on a color wheel. Intermediate colors include blue-green, blue-violet, red-violet, red-orange, yellow-orange, and yellow-green.

Internal-mix airbrush. An airbrush that operates by drawing media inside the airbrush and creating a mixture of media and pressurized gas to produce a fine atomized spray.

Irregular curve. A drafting instrument with a variety of curves, used for drawing noncircular arcs. Also called a *French curve.*

Isometric drawing. An axonometric pictorial drawing in which the axes form 120° angles and measurements along all axes are drawn to the same scale.

Isometric ellipse template. A template of ellipses for isometric drawings.

Isometric hexagon template. A template of hexagons for isometric drawings.

Isometric lines. Horizontal and vertical lines parallel to an isometric axis.

Isometric protractor. A drafting instrument used to measure angles inclined or skewed to the principal isometric planes.

K

Kerning. A design technique that reduces the amount of space between individual letters.

Kicker. A short phrase normally placed near a headline, used to summarize the text or provide clarification of the headline.

L

Landscape orientation. A horizontal orientation in the design layout of a document or publication.

Laser printer. A hard copy computer output device that uses a laser to transfer the image to a light-sensitive drum, which in turn transfers the image to paper using toner.

Layers. In a CAD drawing program, object settings serving a function similar to that of overlaid sheets of paper in a drawing project.

Lead holder. A mechanical pencil.

Lead pointer. A drafting tool used to sharpen, or point, pencil lead.

Leader. Dimensioning object consisting of an arrowhead and angled line connected to a shoulder. The shoulder points to a note or dimension. The angle of the leader is typically 30°, 45°, or 60°.

Leading. A measure of the total line spacing for type, including the point size of the type and the space between lines. See *line spacing.*

Letter spacing. In lettering, the spacing between letters in a word.

Lettering. Placing text characters on a technical drawing.

Lettering machine. A device used to create strips of lettering on clear, adhesive-backed plastic. After the lettering is created, it is pressed in place on the drawing.

Lightness. A color attribute that describes how light or dark a color is.

Line contrast shading. A shading technique that uses object lines of varying thickness to create an impression of different surfaces.

Line gauge. An instrument used to measure type in various units, such as points, picas, and inches.

Line image. A type of image that has a uniform density of color (such as black), rather than a gradation of tone. Also known as *line art*.

Line separation shading. A shading technique that uses slight gaps at the end of object lines to define features that lie behind other features in order to show depth.

Line shading. A special treatment for the object lines of a pictorial to produce shadows and help define shapes.

Line spacing. A measure of the space between lines of type, discounting the point size of the type. When lettering, this is normally equal to one-half the letter height or the full letter height. See *leading*.

Line surface shading. Drawing lines or line patterns on an object's surfaces to create the appearance the object is made of a solid material.

Logo. An identifying symbol for a company or business.

Long axis isometric drawing. An isometric drawing that has one major axis aligned horizontally and the other axes inclined at a 60° angle to horizontal.

M

Magneto-optical (MO) drive. A computer device that writes to an MO disc magnetically and reads the data using a laser.

Maintenance manual. A technical manual that provides instructions for the care or repair of a product.

Master. A reproduction containing the image to be printed.

Material. In a CAD drawing program, a surface finish or texture applied to features of a model in order to define how object surfaces will appear when the scene is rendered.

Mechanical engineer's scale. An open-divided scale using inch measurements to make drawings in whole inches or fractional inches.

Mechanical pencil. Also called a mechanical lead holder, a pencil that accommodates various grades and widths of lead.

Memory. A term used to describe the location where information is stored for use by the computer.

Metric scale. A scale used to make drawings in metric units. Metric scales are displayed on drawings as ratios.

Microprocessors. Small silicon chips used to process data in a computer's central processing unit. Also called *microchips*.

Monitor. A computer output device that displays visual information for the user.

Mouse. A computer input device used to move an on-screen cursor, with buttons used to make selections and initiate commands and functions.

Multiview drawings. Drawings produced in orthographic projection, used to represent different views of three-dimensional objects. Also called *orthographic drawings*.

N

Nameplate. A graphic used on the front page of a publication to identify information such as the name of the publication, the date, and the cost. Also known as the *flag*.

Nib. The tip of a pen.

Nonisometric lines. Lines in an isometric drawing that are not parallel to one of the isometric axes.

Normal features. Features that are parallel to one of the principal orthographic planes.

Normal surfaces. In orthographic projection, surfaces parallel to the projection plane. They are projected as true size. See *inclined surfaces* and *skewed surfaces*.

Normal views. The six perpendicular views (front, right side, left side, top, bottom, and rear) used to completely define a three-dimensional object using orthographic projection. Also called *principal views*.

Novelty. A typeface category that includes decorative or unusual styles of type.

O

Object lines. Thick, dark lines used to represent visible edges of an object. Also called *visible lines*.

Object overlap. The basic principle that objects closer to the viewer will cover or hide objects that are farther from the viewer when the nearer objects are in the line of sight between the viewer and the farther objects.

Object properties. In a CAD drawing program, characteristics of an object, such as the layer setting, line type, line width, and color.

Object snap. In a CAD drawing program, a drawing aid that allows the cursor to "attach" to a specific location on an object, such as the endpoint, midpoint, or center.

Objects. In a CAD drawing program, the elements that make up a drawing, such as points, lines, shapes, dimensions, and text. Also called *entities*.

Oblique drawing. A type of pictorial drawing in which the front face is parallel to the projection plane, and the top and side views are viewed at an oblique angle. A receding axis is used to measure the depth and extends away from the face.

Oblique perspective. A type of pictorial drawing in which the principal planes are inclined to the picture plane and receding axis lines converge at three vanishing points. Also called *three-point perspective*.

Oblique surfaces. In orthographic projection, surfaces that are not perpendicular or parallel to any projection planes. They appear foreshortened in all views. Also called *skewed surfaces*. See *normal surfaces* and *inclined surfaces*.

Old English. A typeface category in which the characters have extra strokes for ornamentation.

One-point perspective. A type of pictorial drawing in which a front view is parallel to the picture plane and receding axis lines converge at a single vanishing point. Also called *parallel perspective*.

Open-divided scale. A scale with which whole values are measured on one side of the zero mark and fractional and decimal values are measured on the other side of the zero mark.

Operating position orientation. The position an object is in when a person is using, controlling, or viewing it in its natural environment.

Operating system. The program that "tells" the computer what to do.

Operation manual. A technical manual that provides instructions on how a product is to be used.

Optical disc. A highly popular, durable computer storage device capable of storing large amounts of data.

Orbit viewing. In a CAD drawing program, a viewing method used to rotate a model dynamically in 3D space.

Orthographic projection. The process of representing several different views of a three-dimensional object to provide a complete description.

Output device. A computer hardware device used to display or produce something generated within the CPU. Monitors, disk drives, and printers are examples of output devices.

Owner's manual. A technical manual that provides operation instructions for a product and other information, such as part specifications and maintenance information.

P

Paint programs. Graphics programs that create images using bitmap graphics.

Parallel perspective. A type of pictorial drawing in which a front view is parallel to the picture plane and receding axis lines converge at a single vanishing point. Also called *one-point perspective*.

Parallel straightedge. A drafting instrument attached to a drawing board, used for drawing horizontal lines and guiding other instruments.

Pattern break. Used to achieve variety, an effect that is created by overlapping an object over a repeating pattern to disrupt the monotonous effect.

Permission form. A written form used to request permission to use copyrighted material from the owner of the property.

Perspective drawing. A type of pictorial drawing in which receding lines converge at one or more vanishing points.

Phantom lines. Thin lines used to indicate alternate positions for moving parts on the object or to indicate repeated details.

Phantom-view drawing. A drawing that uses phantom lines to show an alternate position for a mechanism.

Pica. A standard unit of measurement used in printing, equal to 0.166". There are 12 points in one pica, and six picas are approximately equal to one inch. See *point*.

Picture area. The space within the borders of an illustration or graphic.

Picture plane. A theoretical plane onto which an image is projected.

Pixels. Picture elements that create the image on a computer monitor. Display resolution is measured in pixels.

Plotter. A computer output device used for printing drawings where high-quality line work is needed.

Point. A standard unit of measurement for type and line spacing equal to 0.0138". There are approximately 72 points in one inch, and 12 points are equal to one pica. See *pica*.

Point source light. In a CAD drawing program, a light that projects light rays in all directions in order to illuminate a scene.

Polar coordinates. Coordinates specified as a distance and angle from the origin or a fixed point. Also called *radial coordinates*.

Polygon. A closed shape composed of lines, such as a triangle or rectangle. A *regular polygon* has equal sides and angles.

Polyline. In a CAD drawing program, a continuous line made up of one or more segments.

Portrait orientation. A vertical orientation in the design layout of a document or publication.

Pressure gauge. In reference to airbrushing, an indicator that displays how much pressure is in an airbrush system or pressure tank.

Pressure regulator. In reference to airbrushing, a control used to adjust the pressure of the air supplied to the airbrush.

Principal views. The six perpendicular views (front, right side, left side, top, bottom, and rear) used to completely define a three-dimensional object using orthographic projection. Also called *normal views*.

Process color. Printed color produced through the four-color CMYK printing process.

Processing. With regard to computers, performing functions and operations on data.

Profile views. The right side and left side orthographic views of a three-dimensional object.

Projection plane. An imaginary perpendicular plane onto which an orthographic view of an object is projected.

Projector. An imaginary line that transfers a point on an object to a projection plane. The projector is always perpendicular to the projection plane.

Proportion. The size relationship of one part of an object to the size of the entire object.

Protractor. A drafting instrument used to measure and draw inclined lines and angles.

Public domain. A term used as a classification for original works. Works in the public domain can be freely copied because they have no copyright, or the copyright has expired.

Puck. A computer input device used to select cursor locations on a graphics tablet.

R

Random access memory (RAM). Temporary computer memory storage.

Raster output device. A computer hardware device that produces drawing and text images as series of dots.

Read only memory (ROM). Computer memory that permanently stores instructions necessary to perform tasks.

Read/write head. A part of a computer disk drive, used to store data on and retrieve data from the disk.

Registration. The correct alignment of different colors or images during the printing process.

Registration marks. Circular marks with intersecting lines used to align overlaid image masters in printing applications.

Relative coordinates. Coordinates specified "relative" to the last point drawn. Also called *delta coordinates.*

Relative coordinate system. In a CAD drawing program, a coordinate system created by the user to establish a direction for the X, Y, and Z axes and a specific drawing plane orientation.

Rendering. A pictorial drawing or graphic representation that has the highest degree of realism possible. A rendering is typically produced in color with various effects and background scenery added.

Reverse type. White type set against a colored background.

Reversed axis isometric drawing. An isometric drawing in which the horizontal axes are drawn 30° below horizontal, used to show features on the bottom of an object.

Revolved surface. In a CAD drawing program, a mesh surface created by rotating a profile shape around an axis of revolution.

RGB color model. A color production system that defines colors as percentages of the three additive primaries of light (red, green, and blue).

Rhythm. A design principle that is used to create a sense of movement or repetition.

Roman. A typeface category in which the characters have thin horizontal and vertical strokes at the ends. Also called *serif.*

Rough layout. A shaded sketch approximating the full size of the finished product with objects sketched in proportion and blocked areas representing text. A rough is created during the final solution phase of the design problem-solving process.

Rubber eraser. A drafting instrument useful for removing dark lines, more abrasive than a vinyl or gum eraser.

Ruled surface. In a CAD drawing program, a mesh surface created between two objects that define the two ends.

Runaround. An arrangement for text so that it follows the outline of an irregularly shaped illustration.

Running head. An identifier used to inform the reader of the chapter or section currently being read.

S

Sandpaper pad. A drafting tool used to sharpen, or point, pencil lead.

Sans serif. A typeface category in which the characters do not have serifs.

Saturation. The purity of a color, or absence of gray. Also called *chroma.*

Scale. A device used for measuring distances on drawings.

Scaling. Reducing or enlarging an illustration proportionately for layout purposes.

Scanner. An imaging device used to convert printed images to electronic form.

Scene. In a CAD drawing program, a view of a 3D model specifically set up for rendering.

Schematic drawing. A drawing that shows a detailed diagram of a functional system.

Script. A typeface category in which the characters appear handwritten.

Secondary color. One of the colors formed by mixing two primaries on a color wheel. These include violet, green, and orange.

Section drawing. A drawing that shows internal details of a product that would normally be hidden from view. Also known as a *cutaway.*

Section lines. A series of equally spaced, thin lines drawn at an angle, used to represent a surface that has been cut by a cutting-plane line.

Sectors. In relation to computers, divisions of a track on the surface of a disk.

Semi-sliding sleeve. A fineline pencil tip that exposes lead as you draw, protecting smaller leads.

Serifs. The thin horizontal and vertical strokes at the ends of characters in the Roman typeface category.

Service mark. A trademark used to identify the source of a service.

Set solid. A type specification in which the point size and leading are equal, such as "10/10 Helvetica." Type that is set solid has no room between the descenders and the ascenders on successive lines.

Shading. The process of adding darkness or color to an area of a drawing to produce a visual representation of a surface.

Shadow fill. A shadowing technique that uses a solid fill or darkening effect on surfaces that are not illuminated by either direct or reflected light.

Shadow softening. A shadowing technique used to produce shadows that transition smoothly from dark shading to lighter shading.

Shadowing. Darkening an area of an object that has blocked illumination from the light source.

SI Metric system. The measurement system recognized as the international standard. The basic unit for length is the meter.

Simulated 3D drawing. A computer-generated drawing that gives the illusion of a 3D solid model.

Single-action airbrush. An airbrush that operates by depressing a trigger mechanism in one motion.

Siphon-feed airbrush. An airbrush that draws media by siphon feed or suction from a mounted jar or reservoir.

Skewed surfaces. In orthographic projection, surfaces that are not perpendicular or parallel to any projection planes. They appear foreshortened in all views. Also called *oblique surfaces*. See *normal surfaces* and *inclined surfaces*.

Smudge shading. A shading technique in which a drawing material is applied to media and smeared to create a softly shaded or textured effect.

Snap. In a CAD drawing program, a drawing aid that confines cursor movement to an invisible grid.

Software. A term used to describe the specific instructions for directing the operation of the computer.

Solid model. In a CAD drawing program, a three-dimensional representation with solid material defining the entire mass of an object.

Solid primitives. In a CAD drawing program, predefined solid shapes used in solid modeling applications.

Specular color. Color that is illuminated by light as the light reflects from a surface. Specular color is known as the "highlight" or shiny portion of an object.

Spline. A flexible drafting tool used to draw irregular curves.

Spot color. A type of printed color referring to a solid area of a specific color.

Spotlight. In a CAD drawing program, a light that projects light rays in a cone of illumination.

Spray booth. A closed room or boxlike enclosure with ventilation that allows control of production conditions.

Spubbles. In reference to airbrushing, spurts and bubbles caused when a drop in air pressure occurs and the spray from the airbrush changes from a fine mist to a series of large blobs.

Station point. In perspective drawing, the location of the viewer in relation to the object.

Stencil burner. A special art tool used to burn away material, such as acetate, in order to create a sculpted hole in the material for masking purposes. Also known as a *hot knife*.

Stipple shading. Applying a pattern of dots in an image to create shadows or surface textures.

Storage devices. With regard to computers, hardware devices used to store data.

Stylus. A computer input device used to select cursor locations on a graphics tablet.

Subhead. A heading used to organize the material under a headline into a smaller section.

Substrates. Drawing paper and film.

Subtractive color formation. A method of color formation in which some colors are reflected from a surface and other colors are absorbed or "subtracted." Also called *CMYK color formation.*

Suite. A single software package containing multiple applications.

Surface model. In a CAD drawing program, a three-dimensional representation that uses mesh surfaces to define the exterior surfaces of an object.

Surface shading. The application of special highlights or textures to an object's surfaces.

Surfaces. Design elements used to represent the exterior boundaries of an object.

Swatchbook. A color reference manual or guide with samples used to select colors for printing.

Swivel knife. A razor knife with a blade that can be rotated in the knife handle, used to cut along curved lines.

Syntax. The "grammar" of a computer language.

Systems software. A collection of programs that coordinate the computer, peripheral devices, and software so they all work together correctly.

T

Tabulated surface. In a CAD drawing program, a mesh surface generated from an outline curve along a direction path with a defined length.

Tangent point. A point on an arc or curve that touches another entity at a single point.

Technical illustration. The process of applying graphic skills to produce visual materials that explain or clarify some aspect of a product or process.

Technical manual. A publication produced by a company to provide operating, assembly, installation, or service instructions for a product.

Technical pen. A drawing instrument used to produce a single line width of ink. Ink is supplied to the nib from an ink cartridge.

Template. A drafting instrument with cutouts of shapes or symbols that can be traced onto a drawing using a pen or pencil.

Tertiary color. A color produced by combining a primary and secondary color on a color wheel. Tertiary colors include blue-green, blue-violet, red-violet, red-orange, yellow-orange, and yellow-green.

Textile paints. In airbrushing applications, heavy-bodied media used for buildup and bonding with a fabric surface, such as canvas.

Texture. In a design, an optical simulation of a physical surface texture.

Thermal printer. A hard copy computer output device that uses heat to form an image on paper.

Third-angle projection. An orthographic projection method in which the sides of the object are projected to the sides of an imaginary glass box and toward the viewer. See *first-angle projection.*

Three-point perspective. A type of pictorial drawing in which the principal planes are inclined to the picture plane and receding axis lines converge to three vanishing points. Also called *oblique perspective.*

Thumbnail. A rough draft sketch drawn to record a visual thought or design concept.

Tracing paper. A type of translucent paper used for pencil and ink drawings.

Trackball. A computer input device similar to a mouse, with a ball used to control the on-screen cursor.

Tracks. With regard to computers, concentric, magnetic circles on the surface of a disk.

Trademark. A unique symbol, name, or slogan used to identify the source of a product or service.

Transfer sheets. Commercially prepared sheets containing images, symbols, or screens that are applied to drawings.

Translucent paper. A type of paper, such as tracing paper and vellum, that allows a relatively large amount of light to pass through.

Transparent-view drawing. A drawing that shows the internal features of a product through a transparent outer surface.

Triad. A relationship of colors on a color wheel made up of three equally spaced colors such as red, yellow, and blue.

Triangle. A three-sided drafting instrument used for drawing vertical and inclined lines.

Trimetric drawing. An axonometric pictorial drawing in which all three angles created by the intersections of the axes are unequal, and each axis uses a different scale.

T-square. A drafting instrument consisting of a head and a blade used to draw lines and guide other instruments.

Two-point perspective. A type of pictorial drawing in which the principal planes are inclined to the picture plane and receding axis lines converge to two vanishing points. Also called *angular perspective*.

U

Unity. A design principle that is achieved when many different elements of an illustration are combined into an organized layout, creating a pleasing whole.

US Customary system. The standard system of linear measurement used in the United States. Distance is measured in inches, feet, yards, and miles.

V

Value. The relative lightness or darkness of a color, shade, or tint.

Vanishing point. A point in a perspective drawing where receding lines converge.

Vector graphics. Images created using objects such as lines, circles, and arcs. Also called *object-oriented graphics*.

Vector output device. A computer hardware device that produces drawing and text images as a series of straight lines.

Vellum. Moisture-resistant, treated translucent paper containing oils and waxes to improve quality and durability.

Viewing-plane lines. Lines that indicate the area for which a separate view is drawn. Arrows on the end of the line indicate the direction in which the viewer is looking at the view.

Viewport. In a CAD drawing program, a viewing area containing a preset or user-defined viewpoint.

Vinyl eraser. A drafting instrument useful for removing light lines and smudges, less abrasive than a rubber eraser.

Visible light. The portion of the electromagnetic spectrum that is perceived by human vision.

Visible lines. Thick, dark lines used to represent visible edges of an object. Also called *object lines*.

Visible-object drawing. A pictorial drawing of a product that uses object lines to show the entire part assembled, or each part separately.

W

Watercolor. A type of paint produced by mixing tinted, dry powder with water.

Wooden pencil. The simplest type of pencil, with graphite lead surrounded by wood. Must be sharpened, or pointed, as it becomes dull.

Word spacing. In lettering, the spacing between words, which in practice should be equal to the height of the letters.

World coordinate system. In a CAD drawing program, an absolute coordinate system based on the planes defined by the default X, Y, and Z axes in the Cartesian coordinate system.

X

X coordinate. The horizontal dimension in the Cartesian coordinate system, measured along the X axis.

Y

Y coordinate. The vertical dimension in the Cartesian coordinate system, measured along the Y axis.

Z

Z coordinate. The third dimension in the Cartesian coordinate system, measured along the Z axis and used for three-dimensional objects.

Index